APPLIED QUANTUM MECHANICS

APPLIED QUANTUM MECHANICS

WALTER A. HARRISON

Stanford University

Published by

World Scientific Publishing Co. Pte. Ltd.

5 Toh Tuck Link, Singapore 596224

USA office: 27 Warren Street, Suite 401-402, Hackensack, NJ 07601

UK office: 57 Shelton Street, Covent Garden, London WC2H 9HE

Library of Congress Cataloging-in-Publication Data

Harrison, Walter A. (Walter Ashley), 1930–

Applied quantum mechanics / Walter A. Harrison.

p. cm.

Includes bibliographical references and index.

ISBN-13 978-981-02-4375-3 -- ISBN-10 981-02-4375-8

ISBN-13 978-981-02-4394-4 (pbk) -- ISBN-10 981-02-4394-4 (pbk)

1. Quantum theory. I. Title.

QC174.12 .H377 2000

530.12--dc21 00-043505

British Library Cataloguing-in-Publication Data

A catalogue record for this book is available from the British Library.

Preface

This text was prepared and used in a two-quarter course for graduate electrical-engineering and materials-science students at Stanford University. The aim of the course was to teach those parts of quantum mechanics which an engineer might need or find useful in his profession. To my surprise this made the course almost orthogonal to traditional physics quantum courses, which provide those parts which most physicists feel every student should go through. The analytical solution of the harmonic oscillator states is rarely useful after the course is over. I believe that it is also rare that a solution of Schroedinger's equation is what is needed in engineering activities. For most questions concerning electronic structure of molecules or solids a tight-binding formulation is much more to the point, along with a knowledge of how to obtain the parameters which are needed, and how to calculate properties in terms of them. We have not seen these in other quantum texts. It is also important to have a feeling for when one can use a one-electron approximation and how to include many-particle effects when they are needed. One needs familiarity with perturbation theory and with the variational method, and confidence in the use of Fermi's Golden Rule. One needs the elements of quantum statistical mechanics and I believe also the many other topics one may see from scanning the Table of Contents, including even the elements of the shell model of the nucleus. It is not easy for a student to absorb such a variety of material in a short period, but the more modern approach of learning only that piece of a subject which one needs at the moment is not a viable approach for the fundamental laws which govern physics. Nearly fifty exercises, listed by chapter, are directed at using quantum mechanics for every-day problems, rather than to illustrate features of quantum theory. Solutions are available as a teachers' guide from the publisher. I have found it rather easy to generate problems when the material under discussion really has such wide use.

Graduate students in mechanical engineering, chemistry and chemical engineering, in addition to the electrical engineers and materials scientists, took this course. Most of these had taken no physics nor mathematics courses beyond their sophomore year as undergraduates. For that reason it was essential to include Lagrangian and Hamiltonian mechanics, Chapter 3, and such mathematical techniques as Lagrange multipliers, which physics majors learn only in their third or fourth years. I assumed that a third- or fourth-year undergraduate engineer would also be qualified to take the course, since their physics and mathematics background was the same, but none survived to the end. It appears to me that the graduate engineers have grown in sophistication, partly through other technical courses, to the point that they can deal effectively with such an abstract subject. They of course do it with varying success, but I believe it is so very essential for modern engineers to have a systematic presentation of quantum theory that it was an important experience for all of them. One wonders in particular how a modern materials scientist can obtain a Ph. D. without ever studying the fundamental rules which govern the behavior of materials. Similarly if any engineer needs to work with very small systems, as is increasingly common, he certainly should be able to recognize and deal with quantum effects. Having a "Schroedinger-solving code" on his computer is beside the point.

Although the text is designed for engineers, and engineering backgrounds, it has seemed to me that it might also be useful for physics graduate students who have completed a traditional, more sophisticated, course in quantum mechanics. If that course was light on the approximations which have proven successful for applications, or in dealing with systems which involve many electrons, this text might provide what is needed for that physics student to use the knowledge of quantum theory he has obtained.

We generally use equations which can be evaluated in MKS units, but in the end the energies in atomic systems will be of the order of electron-volts and atomic dimensions are of order Angstroms. Thus it is easiest in all regards to use the composite constants $\hbar^2/m$ = 7.62 eV-Å^2, with m the electron mass, and e^2 = 14.4 eV-Å, with e the electronic charge, so that results are obtained immediately in convenient terms. This is in keeping with almost all treatments of quantum mechanics so that results here can be matched with those in other texts. Then the interaction energy between two electrons a distance r apart is written e^2/r . The main place where the customary units become problematical to one educated with MKS comes with the use of magnetic fields, given here in gauss. Then the parameters needed for evaluation are given explicitly at the beginning of Chapter 22.

There is a central, and hopefully appealing, feature to the text which is not essential to its real goal. That is the assertion that quantum theory follows from a single absolute truth, the wave-particle duality, stated on the first page. The full generality of the statement is only developed as we proceed, and Planck's constant which makes the connection between the two descriptions is obtained from experiment, but no further postulates are required. We do not deduce all of the consequences with elegance and rigor, but believe the basic derivations are all essentially correct. Then if a student is puzzled by some question, such as Schroedinger's cat, he may recognize that if he cannot understand something which follows from the wave-particle duality, it is that duality which he does not understand. He should perhaps address his concern at the source of problem and may not be likely to resolve it by thinking about some remote consequence. Our focus is not on the deep philosophical questions which quantum theory inevitably raises, but it may provide a basis for dealing with them which is appealing to the mind of an engineer.

This view of a single postulate is not apparent in more historical developments where the Pauli Principle, or the Uncertainly Principle, can appear to be independent postulates. This is partly because they initially were, and partly because the teaching of quantum mechanics may be mixed with teaching about the unfinished theory of fundamental particles, which evolved simultaneously. In our view quantum mechanics does not tell us what nature will provide, but does tell us the behavior of anything we can define and specify, either as a particle or as a wave. Similarly the questions of the "true" meaning of measurement and the collapse of wavefunctions do not arise with the pragmatic approach of asking only questions which can be tested experimentally. We see how to ask such questions, and how to answer them. Quantum mechanics cannot do more, nor can any other theory. This reliance on a single postulate can be a comfort to a student who is taken rapidly through an extraordinary range of systems and phenomena in a brief period. Hopefully the net effect is to allow him to recognize when he needs quantum theory, and to know how to proceed when he does.

Walter A. Harrison
Professor of Applied Physics
Stanford University
May, 2000

Contents

I. The Basic Approach

IV. Statistical Physics

V. Electrons and Phonons

I. The Basic Approach

Our goal in this treatment of quantum theory is to provide those aspects of the theory which are most needed for a modern understanding of the world of our experience. Central to this are the many approximations which have turned out to be successful for estimating the properties of matter and predicting the events which occur on a microscopic scale. In that sense the goal is entirely pragmatic, but any study of quantum theory cannot help but raise deep philosophical questions. Perhaps the only real understanding of the theory is through a familiarity with how it works out for many different problems. It seems unlikely that understanding comes from thinking about some complicated extreme case, such as Schroedinger's cat. The approach we take here bypasses much of the question by starting with a single premise of wave-particle duality, from which follows all of quantum theory and its interpretation. Any remaining problems are made moot by the quantum-mechanical view that questions are meaningful only if there is some conceivable experimental way to test the answers.

In keeping with this pragmatic approach we will not emphasize mathematical derivations, nor carry out the detailed analysis of harmonic-oscillator states, hydrogen wavefunctions, or angular momentum, which are part of almost all texts in quantum theory. We treat the simplest case in detail, state the general results, and see that they are the plausible generalizations. We leave the more detailed analysis to other texts. The one we used for the course at Stanford was Kroemer's *Quantum Mechanics for Engineers and Materials Scientists* , which served very well.

Chapter 1. Foundations

1.1 The Premise

For the purposes of this course, quantum mechanics is based upon a single statement, called *the wave-particle duality*, or sometimes *complementarity*, which is :

Everything is at the same time a particle and a wave.

Simply figuring out how this apparently self-contradictory statement can be true will lead us to all of quantum theory. The meaning of "everything" will be made more precise at the beginning of Chapter 3 when we discuss Hamiltonian mechanics. We take the premise itself to be an absolute truth, but we generate approximate ways to deal with it, such as Schroedinger's Equation, or the more approximate tight-binding theory. Within these contexts we can make predictions, and the essential predictions have never been found to be wrong, though the approximations (such as the neglect of relativity for Schroedinger's Equation) make the results approximate.

This premise as applied to light and to electrons is quite familiar. Although light is certainly a wave described by Maxwell's Equations, it is absorbed only in quantum packages, as if it struck photographic film like a bullet. Although electrons are certainly point particles they are diffracted as waves by the grid formed by a crystal lattice. It is also true of a real bullet, and of the beads which make up a necklace. It is even true of the center of gravity of that necklace. It is also true of sound waves and even waves in water. It is also true about a spinning object, such as a top, where the wave-like behavior of the spin limits the states of rotation. The generality of the wave-particle duality will be expressed mathematically, and certainly more precisely, when we discuss Hamiltonian mechanics in Chapter 3.

This premise is a truly remarkable statement, remarkable first because it is so general. It is remarkable second because it has so many consequences. It is remarkable third because it is so difficult to imagine that it could be true. It is remarkable fourth because it is absolutely true; it has never failed an experimental test. It is remarkable finally because although it seems not to be able to answer some questions we would like to ask, these all involve answers for which there is no conceivable test. There is no theory which goes further than quantum theory. Thus it may be as close as we can get to the absolute truth.

At the same time we should point out that this is not the only way to formulate quantum theory. Heisenberg's matrix formulation does not depend upon postulating waves, appears to be totally different, but is in fact mathematically equivalent. We shall see more clearly how this can be when we treat the harmonic oscillator in Chapter 16 using only the fact the operator which represents momentum (Eq. (1.11) in the next section) cannot be interchanged with the position in an equation (in mathematical terms, they do not commute). Then we will obtain results equivalent to what we obtain in Section 2.5 using waves.

1.2 Schroedinger's Equation

We proceed from this premise, by asking how it could be true for a particle, such as an electron with mass m and charge $-e$ which we imagine for simplicity can move along a line in space. We shall see how we can define an average position for a wave, and then match the rate of change with time of this average with the velocity of a particle. If we are to describe a particle by a wave, there must be an amplitude which is a function of position x and time t. In fact a single amplitude, such as $\sin(kx)$, is not enough because it does not tell which way the wave is moving. Thus for a water wave we need not only the height of the water surface, but the velocity of the water motion; for a light wave we need both the electric field and the magnetic field. We require two amplitudes to describe the electron and we choose to make them the real and the imaginary part of a complex amplitude $\psi(x,t)$, as could also be done with a light wave or a water wave.

For a particle we must be able to discuss its position, and the best we can do with that for a wave is to have the amplitude be nonzero only over a limited region, a wave packet as shown schematically in Fig. 1.1a. We

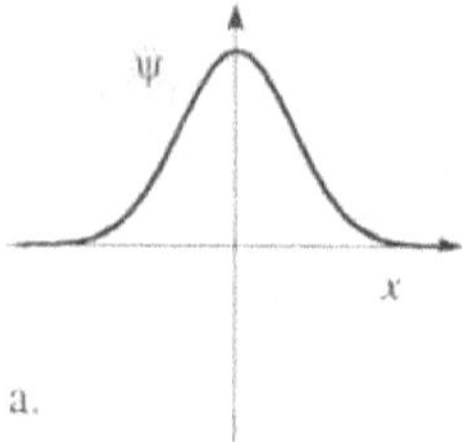

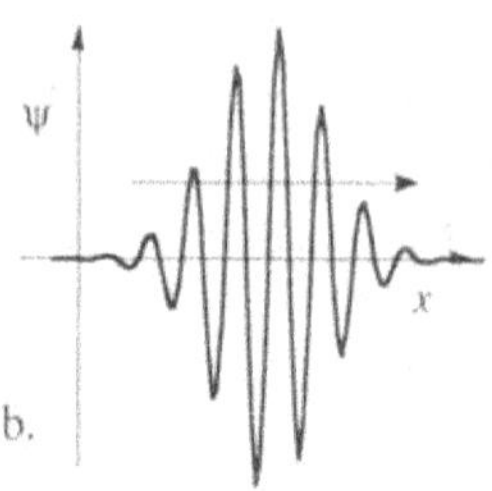

Fig. 1.1. a. The wavefunction $\psi(x)$ for a wave packet is nonzero only over a limited range of position x. It will spread in time. b. A packet multiplied by a plane wave e^{ik_0x} will also move, with velocity $\partial\omega/\partial k$, evaluated at $k = k_0$.

cannot say exactly where the particle represented by such a packet is, but we can specify an average position in terms of the squared amplitude (as we would do for a light wave with an energy density proportional to the square of the electric field plus the square of the magnetic field). In this case the sum of the squares of the real and the imaginary parts of ψ is $\psi^*\psi$ and we would specify the average position by weighting each position in proportion to $\psi^*\psi$ as

$$<x> \equiv \frac{\int \psi^*(x)\, x\, \psi(x)\, dx}{\int \psi^*(x)\psi(x)\, dx} \,. \tag{1.1}$$

This innocent-looking definition of what we will mean by averages, or *expectation values* , will turn out to be very far-reaching and the basic relation of the wavefunction to experiment. It is often interpreted to mean that $\psi^*(x)\psi(x)$ is the probability of finding the particle at x but we really only use the definition of an average from Eq. (1.1).

For discussing the motion of waves it is helpful to go to the familiar plane waves

$$\psi(x,t) = e^{i(kx - \omega(k)t)} \,, \tag{1.2}$$

written in terms of a wavenumber k and the angular frequency $\omega(k)$, which depends upon the wavenumber. We can expand any $\psi(x,t)$, such as the packet shown in Fig. 1.1a, in plane waves, a Fourier expansion of the wavefunction. It may not be necessary to follow the details, but the most familiar such expansion is of a Gaussian packet,

$$\exp[-(\Delta k)^2 x^2] \;\propto\; \int dk\; e^{ikx} e^{-[k/(2\Delta k)]^2} \,, \tag{1.3}$$

centered at the position $x = 0$, as in Fig. 1.1a. Δk is a constant representing the range of wavenumbers included in the expansion. The contribution of any small interval in the integration is a term proportional to Eq. (1.2), at $t = 0$, so each term will evolve with time according to Eq. (1.2).

Of more use will be a packet which is centered around some nonzero wavenumber, k_0. This is accomplished by replacing the Gaussian in Eq. (1.3) by $e^{-[(k-k_0)/2\Delta k]^2}$. The resulting packet is illustrated in Fig. 1.1b (in this case also centered away from $x = 0$). We then introduce the time-dependence factor $e^{-i\omega(k)t}$ from Eq. (1.2) to obtain

$$\psi(x,t) = \int dk e^{i(kx - \omega(k)t)} e^{-[(k - k_0)/2\Delta k]^2}. \tag{1.4}$$

Since the integrand is large only near $k = k_0$, we may approximate $\omega(k')$ by $\omega(k_0) + (\partial\omega / \partial k)(k - k_0)$ and multiply by $e^{i(k_0 - k_0)xt}$. [This "∂" of course means a *partial* derivative. It is no different from the usual derivative here because ω depends only upon the one variable, k.] Then we find that the packet is given by

$$\psi(x,t) = e^{i(k_0 x - \omega(k_0)t)} \int dk\, e^{i(x - (\partial\omega/\partial k)t)(k - k_0)} e^{-[(k-k_0)/2\Delta k]^2}. \tag{1.5}$$

We may change to a dummy variable $\kappa = k - k_0$ and we see that the integral itself is exactly the integral in Eq. (1.3) except that x is replaced by $x - \partial\omega/\partial k\, t$ so the center of the packet is not at $x = 0$, but at $x = \partial\omega/\partial k\, t$. Thus the packet is moving with velocity

$$v = \left.\frac{\partial\omega}{\partial k}\right|_{k_0}, \tag{1.6}$$

evaluated at $k = k_0$ as indicated. It is also multiplied by the plane wave $e^{i(k_0 x - \omega(k_0)t)}$.

This is a familiar result from any wave theory that the group velocity of a wave is given by the derivative of the frequency (in radians per second) with respect to wavenumber. This is the velocity we must associate with the particle in a state with wavenumber k_0. It is the rate that $\langle x \rangle = \int \psi(x,t)^* x\, \psi(x,t)dx / \int \psi(x,t)^* \psi(x,t)dx$ changes with time. The factors $e^{i(k_0 x - \omega(k_0)t)}$ and its complex conjugate cancel in the integral and do not affect the result. The *phase velocity*, ω/k, is of no physical significance and in fact we shall see that simply changing our zero of energy (or adding an unphysical constant potential) changes that phase velocity.

We now return to the particle point of view in order to identify properties of the wave with those of the particle. In classical (nonrelativistic) mechanics the energy of a particle in free space is $E = {}^1/_2 mv^2$, so the velocity is given by $(1/m)\,\partial E/\partial v$ or $\partial E/\partial p$ where the momentum is given by $p = mv$. This becomes clear as the natural and general choice when we review classical Hamiltonian mechanics in Chapter 3, where the Hamiltonian H is the energy written in terms of momentum and position, here $H = p^2/2m$. In any case, it must be true for a classical particle in free space that the velocity is given by

$$v = \frac{\partial H}{\partial p}. \tag{1.7}$$

Then we can only identify the wave description, Eq. (1.6), and the particle description, Eq. (1.7), if the energy H is a constant times the frequency and the momentum is the same constant (which we call Planck's constant $\hbar$) times the wavenumber,

$$H = \hbar\omega, \qquad p = \hbar k. \tag{1.8}$$

The value of Planck's constant must be determined from experiment, and it will be universal for *all* waves since we shall find that when different particles interact, or scatter from each other, the sum of wavenumbers afterward equals that before. If momentum $p = \hbar k$ is to be conserved, $\hbar$ must be the same for all waves. It is, in electron-volt seconds,

$$\hbar = h/(2\pi) = 6.6\times10^{-16}\ \text{eV-sec}. \tag{1.9}$$

The bar on the h indicates the division of a Planck's constant h, defined earlier, by 2π. Almost always our calculations will lead us to values which depend upon combinations of constants which are more useful if given in terms of electron-volts (eV) and Angstroms (Å),

$$\frac{\hbar^2}{m} = 7.62\ \text{eV-Å}^2, \qquad e^2 = 14.4\ \text{eV-Å}, \tag{1.10}$$

with m the mass of the electron and e the magnitude of its negative charge.

We developed the relations, Eq. (1.8), to relate the properties of a free particle to a plane wave $e^{i(kx - \omega(k)t}$, but to do that we needed to consider more general wavefunctions which were a combination of such plane waves, as in Eq. (1.4). As we continue with these more general wavefunctions, we may represent the momentum by an operator,

$$p = \frac{\hbar}{i}\frac{\partial}{\partial x} \tag{1.11}$$

and the energy by an operator,

$$H = -\frac{\hbar}{i}\frac{\partial}{\partial t} \tag{1.12}$$

since these are the operators, operating on the wavefunction which contains a sum of different plane waves, which will multiply each term by the value of momentum or energy appropriate to that plane wave. Thus to obtain the expectation value of the momentum of a particle with a wavefunction $\psi(x,t)$ we write in direct analogy with Eq. (1.1)

$$\langle p \rangle \equiv \frac{\int \psi^*(x,t) \frac{\hbar \partial}{i \partial x} \psi(x,t) dx}{\int \psi^*(x,t)\psi(x,t)\, dx} . \tag{1.13}$$

The expectation value of the momentum is the weighted average over that for all of the plane waves making up the wavefunction. Similarly the expectation value of the energy of the wave is

$$\langle H \rangle \equiv \frac{\int \psi^*(x,t) \frac{-\hbar \partial}{i \partial t} \psi(x,t) dx}{\int \psi^*(x,t)\psi(x,t)\, dx} . \tag{1.14}$$

These relations will turn out to have far-reaching consequences.

Once we are discussing wavefunctions which are not plane waves, we can also discuss systems for which there is a potential energy for the particle, $V(x)$, which is nonzero. Clearly the expectation value of the potential energy is

$$\langle V(x) \rangle \equiv \frac{\int \psi^*(x)\, V(x)\, \psi(x) dx}{\int \psi^*(x)\psi(x)\, dx} . \tag{1.15}$$

We now have two operators which represent the energy of the electron, $H = -(\hbar/i)\partial/\partial t$ and $H = p^2/2m + V(x)$, with the momentum operator given by Eq. (1.11). We may combine these to give an equation which will tell us the evolution of the wavefunction with time,

$$-\frac{\hbar}{i}\frac{\partial}{\partial t}\psi(x,t) = H\,\psi(x,t) = -\frac{\hbar^2}{2m}\frac{\partial^2\psi(x,t)}{\partial x^2} + V(x)\psi(x,t)\,, \tag{1.16}$$

which is called the *Schroedinger Equation* for a particle of mass m. It guarantees that the time-dependence of the expectation value of any operator is correctly given

We seem to have derived the equation, but really we have simply sought to find what must be true if the electron is to behave both as a particle and as a wave. That premise is exactly true, but we have made a number of approximations or guesses on the way. For one thing, we assumed nonrelativistic dynamics and if the electron energy is large enough we must certainly use special relativity to describe the dynamics. For a second, we needed at least two components (real and imaginary) for the wavefunction, but it turns out that the electron has a spin, and to describe the electron fully requires four components. These two features turn out to be intimately related. When Dirac (1926) sought to invent an equation for relativistic particles he found he needed four components, and the resulting particle showed an intrinsic angular moment of $\hbar/2$ which *is* the electron spin. The corresponding Dirac theory is a more complete theory of the electron, but we will not need the extra complexity and will proceed with the simpler Schroedinger representation with spin added as a separate feature.

While we accept these approximate features of the Schroedinger representation of the particle, we should point out that it is extraordinarily general. Though we invented it for an electron, it applies also to a proton with a different m , or a neutron, or an atom, a molecule, or a solid. It will apply not only to the translational motion of that solid but will generalize to the rotational motion. It also generalizes to a light wave, with mass equal to zero, to a sound wave, or to a water wave. It generalizes to everything. Again the full significance of this generality will only be clear when we present Hamiltonian mechanics in Chapter 3. There we will see for example that it generalizes to the center of gravity of an object. If that object is a doughnut, the center of gravity lies in a region of space where there is no dough. Thus where the corresponding wavefunction is largest, there may be no likelihood of finding any real material. Clearly, then, we have also invented this wavefunction. It has no independent existence, cannot be detected, and yet all of the predictions which we make using it will be correct.

1.3 Light Waves

Before going on with the consequences of our formulation, it may be helpful to redo it for light waves, which will also behave as a particle, but one without mass. It is again requiring the consistency of the wave and particle picture, but in this case it is the wave whose properties we know from classical physics.

To specify such a wave we will need to give the electric field, which requires three components (x, y, and z components), given as a function of position and time. That is not enough. Just as one function (the real part of ψ) was not enough to tell which direction the electron was moving, we must also specify the magnetic field as a function of position and time. The two are related through Maxwell's Equations. Later in this text it will be important to treat these fields in terms of a scalar potential $\phi(\mathbf{r},t)$ and a vector potential $\mathbf{A}(\mathbf{r},t)$, so it may be best to do that here for this very simple analysis. With no charges present, the vector potential is enough and in terms of it the electric field $\mathbf{E}$ and the magnetic field $\mathbf{H}$ are given by

$$\mathbf{E} = -\frac{1}{c}\frac{\partial \mathbf{A}}{\partial t}, \qquad \mathbf{H} = \nabla\times\mathbf{A}. \tag{1.17}$$

Then for a plane wave propagating in an x- direction we shall write

$$\mathbf{A}(\mathbf{r},t) = \mathbf{A}_0\,[e^{i(kx - \omega(k)t)} + e^{-i(kx - \omega(k)t)}], \tag{1.18}$$

where the amplitude $\mathbf{A}_0$ might lie in a z-direction and the direction of propagation is specified by the sign of ω. We have chosen to use complex exponentials because that will be the most convenient way to proceed when we treat light waves more completely later. The two terms are necessary because the fields from Eq. (1.17) are real quantities. Using Eq. (1.17) we see that the electric field in the z-direction is $-(2\omega A_0/c)\sin(kx - \omega(k)t)$ and the magnetic field lies in the y-direction and is $2qA_0\sin(kx - \omega(k)t)$.

If this wave is to be regarded as a particle, or a collection of particles, these particle must have, according to Eq. (1.8), energy $\hbar\omega$ and momentum $\hbar k$. We shall see this more explicitly in Chapter 18 when we treat the light wave as a harmonic oscillator. The relation between the frequency ω and the wavenumber k is $\omega = ck$ which gives us the counterpart of our starting description of the electron, $H = \mathbf{p}^2/2m$. It is $H = cp$ for the light particle, or *photon*. Substituting Eqs. (1.11) and (1.12) gives us the counterpart of the Schroedinger Equation, which for our wave is

$$i\hbar \frac{\partial A}{\partial t} = -i\hbar\ c \frac{\partial A}{\partial x} \ . \tag{1.19}$$

The $i\hbar$ cancels so no $\hbar$ appears in the wave equation. A more familiar form comes from writing $H^2 = c^2p^2$, which gives

$$\frac{\partial^2 A}{\partial t^2} - c^2 \frac{\partial^2 A}{\partial x^2} = 0. \tag{1.20}$$

This is exactly the result of writing one of Maxwell's Equations , $\nabla\times\mathbf{H} = -(1/c)\ \partial\mathbf{E}/\partial t$ (if there are no currents present), combined with Eqs. (1.17). $4\pi c j_z$ would appear in place of the zero on the right in Eq. (1.20) if there were currents present, with j_z the current density in the z-direction.

The treatment of light and of electrons is entirely parallel, as it must be in accordance with the wave-particle duality. For classical physics it is the wave description of light which is familiar and the particle picture of the electron, but both descriptions are appropriate in both cases. Different kinds of systems have different dynamical relations between momentum and energy and correspondingly different wave equations. Given the equations for light, we can introduce a refractive index which varies with position, and therefore a $c(\mathbf{r})$ and study the dynamics of the photons, or refraction of the light.

1.4. New Meaning for Potentials

The vector potential **A** which we introduced in Section 1.3 is an invention, just as the classical electrostatic potential ϕ , is an invention. In classical physics the vector potential only has meaning through the defining equations, Eqs. (1.17), which relate the observable fields to it. If we add a constant to the vector potential (or to the electrostatic potential) it does not change the fields and we regard it as a simple definition, like the definition of an origin to a coordinate system, x, y, z . It is playing for the photon the role played by the wavefunction ψ which we also regarded as an invention. It may not be surprising that in quantum mechanics these electrostatic and vector potentials take on real new meaning.

This meaning is associated with the Aharanov-Bohm Effect (Aharanov and Bohm (1959)). They proposed two experiments, which appeared to be paradoxes. One is for an electron wave, illustrated in Fig. 1.2. We imagine an electron packet moving from the left, and then being split into two packets, which finally recombine on the right and produce a diffraction pattern on a luminescent screen to the right, just as light - or light wave packets - will form a diffraction pattern with constructive and destructive interference on a film in a two-slit experiment. Now we add two Faraday cages, conducting cages shown by dashed lines, and while the packets are entirely within the cages we apply a potential difference between them. No electric field is seen by either packet; their relative potential is simply shifted. However, according to the Schroedinger Equation, Eq. (1.16), this constant shift will advance the phase of one wavepacket [$i\hbar\,\partial\psi/\partial t = H\psi$ means that the rate of change of the phase is equal to the energy divided by $\hbar$ as we shall see more completely in Eq. 1.22).] relative to the other. If we keep it on long enough to shift the relative phase by π, and then again put the potential difference to zero, the points on the screen at which constructive and destructive interference occurs will be interchanged. The potential is removed before the packets reach the cage surfaces but the interference pattern is modified even though no field has ever been felt by the packets. In this sense the potential, or potential difference, has taken on new physical meaning with measurable consequences. The electrostatic potential which was invented to describe electric fields obtains new meaning in quantum mechanics. It is natural at first to try to dismiss the paradox by saying that there will be small leakage fields within the cages, but that is not the point. Effects such as that can be made as small as one likes and the physical consequences of the phase shift remain large.

There is an important message from this exercise. Once one is sure that the argument is correctly made, there is no real need to test it experimentally, any more than one needs to test a proposed perpetual motion machine once one sees that it violates the first law of thermodynamics. One does better to adjust one's intuition. Our feeling that only the electric field has consequences is generalized here to an electron which in some sense is in two places, and then the potential difference between the two places has consequences. Considerable experimental effort goes into displaying the

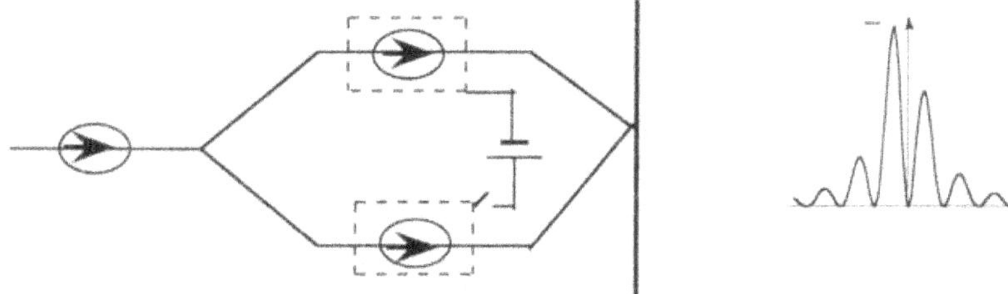

Fig. 1-2. The Aharanov-Bohm Paradox. An electron packet from the left is split into two packets, which pass through two Faraday cages. A potential difference applied to the cages, while the packets are in them, will advance the phase of one packet relative to the other and shift the diffraction pattern, though no field ever exists where the wavefunction is nonzero.

surprising consequences of quantum theory, but it is clear that one is not "testing" quantum mechanics, but testing one's understanding of quantum mechanics. We can think of the starting postulate as an absolute truth, certainly on the scale of the truth of conservation of energy.

The second Aharanov-Bohm Paradox split an electron packet so that the packets went on opposite sides of a long coil containing a magnetic field. The field in such a long coil is contained entirely within the coil so that neither packet ever feels a magnetic field. However, the vector potential is nonzero outside the coil and its presence will also shift the fringes. This "fictitious" vector potential has physical consequences. Similarly our fictitious wavefunctions will have important physical consequences which we explore in this text.

1.5 Measurement

The discussions above have touched on the question of measurement, which receives considerable attention from physicists. Our more pragmatic view is based upon Eq. (1.1) which tells us how the average of many measurements (in this case of position x) is predicted using $\psi(x)^*\psi(x)$. Quantum mechanics can tell us what sets of circumstances are consistent with each other. It tells us what we will see on the screen in the experiment shown in Fig. 1-2. From a practical point of view, that is what is needed. We should not speculate "which path the electron followed". We could set up another experiment which would also detect which way it went, and we would predict, and find, that the interference pattern would disappear, as we shall see in detail in Section 23.4.

People have sought ways to avoid the problem, as in the consideration of fringing fields discussed above for the Aharanov-Bohm experiment, by thinking of many electrons interfering with each other. However, that experiment can be done with so few electrons passing through per second that there is almost no chance of two electrons being in the apparatus at once, and the same result is obtained. It is again certainly better to adjust our intuition to fit the truth, rather than the other way around.

We may make a classical analogy as to how quantum mechanics tells us what is consistent, though it may or may not be helpful to follow such an analogy. Imagine walking past the window of a pool hall and noting a tall and a short man playing pool. The tall man is about to hit the cue ball aimed at the five ball. You estimate that it is an easy shot and the five ball should go in the corner pocket. From where the cue ball will then be he will probably choose to put the three ball in the side pocket. After you pass the window you recognize that maybe the five ball will *not* go in and an entirely new scenario arises. The short man will pick up his cue and will probably seek to put the six ball in the other side pocket, etc. You could carry the first scenario, and the second scenario, as far as you like (with decreasing certainty of the details) and what you are really doing is determining sets of circumstances which are consistent with each other, but inconsistent with the circumstances of another scenario.

If you turn around and pass the window again, you may note that the five ball is still on the table: *scenario one has become irrelevant.* Indeed the short man is standing at the table, shooting at the six ball.

The scene is obviously chosen to indicate that the well-known "collapse of the wavefunction", which is supposed to occur when someone makes a measurement, is not a quantum phenomenon, but one of everyday classical experience. Schroedinger's cat should not to be of concern. More importantly, this analogy sets the stage: all we can do in quantum mechanics is to estimate sets of consistent circumstances, or scenarios, and the likelihood of each occurring. When we do an experiment, we eliminate - or make irrelevant - a large number of other scenarios. No other theory may ever do more than that.

1.6 Eigenstates

We have found that the observables position, momentum, and energy, are represented by operators on the wavefunction, and that a statistical average of measurements of such an observables O for a given wavefunction is given by $\langle O\rangle = \int\psi^*O\psi\, d^3r / \int\psi^*\psi\, d^3r$ as in Eqs. (1.1) and Eqs. (1.13) through (1.15), but now written for wave functions in three-dimensions, $\psi(\mathbf{r},t)$. The d^3r indicates a volume integral. We are thinking of electrons, but this applies to the energy-density for light waves which can produce diffraction patterns such as we observed for electrons on the screen in Fig. 1.2. The mathematical consequences of this statement about statistical averages are extraordinary, and we turn to them next.

We do not focus on the mathematical details until we need them, but must mention that we always assume some set of boundary conditions on all wavefunctions, such as the condition that the wavefunction be nonzero only inside some surface, and therefore zero on the surface. It is also necessary that the operators be *Hermitian*, which means that for any two wavefunctions they satisfy $\int\psi_1^*(\mathbf{r})O\psi_2(\mathbf{r})d^3r = \{\int\psi_2^*(\mathbf{r})O\psi_1(\mathbf{r})d^3r\}^*$, which will always be true here and is readily verified for the operators we have introduced, using the boundary condition such as we just gave (partial integrations are required to prove it for the momentum and energy operators).

We now state a mathematical fact that for any such operator O, with appropriate boundary conditions, there exist ***eigenstates***, functions which satisfy $O\psi_j(\mathbf{r}) = \lambda_j\psi_j(\mathbf{r})$. They are analogous to normal modes of a violin string, and we shall follow that analogy rather than providing the detailed proofs. The eigenvalues λ_j will be real numbers. The most important such operator for us is the *Hamiltonian* operator representing the energy, for which the condition is

$$H\psi_j(\mathbf{r}) = \varepsilon_j\psi_j(\mathbf{r}). \tag{1.21}$$

We see that this looks like the Schroedinger Equation, Eq. (1.16), with $-\hbar/i\ \partial/\partial t$ replaced by the energy eigenvalue, ε_j. For this reason it is also called the *time-independent Schroedinger Equation*. In fact, any wavefunction which satisfies Eq. (1.21) can be seen from Eq. (1.16) to have a very simple time dependence given by

$$\psi_j(\mathbf{r},t) = \psi_j(\mathbf{r})e^{-i\varepsilon_j t/\hbar}. \tag{1.22}$$

In this way also this is closely analogous to the normal modes of a violin string, which are distortions which exactly retain their shape, but change in phase, or amplitude, with time as $\cos(\omega t + \delta)$. The displacements in the normal modes of a string of length L can be written $u_n(t)\sin(xn\,\pi/L)$ with n any integer, x the distance along the string, and $u_n(t)$ the amplitude for the n'th mode, varying with time as $u_n(t) = u_n\cos(\omega_n t + \delta_n)$.

It is best always to *normalize* our wavefunctions,

$$\int \psi^*(\mathbf{r},t)\psi(\mathbf{r},t)\, d^3r = 1. \tag{1.23}$$

This is called the *normalization condition* and sets the scale of the wavefunction so that $\psi^*(\mathbf{r},t)\psi(\mathbf{r},t)$ is the *probability density.* Then $\psi^*(\mathbf{r},t)\psi(\mathbf{r},t)d^3r$ is the probability that the electron with that wavefunction will be found in the small volume d^3r at the position $\mathbf{r}$ at the time t. The probability density may of course shift with time, but if the wavefunction is normalized at one time, it will remain normalized. This is obvious for an energy eigenstate from Eq. (1.22), and follows in general from substituting the time dependence of each factor in Eq. (1.23) from the Schroedinger Equation, Eq. (1.16), and using the Hermiticity condition on the Hamiltonian. This normalization does not change any of the results we obtained earlier but simplifies the formulae for the expectation values in Eqs. (1.1), (1.13), (1.14) and (1.15) by making the denominator unity.

The set of all eigenstates for an operator (and the Hamiltonian operator in particular) form a complete set; any function satisfying the same boundary conditions can be expanded in this set. Further, these eigenstates are orthogonal to each other, written in terms of a *delta function* as

$$\int \psi_j{}^*(\mathbf{r})\psi_i(\mathbf{r})\, d^3r = \delta_{ij}\,, \quad \text{with } \delta_{jj} \equiv 1,\ \delta_{i\neq j} \equiv 0. \tag{1.24}$$

The $i = j$ equation is the normalization condition. Again, this is just as for a normal mode where any displacement of the string, $\delta z(x)$, which is zero at both ends, can be expanded as $\delta z(x) = \Sigma_n u_n \sin(xn\pi/L)$. This is one of the principal uses of the normal modes. If one determines the frequency ω_n of each normal mode, one can expand the displacements at time $t = 0$ in normal modes and determine the future displacements directly. [Actually, one must expand both the displacements and the velocities in normal modes, by writing $\delta z(x,t) = \Sigma_n u_n \sin(xn\pi/L) \cos(\omega_n t + \delta_n)$ and fitting δ_n and u_n to the displacement and velocity at time $t = 0$. The same form then gives the displacement and velocity at any later time. This is the counterpart of expanding the real and imaginary part of the wavefunction in terms of energy eigenstates. For the analogy we do not need this, but we return to such an analysis when we treat lattice vibrations in solids in Chapter 15. Also at that time we shall discuss normalization for the normal modes.] Similarly we can expand electronic wavefunctions in energy eigenstates and immediately obtain the expansion for future times.

If the wavefunction of an electron $\psi(\mathbf{r})$ is an eigenstate of the Hamiltonian, then we know that if we measure the energy we will, on the average, obtain

$$\langle H \rangle = \int \psi_j{}^*(\mathbf{r}) H \psi_j(\mathbf{r})\, d^3r = \varepsilon_j\,. \tag{1.25}$$

But even more importantly, the mean-square deviation from that average,

$$\langle (H-\varepsilon_j)^2 \rangle = \int \psi_j{}^*(\mathbf{r})(H-\varepsilon_j)^2 \psi_j(\mathbf{r})\, d^3r \equiv 0, \tag{1.26}$$

which we obtain by expanding the expression in parentheses and using Eq. (1.21) to evaluate each term. This means that we would *always* measure exactly that energy ε_j. In just this way if we use a mode analyzer to determine the frequency of a string vibrating in a single normal mode, we obtain only the single mode frequency. For the electron we say that the electron is in an *energy level*, or that it occupies an energy eigenstate.

More generally, an electron can have any wavefunction, and if we expand it in energy eigenstates, we may obtain the time dependence as

$$\psi(\mathbf{r},t) = \Sigma_j \, a_j\psi_j(\mathbf{r})e^{-i\varepsilon_j t/\hbar} \, . \tag{1.27}$$

If we measure the energy at some time t we obtain on average

$$\begin{aligned} \langle H\rangle &= \int\psi^*(\mathbf{r},t)H\psi(\mathbf{r},t)\,d^3r = \Sigma_{i,j}a_j^*a_i\int\psi_j^*(\mathbf{r})H\psi_i(\mathbf{r})\,d^3r \\ &= \Sigma_j \, a_j^*a_j \, \varepsilon_j \, . \end{aligned} \tag{1.28}$$

In the last step we wrote $H\psi_i(\mathbf{r})$ as $\varepsilon_i\psi_i(\mathbf{r})$, took the ε_i out of the integral, and used Eq. (1.24) to eliminate all terms with i differing from j.

That is, we can say that we obtain the average of the energy eigenvalues, weighted by the *probability* $a_j^*a_j$ that the j'th state is occupied. We can say more than that. It is easy to show that we will always measure exactly one of the eigenvalues, and so that $a_j^*a_j$ really is the probability of finding the electron with exactly that energy ε_j. [A way to show this is to evaluate the product $\langle\Pi_i(H-\varepsilon_i)^2\rangle$ using a wavefunction $\Sigma_j \, a_j\psi_j(\mathbf{r})$ to see that there is one factor of zero for every term, but if we ever measured an energy H different from all eigenvalues that product would be nonzero.] This is the same again as for a violin string. For an arbitrary vibration, a mode analyzer will detect vibrations at each of the normal-mode frequencies, but none between. Electrons only appear in energy levels because they are like other waves. $\psi(\mathbf{r})$ can be anything, but if we make a measurement we only find it in one of the eigenstates of the operator for the variable we are measuring.

1.7 Boundary Conditions

We return briefly to boundary conditions which we apply to the wavefunctions. Up till now we have simply limited ψ to one region of space by requiring the wavefunction to go to zero on the surrounding surface. These are called *vanishing boundary conditions* . For one dimension they are like those on the violin string, for which the displacements must vanish at the two ends, at the stock and the bridge. Often a more useful set are *periodic boundary conditions* , which in one dimension requires that both the value and the slope be the same at the two ends, as illustrated in Fig. 1.3. This corresponds to bending the line on which an electron moves into a ring and requiring a continuous (or single-valued) wavefunction and no cusps (which would correspond to infinite kinetic energy, $-(\hbar^2/(2m))\partial^2\psi(x)/\partial x^2$,

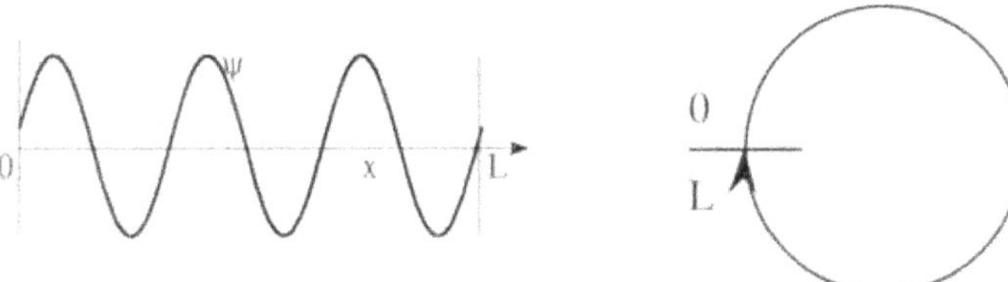

Fig. 1.3. Periodic boundary conditions on a wavefunction require that $\psi(L) = \psi(0)$ and $\partial\psi(x)/\partial x\,|_L = \partial\psi(x)/\partial x\,|_0$. It corresponds to bending the system into a ring of circumference L.

locally). This is directly generalized to a three-dimensional box of dimensions L_x, L_y, and L_z, by requiring these same conditions on opposite faces of the box. Some video games have such boundary conditions such that an object leaving one side of the screen appears on the other. It allows us to use plane waves as in Section 1.2. It is important that boundary conditions be placed on the outside of the system, but then most properties are quite insensitive to which set are used. Periodic boundary conditions eliminate the surface of the system; the boundary planes are no different from other planes in the system. They are usually appropriate unless we really are interested in the properties of a surface.

1.8 Sound Waves

Application of boundary conditions is familiar for sound waves, and many other problems. In some ways sound waves, in gases and in solids, are easier to think about than light waves, or wavefunctions, and we might introduce them at the outset so that we can use them for illustration. They are of course subject to the wave-particle duality as is everything else, and we shall see how they are treated as particles, phonons, in Chapter 16.

In a compressional wave propagating along an x-axis we may write displacements $u(x, t)$ of the medium in the x-direction as

$$u(x, t) = u_0(e^{i(qx - \omega t)} + e^{-i(qx - \omega t)}) \qquad (1.29)$$

as for light in Eq. (1.18). This gives rise to a dilatation (a local fractional volume change) equal to the first derivative of $u(x, t)$ with respect to x and a change in local pressure equal to the bulk modulus B times the negative of that local dilatation. That change in pressure exerts a force per unit area on a slab of the medium of thickness δx equal to δx times the negative of the derivative of that pressure with respect to x, or $-Bq^2u(x, t)\,\delta x$. That force per unit area is equal to the acceleration of that slab, $\partial^2 u(x, t)/\partial t^2$, times its mass per unit area $\rho\delta x$ so

$$\rho\frac{\partial^2 u(x,t)}{\partial t^2}\delta x = B\frac{\partial^2 u(x,t)}{\partial x^2}\delta x$$
$$-\omega^2\rho u(x, t)\,\delta x = -Bq^2u(x, t)\,\delta x. \tag{1.30}$$

or

$$\omega = \sqrt{\frac{B}{\rho}}\,q\,. \tag{1.31}$$

This corresponds to a speed of sound equal to $\sqrt{B/\rho}$. For an ideal gas, the adiabatic $B = {}^5/_3\,P$, with P the pressure (e., g., Kittel and Kroemer (1980) p. 433). Sound actually cannot propagate in an ideal collisionless gas, but if we assume enough collisions to reduce the mean free path well below the sound wavelength we might still use the ideal gas value for B. Then the speed of sound is $\sqrt{5/3}$ times the root-mean square thermal velocity of atoms or molecules in this classical ideal gas.

If a gas is confined to a pipe with closed ends we could construct normal modes using vanishing boundary conditions on the displacements $u(x, t)$ at the ends. If it was open at both ends we could use vanishing boundary conditions on the pressure at both ends. These modes of frequency ω will have quantized energy in units of $\hbar\omega$ and will be absorbed as discrete phonons, in spite of our classical-mechanical, granular view of their nature, emphasizing again the extraordinary generality of the wave-particle duality.

Chapter 2. Simple Cases

Having set up the general rules for quantum mechanics, we turn to the simplest systems, the simplest of all being a free electron, as discussed in Section 1.1.

2.1 Free Electrons in One Dimension

We begin with an electron moving along a line of length L, and find the energy eigenstates. We take the potential $V(x) = 0$ so the energy-eigenvalue equation, Eq. (1.21), becomes

$$\frac{-\hbar^2}{2m}\frac{\partial^2\psi}{\partial x^2} = \varepsilon\,\psi\,. \tag{2.1}$$

The general solution is of the form $\psi(x) = ae^{ikx} + be^{-ikx}$ with energy $\varepsilon_k = \hbar^2k^2/2m$. If we apply vanishing boundary conditions at $x = 0$ and L, the eigenstates become $\sqrt{2/L}\,\sin kx$ with kL an integral multiple of π . The amplitude $\sqrt{2/L}$ gives a normalized state, satisfying Eq. (1.24). These eigenstates illustrate all of the features which we discussed in Chapter 1. The integer zero, corresponding to $k = 0$ is not allowed since it gives a wavefunction equal to zero, and that is not a state.

We may alternatively apply periodic boundary conditions on the same line. Then $\sqrt{1/L}\,e^{\pm ikx}$ are eigenstates with kL equal to an integral multiple of 2π. It is an important point that any linear combination of e^{ikx} and e^{-ikx} (which can again be normalized) is also an eigenstate if its kL is an integral multiple of 2π. In general if we have two eigenstates of the same energy, any linear combination of them is also an eigenstate, which follows immediately from the eigenvalue equation, Eq. (1.21). For periodic boundary conditions, the state with $k = 0$ is $\psi(x) = \sqrt{1/L}$ and is an allowed state.

We have sketched the energy $\varepsilon_k = \hbar^2k^2/2m$ in Fig. 2.1 and indicated the states satisfying periodic boundary conditions by diamond solid dots, continuing to arbitrarily large positive and negative values of k. The states satisfying vanishing boundary conditions are indicated by crosses. No states are indicated with negative k for vanishing boundary conditions since such states are the same as the corresponding states with positive k, though perhaps with a different sign of the normalization constant. The two are not orthogonal to each other and are not to be considered different states. However these states with vanishing boundary conditions are spaced half as far apart in energy so that over a sizable energy range there are approximately the same number of states.

Each state can accommodate one or two electrons (of opposite spin if two) when many electrons are present, as we shall see for atoms in Section 4.2 and prove in general in Section 10.5. For a finite length L and a finite number of electrons, the total energy, obtained as the sum of the energies of the occupied states, will depend upon which boundary conditions are

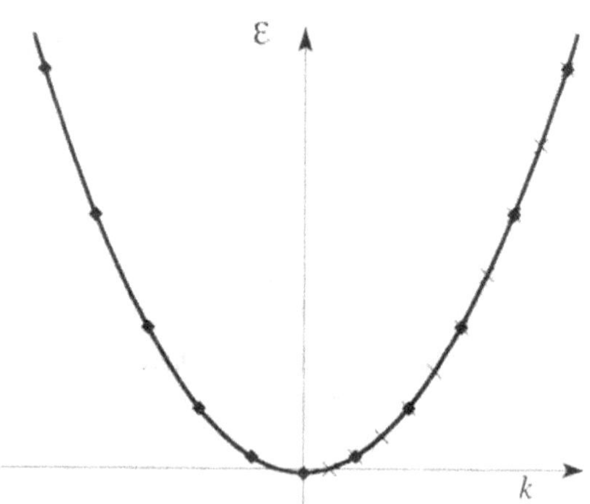

Fig. 2.1. The energy, as a function of wavenumber, for an electron (or other particle) moving in one dimension. The states at wavenumbers indicated by a diamond satisfy periodic boundary conditions. Those with an × satisfy vanishing boundary conditions, and changing the sign of k does not yield a different state.

utilized. This is illustrated by the π-states in a benzene ring in Problem 2.1. These states, which will be discussed in detail in Section 6.1 in the text, are approximated as six free electrons in a ring 15.7 Å in circumference. The benzene ring is broken in Problem 2.1 so that it becomes a line of the same length, with vanishing boundary conditions, and the change causes an increase in energy which may be thought of as an estimate of the energy to break a π-bond.

2.2 Free Electrons in Three Dimensions

The one-dimensional system just discussed generalizes directly to three dimensions. The wavefunction is now of course a function of the coordinates x, y, and z. With $V(\mathbf{r}) = 0$ (or constant), the energy eigenvalue equation, $H\psi(\mathbf{r}) = \varepsilon\psi(\mathbf{r})$, is

$$\frac{\hbar^2}{2m}\left(-\frac{\partial^2}{\partial x^2}-\frac{\partial^2}{\partial y^2}-\frac{\partial^2}{\partial z^2}\right)\psi(x,y,z)=\varepsilon\psi(x,y,z)\,. \tag{2.2}$$

We may make an important general point about such an equation. When we seek an eigenstate of many variables, $\psi(x_1, x_2, ...)$, and the Hamiltonian may be written as a sum of individual Hamiltonians, one for each variable, $H(x_1, x_2, ...) = H_1(x_1) + H_2(x_2) + ...,$ then we can obtain a product solution,

$$\psi(x_1, x_2, ...) = \psi_1(x_1)\psi_2(x_2)... \tag{2.3}$$

with each of the factors obtained from its own eigenvalue equation,

$$H_1(x_1)\psi_1(x_1) = \varepsilon_1\psi_1(x_1), \text{ etc.,} \tag{2.4}$$

and the eigenvalue for the state $\psi(x_1, x_2, ...)$ is the sum of the individual eigenvalues,

$$\varepsilon = \varepsilon_1 + \varepsilon_2 + ... \tag{2.5}$$

This important result is readily verified by substituting the product form into the multidimensional eigenvalue equation with the eigenvalue from Eq. (2.5) on the right. Then for the term on the left with $H_1(x_1)$ and the term with ε_1 on the right, all other factors ψ_2, ψ_3, etc., may be canceled. Thus if Eqs. (2.4) is satisfied, we have shown the equality term-by-term.

One very important consequence of such a factorization is that if we consider a system of many electrons, each moving independently in a potential field, the many-electron wavefunction can be factored into one-electron wavefunctions, and the energy eigenvalue for the system is equal to the sum of the energies of the individual electrons. This can never really be exactly true since the electrons interact with each other through terms in the Hamiltonian such as $e^2/|\mathbf{r}_1 - \mathbf{r}_2|$ which are not separable as individual Hamiltonians. However, this *one-electron approximation* of assuming a separation is the fundamental approximation for almost all treatments of many-electron systems and will be heavily used in this book. We in fact already made it in discussing the benzene molecule of Problems 2.1 and 2.2, where we took the energy of the π-electrons to the be sum of their individual energies, and this approximation is basic to our understanding of electronic structure.

We apply periodic boundary conditions on a box, as shown in Fig. 2.2, and for the free-electron gas in three dimensions this factorization allows us to generalize the one-dimensional energy eigenstates immediately as

$$\psi(x,y,z) = \frac{1}{\sqrt{L_x L_y L_z}} e^{ik_x x} e^{ik_y y} e^{ik_z z} = \frac{1}{\sqrt{\Omega}} e^{i\mathbf{k}\cdot\mathbf{r}} , \tag{2.6}$$

where Ω is the total volume of the system. The energy eigenvalue of each is $\varepsilon_k = (\hbar^2/(2m))(k_x{}^2 + k_y{}^2 + k_z{}^2) = \hbar^2 k^2/(2m)$. The periodic boundary conditions restrict the wavenumbers, as in one dimension, to

$$k_x L_x = 2\pi n_x, \quad k_y L_y = 2\pi n_y, \quad k_z L_z = 2\pi n_z. \tag{2.7}$$

This makes a grid of allowed wavenumbers, with each grid cell size equal to

$$\delta k_x \delta k_y \delta k_z = \frac{2\pi}{L_x}\frac{2\pi}{L_y}\frac{2\pi}{L_z} = \frac{(2\pi)^3}{\Omega} . \tag{2.8}$$

For macroscopic systems the spacing of successive grid points is so small that the allowed wavenumbers form an almost continuous set. Had we similarly taken a much larger L for Fig. 2.1, the points representing allowed

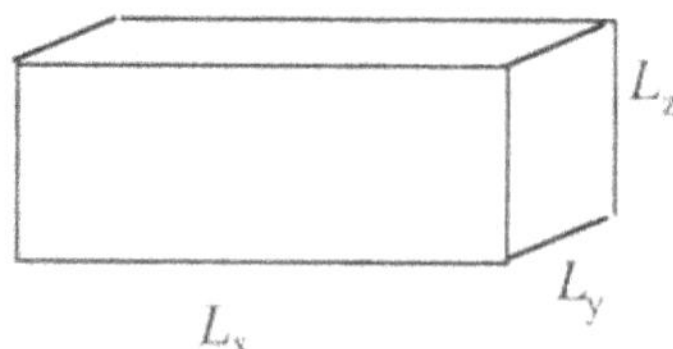

Fig. 2.2. Periodic boundary conditions are applied on a large box, so that each component of the wavenumber must be an integral multiple of 2π divided by the corresponding dimension.

states would have been very much closer to each other. When there is such an almost continuous variation from state to state, a band of energies is formed, and the ε_k when it arises in a solid is called an *energy band.*

The number of states in a volume of wavenumber space d^3k is then obtained by dividing that volume by the cell volume, as

$$\delta N = \frac{\Omega}{(2\pi)^3}\, d^3k. \tag{2.9}$$

Thus if we had N free electrons confined to a box of volume Ω, the lowest-energy state would place two electrons (one of each of the two spin states, as we shall prove in Chapter 10) in every state with wavenumber less than some k_F (the *Fermi wavenumber*) and

$$\frac{2\Omega}{(2\pi)^3}\frac{4\pi k_F{}^3}{3} = N \quad \text{or} \quad k_F{}^3 = 3\pi^2\frac{N}{\Omega}, \tag{2.10}$$

with N/Ω the electron density. (As suggested by the form, this is independent of size or shape of the box). The surface dividing the regions of wavenumber space between occupied and empty states is called the *Fermi surface* , and in this case is a sphere. In Problem 2.3, we obtain the radius, k_F, for that sphere, and the Fermi energy $E_F = \hbar^2 k_F{}^2/2m$, for Na, Mg, and Al, which have 1, 2, and 3, respectively, free electrons per atom. It is a central number determining the electronic properties of these simple metals.

Another important quantity is the *density of states,* $n(\varepsilon)$ or $D(\varepsilon)$, the number of states per unit volume and per unit energy (including two spin states for each wavenumber . We calculate it as twice the number of allowed wavenumbers in a spherical shell of thickness δk, or energy interval $\delta\varepsilon = \delta k/(\partial\varepsilon_k/\partial k)$. It is $\delta N = 2[\Omega/(2\pi)^3]4\pi k^2\delta k = 2[\Omega/(2\pi)^3]4\pi k^2\delta\varepsilon/(\hbar^2 k/m)$, corresponding to a density of states,

$$n(\varepsilon) = \frac{\delta N}{\Omega\delta\varepsilon} = \frac{1}{2\pi^2}\left(\frac{2m}{\hbar^2}\right)^{3/2}\sqrt{\varepsilon}\,. \tag{2.11}$$

This density of states, evaluated at the Fermi energy, will determine, for example, the electronic specific heat of the metal. In Problem 2.4 we redo this derivation for electrons in a plane to get the density of states (again including spin) per unit energy and per unit *area* .

We can also of course obtain the total kinetic energy, per unit volume, of the electron gas by integrating $\int \varepsilon\, n(\varepsilon)\, d\varepsilon$ from zero to the Fermi energy.

The integrand there is proportional to the $^3/_2$ power of energy while the integrand to obtain the total number of electrons is proportional to the $^1/_2$ power of energy, so when we obtain the average energy per electron the integrations give factors of $^5/_2$ and $^3/_2$ and the average energy per electron is

$$<\varepsilon_k> = \frac{E_{tot}}{N} = \frac{3}{5} E_F. \tag{2.12}$$

We have used periodic boundary conditions, as is usual, for discussing the free-electron gas. It eliminates the effects of the surfaces and our resulting parameters depend only upon volume. We can also apply vanishing boundary conditions to the box shown in Fig. 2.2. As in the one-dimensional case this decreases the spacing between allowed wavenumbers, but only allows positive k_x, k_y, and k_z so the Fermi energy, the density of states, and total energy come out approximately the same when the system is large. If we proceed more carefully we will obtain additional terms, as we did in Problem 2.2 for the one-dimensional case, which are proportional to the area of the box and can be interpreted as contributions to the surface energy. We consider such terms next.

2.3 Quantum Slabs, Wires, and Dots

Periodic boundary conditions allowed us to eliminate the effects of surfaces, but sometimes we are interested in the surfaces themselves or wish for example to discuss very thin sheets of material. Then we might take L_z very small and take vanishing boundary conditions at $z = 0$ and L_z, but take the other two dimensions large and use periodic boundary conditions on the lateral surfaces enabling us to eliminate any effects of the lateral surfaces. We then can construct free-electron states which are of the form

$$\psi(x,y,z) = \sqrt{\frac{2}{L_x L_y L_z}}\, e^{ik_x x} e^{ik_y y} \sin(n\pi z/L_z)\,. \tag{2.13}$$

This is again normalized and we think of L_z as quite small but the other dimensions large. The energies of the states are given by

$$\varepsilon_k = \frac{\hbar^2\pi^2}{2mL_z^2} n^2 + \frac{\hbar^2(k_x^2 + k_y^2)}{2m}\,. \tag{2.14}$$

With L_z small this is ordinarily described as a set of *sub-bands*, numbered by n , each having a two-dimensional dispersion in the y- and z-directions. We note that if L_z is small, and there are not too many electrons, all of the occupied states will lie in the first sub-band. Such a system really becomes a two-dimensional system as we treated in Problem 2.4. In quantum mechanics often the motion in some of the dimensions is pushed to such high energies that those dimensions become irrelevant and we have truly lower-dimensional systems. In classical physics a gas of atoms will have the same energy and specific heat even if it is confined to a very thin slab. We see in Problem 2.4 that the density of states retains a constant value at low energies, in contrast to Eq. (2.11). We shall see in Section 18.4 how this is used in solid-state lasers to enhance the laser performance.

At intermediate thicknesses the total density of states for a free-electron gas, defined for Eq. (2.11), is the sum of the density of states for each of the subbands, each subsequent subband giving a constant density of states beginning at successively higher energies, as illustrated in Fig. 2.3. Note the sum of these matches the free-electron density of states of Eq. (2.11) just after each rise. If we doubled the slab thickness a new step would arise between each old one, all being half as high, and this would again be true. In the limit of very thick slabs the slab curve approaches the free-electron curve, as we would expect.

A bulk metal, corresponding to a very thick slab, as in Problem 2.3, has a particular Fermi energy as calculated. We might mark that on the parabola in Fig. 2.3. If this same metal were deformed into a thin slab, with the density of states represented by the stepped function in Fig. 2.3, the electrons could no longer be accommodated below that same Fermi energy because the density of states is generally below the bulk value, so we would need to

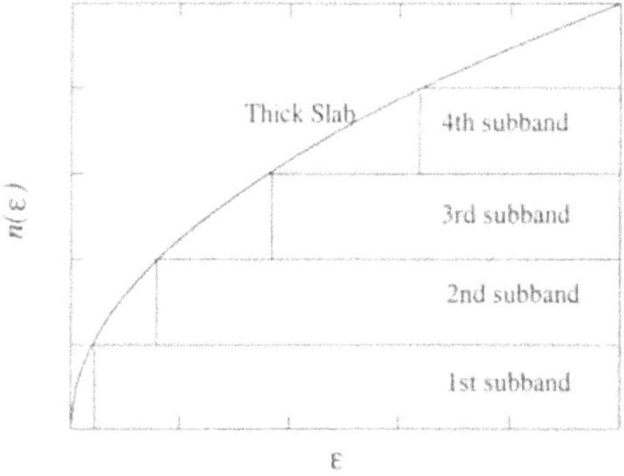

Fig. 2.3. The density of states for a slab is a sum of the densities of states for each subband, giving a stepped density of states. As the slabs are made thicker the steps are more frequent and not so high, approaching closer to the parabolic density of states shown for a very thick slab, given in Eq. (2.11).

fill the states to a higher Fermi energy, and the energy of the electron gas would be higher, just as it was when we cut the benzene ring in Problem 2.2. A calculation of the energy for a free-electron gas in such a slab at intermediate thicknesses, L_X , was made for example by Edwards and Mathon (1991), Edwards, Mathon, Muniz, and Phan (1991), and more recently by Harrison (1999). It is important in such a calculation to hold the number of electrons fixed as we change the boundaries, which will mean that the Fermi energy at which we stop filling will vary with thickness L_Z, but will always be higher than for the thickest slabs.

Holding the number of electrons per unit volume fixed, by increasing the lateral area in proportion to $1/L_Z$, we obtain the total energy as a function of thickness L_Z. The term in the energy independent of L_Z is the $\langle\varepsilon\rangle$ of Eq. (2.12) times the number N of electrons. The term proportional to lateral area (and thus to $1/L_Z$) can be divided by two and each half associated with a surface energy; it turns out to be given by given by $k_F{}^2E_F/(8\pi)$. When we treat electronic states later in terms of tight-binding theory we shall see that it is appropriate to set the vanishing boundary condition at a full interplanar atomic distance s beyond the last plane of atoms, rather than at half that distance which we tacitly assumed here by holding the lateral area times L_Z fixed. Then there is a reduction in the energy from this relaxation of the surface condition leading to a surface energy of (Harrison (1999), p. 721)

$$E_{surf.} = \frac{k_F{}^2E_F}{8\pi}\left(1 - \frac{8k_Fs}{15\pi}\right). \tag{2.15}$$

We may also subtract this surface energy, and the bulk energy of Eq. (2.12), from the total energy, leaving only terms which drop off faster than $1/L_Z$ at large L_Z . We may associate this remainder with an interaction between opposite surfaces. Such an interaction is shown in Fig. 2.4, showing the oscillations which arise as successive sub-bands cross the Fermi energy.

These turn out also to be important and were the reason why Edwards and coworkers carefully calculated the surface energy. An oscillatory coupling between the magnetization of two iron crystals, through a simple metal such as copper, had been observed by Bennett, Schwarzacher, and Egelhoff (1990) and by Parkin, Bhadra, and Roche (1991). We shall clarify some of the concepts needed later, but the effect can be understood as arising from a ferromagnetic metal such as iron having electrons only of one spin moving in a [100] direction (parallel to a lattice cube edge) at the Fermi energy. Then if the iron on both sides of a copper slab with (100) orientation (perpendicular to a [100] direction) has parallel magnetization, the electrons of the opposite spin are confined to the copper slab giving an energy varying as in Fig. 2-4. If the magnetization in the two iron crystals is antiparallel, electrons of one spin spread out far to the right and of the other spin far to the left and no such oscillatory interaction arises. Only integral numbers of copper planes can occur so only the integral values of N_p in Fig. 2.4 arise and the oscillation is "aliased" with the lattice spacing to produce the long period shown by the points in the figure and by the experiments.

Such an oscillatory coupling can be very useful technologically. If we construct a system with thickness such that the energy in Fig. 2.4 is positive, the magnetization of the iron crystals will tend to be antiparallel. Then the resistance of the system parallel to the planes will be high because almost all electrons spend time in both the highly conducting copper and poorly conducting iron. If, however, a magnetic field is applied which aligns the magnetization on both sides, electrons of one spin are confined to the highly-conducting copper and provide an electrical short which greatly reduces the resistance, an effect in this case called a *Giant Magnetoresistance* . Such a magnetoresistance can be used, for example, in a device to read magnetic data stored on disks.

Rather than constructing a slab, as above, we might construct a system with vanishing boundary conditions on two small dimensions, L_x, and L_y, and periodic boundary conditions on one very long dimension, L_z, as

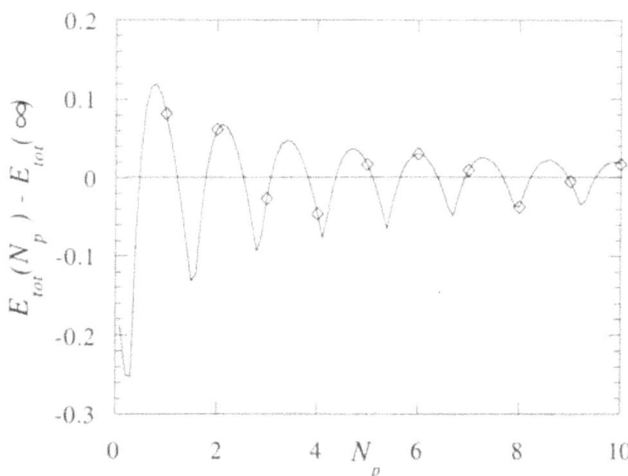

Fig. 2.4. A plot of the energy per unit area (in units of $E_F k_F^2/8\pi$) of an electron gas, as a function of the thickness of the slab, with the value for infinite spacing subtracted. Parameters were taken for copper and thickness given as the number N_p of (100) copper atomic planes. Only integral values, shown by diamonds, are observable. (After Harrison (1999))

illustrated in Fig. 2.5. This would be called a *quantum wire.* The energy eigenvalues are clearly given by

$$\varepsilon_k = \frac{\hbar^2\pi^2}{2mL_x^2} n_x^2 + \frac{\hbar^2\pi^2}{2mL_y^2} n_y^2 + \frac{\hbar^2 k_z^2}{2m}, \tag{2.16}$$

with the integers n_x and n_y specifying the subband. The lowest sub-band has $n_x = n_y = 1$ and is the lowest band shown to the right in Fig. 2.5. There are two sub-bands, $n_x = 1$, $n_y = 2$ and $n_x = 2$, $n_y = 1$, *degenerate* (meaning states of the same energy) if $L_x = L_y$, also shown, and sub-bands of larger energy. Electrons in each behave as electrons in one-dimension, with the properties as we obtained in Section 2.1.

One property of a one-dimensional metal of current interest is its "quantized conductance", discussed, for example, by Landauer (1989). It is readily understandable in terms of the sub-bands of Fig. 2.5. We imagine a quantum wire as to the left in Fig. 2.5 with large metal crystals (three-dimensional electron gases) at each end. Then an electron reaching one end of the quantum wire can escape into the metal crystal and we assume for the moment that the transmission of the junction is one, so it will escape in all cases. (We return in a moment, and in Problem 2.5, to the case of partial transmission.) The lowest-energy state of this composite system, called the *ground state,* will have all levels filled to a Fermi energy, some height in the bands to the right in Fig. 2.5. The electrons in the wire at any energy below the Fermi energy will be flowing into the metal crystals, but they are

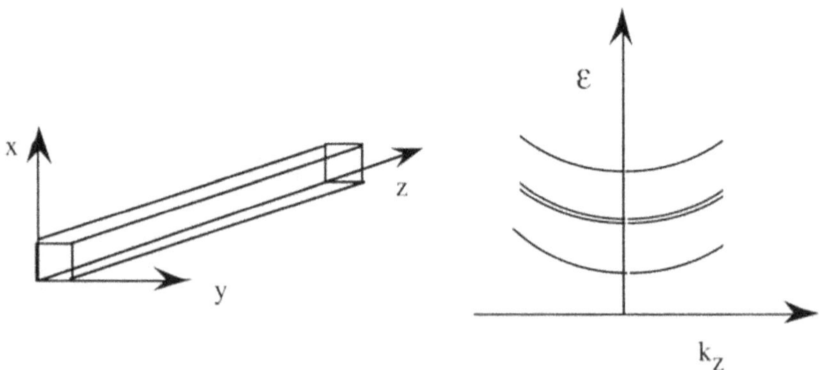

Fig. 2.5. A quantum wire has vanishing boundary conditions on two small dimensions and periodic boundary conditions on a large one, as on the left. Each sub-band corresponds to fitting an integral number of half-wavelengths to L_x and to L_y , and then the energy varies almost continuously with k_z, as shown to the right.

certainly flowing also into the wire at the same rate since there is no current flow in this ground state.

We may immediately calculate these two identical rates of flow by a scheme such as we will need to use many times in this text. We sum over the electron states with positive k_z in this energy range, with each contributing a current equal to $-e$ times the rate it strikes the end, equal to its velocity [$(1/\hbar)\partial\varepsilon_k/\partial k_z$ from either Eq. (1.6) or (1.7) using Eq. (1.8)] divided by the length of the wire . We double this to include electrons of both spins. This is current given by the first form in

$$\delta J = 2\sum\nolimits_{k_z}(-e)\frac{1}{\hbar}\frac{\partial\varepsilon}{\partial k_z}\frac{1}{L_z} = 2\frac{L_z}{2\pi}\int dk_z\,(-e)\frac{1}{\hbar}\frac{\partial\varepsilon}{\partial k_z}\frac{1}{L_z} = \frac{-2e}{2\pi\hbar}\int d\varepsilon\,. \qquad (2.17)$$

The sum is just over the wavenumber in the desired energy range. We next convert the sum to an integral over wavenumber, divided by the spacing (from Eq. (2.7)), $2\pi/L_z$ between subsequent states. If the spacing is small enough (or L_z large enough), this will equal the sum and the energy range can be small enough that the variation of speed is negligible. However, because of the factor $\partial\varepsilon_k/\partial k_z$, this becomes an integral over energy and there are no remaining energy-dependent factors. It is customary to write the denominator as $h = 2\pi\hbar$ (Eq. (1.9)), one of the few times that the constant h arises.

We now raise the energy of the electrons in the metal on the left very slightly and lower the energy of the electrons in the metal on the right very slightly. This corresponds to applying a voltage equal to the resulting energy difference (divided by the electronic charge $-e$). With our transmission of unity the metal to the left is supplying electrons moving to the right in the wire at energies below its Fermi energy, but the metal to the right is supplying electrons to the wire only below the Fermi energy on the right (which is lower), *if the Fermi energy lies in the quantum-wire subband we are considering*. The Fermi energy on the left is higher than that on the right by the applied voltage ϕ times the electron charge $-e$ so from Eq. (2.17) there is a net current in the wire due to the applied voltage of

$$\delta J = \frac{2e^2}{2\pi\hbar}\,\phi = \frac{2e^2}{h}\,\phi\,, \qquad (2.18)$$

which is to be added for every sub-band at the Fermi energy. This is a conductance contribution $\delta C = 2e^2/h$ for every sub-band at the Fermi energy. For a thick wire there will be very many bands below the Fermi energy determined from Eq. (2.10).

If we shrink a wire by pulling on it so that it necks down as shown in the insert in Fig. 2.6, the number of sub-bands below the fixed Fermi energy will decrease. It is the number of n_x, n_y pairs leading to an energy from Eq. (2.16) (with $k_z = 0$) which lie below the Fermi energy. That number is seen to be in proportion to the cross-sectional area using an argument parallel to that which led to Eq. (2.10). Thus the conductance will decrease in proportion to the cross-section, L_xL_y , as we would expect, due to the decrease of the number of contributing sub-bands. However, it is decreasing *in discrete steps, which becomes observable when the cross-section is very* small, just before breaking. Thus the conductivity varies as shown in Fig. 2.6. Such steps of approximately $2e^2/h$ in the conductivity were indeed observed by Costa-Krämer, Garćia, Garćia Mochales, and Serena (1995).

In this treatment we assumed 100% transmission at the ends of the quantum wire. We may correct for this by noting that the current J entering from the left would be reduced by a factor of the transmission, T, but then there would also be a reflection of that current from the right, giving a current to the left of $JT(1 - T)$ which subtracts from the initial current JT. We should subtract that, but add the current which arises when the reflected current strikes the left end, $JT(1 - T)^2$, etc. We obtain an infinite series for the current given by

$$J_{\text{net}} = JT[\,1 - (1 - T) + (1 - T)^2 - (1 - T)^3 + ...] = \frac{JT}{2 - T}\,, \qquad (2.19)$$

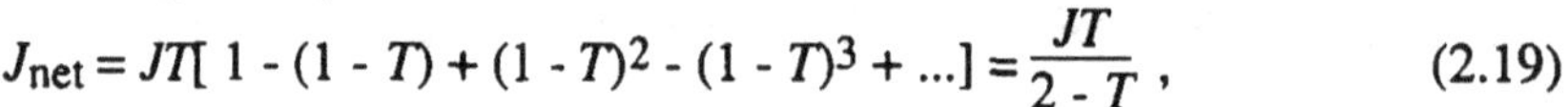

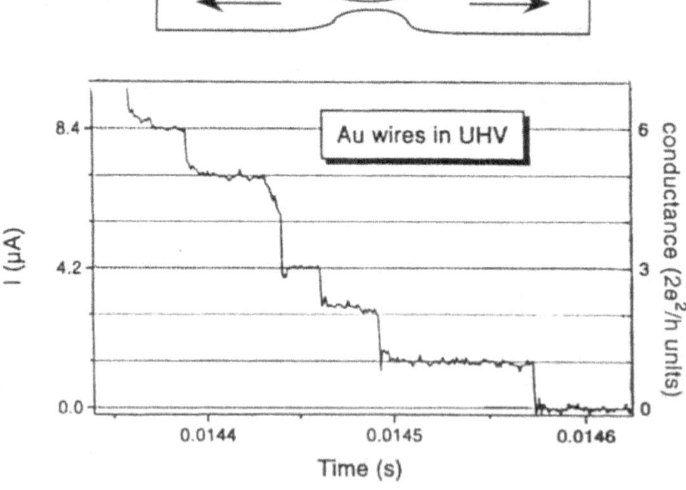

Fig. 2.6. A gold wire is slowly stretched, as shown above, necking down to a small cross-section. The conductivity decreases in proportion to the cross-sectional area, by steps of $2e^2/h$ as conducting sub-bands rise above the Fermi energy. As it is about to break, the individual steps can be seen. From Costa-Krämer, Garćia, Garćia Mochales, and Serena (1995).

where in the last step we used $1/(1+x) = 1 - x + x^2 - x^3 + ...$

Note that the result is independent of the length of the quantum wire. The fact that the observed steps were close to $2e^2/h$ indicates that the transmission is indeed near one, as one might expect if the taper is slow on the scale of an electron wavelength. Such sums of infinite series are frequently useful. Problem 2.5 is a redoing of this calculation when the transmission is different on the right from on the left. In any case the fact that the experimentally observed steps in Fig. 2.6 are so close to the theoretical $2e^2/h$ indicates, as we indicated, that the transmission in the experiments was very close to one.

Note that this analysis has predicted a minimum conductivity of $2e^2/h$, corresponding to a maximum resistance of

$$R_{\text{max.}} = \frac{h}{2e^2} = \frac{2\pi\times 6.6\times 10^{-16}\,\text{eV-sec}}{2\times 1.6\times 10^{-19}\text{Coulomb}\times e}$$
$$= 12.9\times 10^3 \frac{\text{sec.-Volt}}{\text{Coulomb}} = 12.9 \text{ kilo-ohms.} \tag{2.20}$$

Above that, the resistance becomes infinite; it is insulating. This concept of maximum resistance comes up in other circumstances, but we note that it is predicted to be higher than this value when the transmission is less than one. Quantized conductance is an important concept in mesoscopic devices.

We might go on to a quantum wire which was also very short, so that the energy steps were large for all three directions of motion. This is called a *quantum dot*, and has discrete states as does an atom. We consider these in the context of a spherically symmetric well in the following section.

2.4 Circularly and Spherically-Symmetric Systems

Spherically-symmetric systems are of particular importance in quantum mechanics, partly because they include atoms. The mathematics is simpler for *circularly-symmetric*, two dimensional systems, so we consider those and

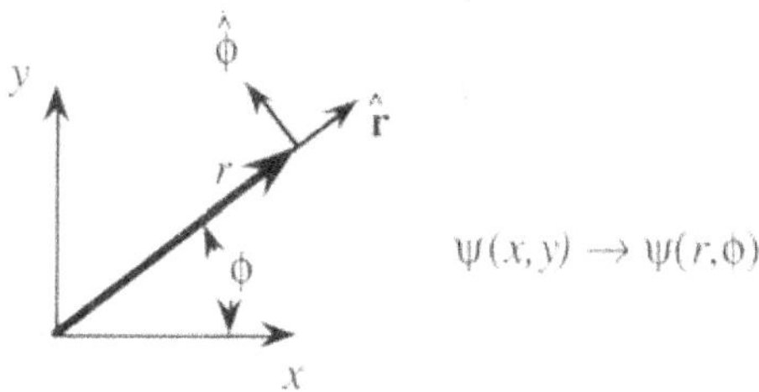

Fig. 2.7 A radial and angular coordinate system appropriate to circularly-symmetric systems in a plane.

then write the direct generalization to three dimensions.

For such a circularly-symmetric system in a plane we take a polar coordinate system, as shown in Fig. 2.7, with a radial distance r in the direction $\hat{\mathbf{r}}$ (a unit vector in the radial direction), and an angle ϕ measured from the x-axis. We write a unit vector $\hat{\phi}$ in the direction of increasing ϕ.

Then we specify the wavefunction as a function of r and ϕ rather than x and y. The gradient vector of any function can also be written in the new coordinate system,

$$\nabla\psi = \frac{\partial\psi}{\partial x}\,\hat{\mathbf{x}} + \frac{\partial\psi}{\partial y}\,\hat{\mathbf{y}} \rightarrow \frac{\partial\psi}{\partial r}\,\hat{\mathbf{r}} + \frac{1}{r}\frac{\partial\psi}{\partial\phi}\,\hat{\phi}\,. \tag{2.21}$$

Similarly the Laplacian is given by

$$\nabla^2\psi = \frac{1}{r}\frac{\partial}{\partial r}r\frac{\partial\psi}{\partial r} + \frac{1}{r^2}\frac{\partial^2\psi}{\partial\phi^2}\,. \tag{2.22}$$

Even with a circularly-symmetric potential $V(r)$ this does not separate the Hamiltonian into terms depending only upon r and terms depending only upon ϕ, which we used to factor the wavefunction in Section 2.2. However, it can be verified that we can obtain solutions of the form $R_{\mathrm{m}}(r)Y^{\mathrm{m}}(\phi)$, with $Y^{\mathrm{m}}(\phi) = e^{i\mathrm{m}\phi}/\sqrt{2\pi}$ for $m = 0, \pm1, \pm2, \ldots$, and with the corresponding radial function satisfying

$$-\frac{\hbar^2}{2m_e}\frac{1}{r}\frac{\partial}{\partial r}r\frac{\partial R_{\mathrm{m}}(r)}{\partial r} + \frac{\hbar^2 m^2}{2m_e}\frac{R_{\mathrm{m}}(r)}{r^2} + V(r)R_{\mathrm{m}}(r) = \varepsilon\, R_{\mathrm{m}}(r)\,. \tag{2.23}$$

We have written the electron mass as m_e, to distinguish it from the integer m which traditionally describes the angular wavefunction. We have written the angular part with a factor $1/\sqrt{2\pi}$ so that $\int Y^{m*}(\phi)Y^{\mathrm{m}}(\phi)\,d\phi = 1$ and the normalization of the full wavefunction is accomplished if $\int R_{\mathrm{m}}(r)^*R_{\mathrm{m}}(r)\,dr = 1$.

We may see that these are also eigenstates of the angular-momentum operator L, which may not be surprising in a circularly-symmetric system. The angular moment is the radial distance r times the tangential momentum, $(\hbar/(ir))\partial/\partial\phi$ as seen from Eq. (2.21). The angular momentum eigenvalue equation becomes

$$L\,\psi(r,\phi) = r\frac{\hbar\partial}{ir\partial\phi}\psi(r,\phi) = \frac{\hbar}{i}\frac{\partial}{\partial\phi}\psi(r,\phi) = m\hbar\,\psi(r,\phi)\,, \tag{2.24}$$

where the $R_m(r)$ factors cancel immediately so that the $Y^m(\phi)$ become eigenfunctions of the operator, with eigenvalue $m\hbar$.

Thus we have simultaneously eigenstates of angular momentum and energy. This turns out always to be possible if the two operators commute, $LH - HL = 0$. For this case it is exactly the requirement that the Hamiltonian be circularly symmetric, $\partial H/\partial\phi = 0$. We number the eigenstates by their angular momentum index m and then determine the energy from the *radial eigenvalue equation* in two dimensions, Eq. (2.23). There will be many solutions, subject to any given boundary conditions, and these would be numbered by another index n. Note that the radial equation contains what is called a *centrifugal potential* , the second term. It arises from the angular kinetic energy and, by diverging at small r , forces the wavefunction to go to zero if $m \neq 0$. It will also clearly lead to an additive term in the energy ε from that angular kinetic energy. All of these features will generalize to the case of spherical symmetry. We may also note that we could generalize this case to a case of cylindrical symmetry, with eigenstates of the form $\psi(r,\phi,z) = R_m(r)(e^{\pm im\phi}/\sqrt{2\pi})Z(z)$ and if the potential does not vary in the z-direction, $Z(z)$ can be taken of the form $e^{ik_zz}/\sqrt{L_z}$.

Eq. (2.23), when $V(r)$ is equal to zero, is the equation for free particles written in cylindrical coordinates. It is a form of Bessel's equation (e. g., Mathews and Walker (1964), 171ff) and the solutions $R_m(r)$ are Bessel functions of integral order $J_m(kr)$, with $\varepsilon = \hbar^2k^2/(2m_e)$. There is less occasion to use them in quantum mechanics than the spherical Bessel functions which are solutions of the similar Eq. (2.31) which we shall come to. However, they are used here in Problem 2.6 in the discussion of quantum wires with a circular cross-section. In such problems with cylindrical symmetry one can utilize a mathematical text, such as Mathews and Walker (1964), which gives their properties. We do this in Problem 2.6.

We go now to spherical coordinates, in close parallel with the circular system and illustrated in Fig. 2.8. Relative to a Cartesian system (x, y, z), r is the distance from the origin, the angle θ is measured from the z-axis, and ϕ is the azimuthal angle of the plane of **z** and **r** relative to the x-axis. In place of Eq. (2.22), the Laplacian is now given by

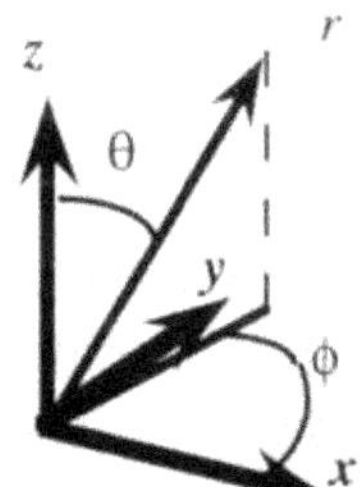

Fig. 2.8. Spherical coordinates relative to a Cartesian system.

$$\nabla^2 = \frac{1}{r^2}\frac{\partial}{\partial r} r^2 \frac{\partial}{\partial r} + \frac{1}{r^2}\left(\frac{1}{\sin\theta\partial\theta}\left(\sin\theta\frac{\partial}{\partial\theta}\right) + \frac{1}{\sin^2\theta\partial\phi^2}\right) \tag{2.25}$$

As for the circular case if the potential is independent of angle we can factor the wavefunction into a radial and an angular factor,

$$\psi(r,\theta,\phi) = R_l(r) Y_l^m(\theta,\phi), \tag{2.26}$$

with the angular factor in this case called a *spherical harmonic* , normalized as $\int d\Omega Y_l^{m*} Y_l^m = \int \sin\theta \, d\theta \int d\phi Y_l^{m*}(\theta,\psi) Y_l^m(\theta,\phi) = 1$. These are given by

$$Y_l^m(\theta,\phi) = \pm\sqrt{\frac{2l+1}{4\pi}\frac{l-|m|}{l+|m|}} P_l^m(\cos\theta) e^{im\phi} \ , \tag{2.27}$$

specified by two integers, $l = 0,1,2,...$ $m = -l, -l+1, -l+2,...,l$. The leading sign is plus unless m is an odd positive integer, then minus. The P_l^m are associated Legendre functions, which we shall supply when needed. As for the circular case, these are also angular-momentum eigenstates. The angular momentum around the z-axis L_z is given by $m\hbar$ as for the circular case. The total angular-momentum squared, $L_x^2 + L_y^2 + L_z^2$, eigenvalues are $l(l+1)\hbar^2$, and this limits the component which can appear along the z-axis; $l(l+1)\hbar^2$ is always greater than $m^2\hbar^2$. Thus states in spherical systems can be chosen to be eigenstates of energy, of total angular-momentum-squared, and of component along some chosen z-axis. For a particular energy and l there will always be $2l + 1$ states of that same energy, with varying m corresponding to $2l+1$ different orientations of angular momentum. We shall discuss these more completely in Section 16.3.

States of $l = 0$ are called s-states ("sharp" from atomic spectra data). They are spherically symmetric and we can associate the "s" with spherical. Only one m-value is allowed and the spherical harmonic is

$$Y_0^0(\theta,\phi) = \sqrt{1/(4\pi)} \ , \tag{2.28}$$

normalized as indicated above as $\int \sin\theta \, d\theta \int d\phi \, Y_0^{0*} Y_0^0 = 1$. We will frequently have occasion to sketch composite states made of a number of atomic orbitals, and then it is convenient to sketch s-states as a circle.

States of $l = 1$ are called p-states ("principal"). That for $m = 0$ is given by $Y_1^0(\theta,\phi) = \sqrt{3/(4\pi)} \cos\theta = \sqrt{3/(4\pi)} \, z/r$. Those for $m = \pm 1$ are given by $\sqrt{3/(8\pi)} \sin\theta \, e^{\pm im\phi}$ but for most applications it is more convenient to take combinations, $(Y_1{}^1 \pm Y_1{}^{-1})/\sqrt{2}$, which are also eigenstates so that the three normalized "cubic harmonics" for the p-states are

$$\sqrt{\frac{3}{4\pi}}\frac{x}{r},$$

$$\sqrt{\frac{3}{4\pi}}\frac{y}{r}, \qquad (2.29)$$

$$\sqrt{\frac{3}{4\pi}}\frac{z}{r}.$$

The first two are p-states with zero angular momentum around x- and y-axes, respectively. They are analogous to three components of a polar vector and we may think of the "p" as standing for "polar". They are zero in the central plane perpendicular to their axes, positive on one side of the plane and negative on the other.

The wavefunctions corresponding to energy eigenstates are spherical functions $R_l(r)$ times these angular functions. When we construct states which are mixtures of these it will be useful to use Dirac notation for the states. The |s> represents an s-state, a spherically symmetric function and the circle we sketch for it corresponds to a contour of constant probability density. For the p-states $|p_x\rangle$, $|p_y\rangle$, and $|p_z\rangle$ a contour of constant probability density will be a pair of closed surfaces, one where the wavefunction is positive and one where it is negative. They are ordinarily sketched as in Fig. 2.9, with the sign of the wavefunction indicated as shown. The s-state is ordinarily taken to have positive wavefunction at large distance.

States with $l = 2$ are d-states ("diffuse") and the cubic harmonics can be chosen to have symmetry (not normalized) of

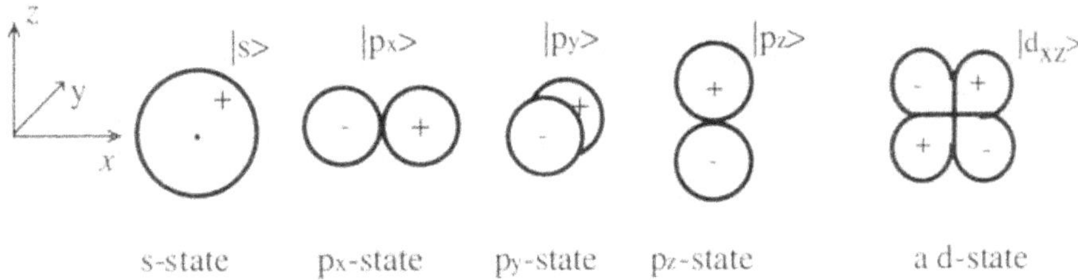

Fig. 2.9 The symbols which are used to represent the s-, p-, and d-states of the atom. The Dirac notation, |s> , etc., is described in Section 5.1.

$$\frac{yz}{r^2}, \ \frac{zx}{r^2}, \ \frac{xy}{r^2}, \ 3\frac{z^2}{r^2} - 1, \ \text{and} \ \frac{x^2 - y^2}{r^2} \ . \tag{2.30}$$

The second of these is illustrated in Fig. 2.9 to the far right. These will be of less interest to us here than the s- and p-states.

For each l the *radial time-independent Schroedinger Equation* can be written (returning to $m = m_e$ for the electron mass)

$$-\frac{\hbar^2}{2m}\frac{1}{r^2}\frac{\partial}{\partial r}r^2\frac{\partial R_l(r)}{\partial r} + \frac{\hbar^2 l(l+1)}{2mr^2}R_l(r) + V(r)R_l(r) = \varepsilon R_l(r) \ . \tag{2.31}$$

As for the circular case, the second term, $\hbar^2 l(l+1)/(2mr^2)$ is called a centrifugal potential and arises from angular kinetic energy. Note that it is related to the angular momentum squared as $L^2/(2mr^2)$, as one might have guessed. The $R_l(r)$ will ordinarily be real and are normalized as $\int dr\, r^2\, R_l(r)^2 = 1$, since then with $\int d\Omega Y_l^{m*} Y_l^m = 1$ we satisfy $\int d^3r\, \psi(r,\theta,\phi)^*\, \psi(r,\theta,\phi) = 1$.

For solving this equation, it can be simplified by defining $\chi(r) = rR_l(r)$, normalized as $\int \chi^2 dr = 1$. If we substitute for $R_l(r)$ in Eq. (2.31) we obtain

$$-\frac{\hbar^2}{2m}\frac{\partial^2 \chi}{\partial r^2} + \frac{\hbar^2 l(l+1)}{2mr^2}\chi + V(r)\chi = \varepsilon\chi \tag{2.32}$$

We will extensively use Eqs. (2.31) and (2.32) in this text, but at this point make only the simplest application, that to free electrons in a spherical well.

We take the potential $V(r) = 0$ for $r < R$, $V(r) = \infty$ for $r > R$, so that at this outer radius $\chi(R) = 0$. Also, since $\chi = rR_l$ we also have $\chi(0) = 0$. For $l = 0$, we have simply

$$-\frac{\hbar^2}{2m}\frac{\partial^2 \chi(r)}{\partial r^2} = \varepsilon\, \chi(r) \ . \tag{2.33}$$

The general solutions are $\chi(r) = A \sin(kr) + B \cos(kr)$, with $\varepsilon = \hbar^2 k^2/(2m)$. To satisfy the boundary conditions, B=0, $kR = n\pi$. The resulting radial wavefunction (not normalized) is

$$R_0(r) = \frac{\chi(r)}{r} = \frac{\sin kr}{kr} = j_0(kr) \ . \tag{2.34}$$

It is called the *spherical Bessel function* of order $l = 0$. (It is regular at r = 0. There are also $n_l(r) \propto 1/(kr)^{l+1}$ diverging as $r \to 0$; e. g., Schiff (1968), 84ff, but we will not need them.) For higher l there is an additional term on the left in Eq. (2.33), $\hbar^2 l(l+1)\chi(r)/(2mr^2)$, and the solutions which are regular at the origin, $j_l(kr) \propto (kr)^l/(1\cdot3\cdot5\cdot...\cdot(2l+1))$ at small r, are spherical Bessel functions of higher order. At large distances they vary as $\sin(kr - l\pi/2)/(kr)$. It can be confirmed that these same formulae apply to the $l = 0$ case. For $l = 1$ the spherical Bessel function is

$$\frac{\chi_1(r)}{r} = j_1(kr) = \frac{\sin kr}{(kr)^2} - \frac{\cos kr}{kr} . \tag{2.35}$$

$j_2(kr)$ will have three terms, etc. These are all analogous to the $\sin kx$ and $\cos kx$ solutions we found for free electrons moving in one dimension. We obtain energy eigenstates by adjusting the coefficients to satisfy the boundary conditions and normalization.

In particular, if we consider free electrons, confined to a spherical box of radius R, we find solutions by requiring the $j_l(kR) = 0$, which will give a set of values of k for each l. The first three for $l = 0$ and the third state for $l = 1$ are illustrated in Fig. 2.10. As will always be the case for spherical potentials, the lowest state of any l will have no *nodes*, radii at which $\psi = 0$ as for the nodes in a vibrating string, except possibly at $r = 0$ or at the upper limit. Each successive n-value is a state with an additional node. The states can be normalized by a scale factor A such that $A^2\int dr\, r^2 j_l(kr)^2 = 1$. In

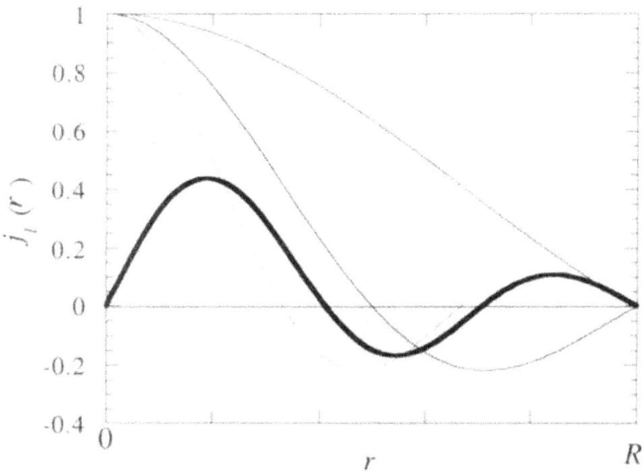

Fig. 2.10. The three lowest-energy states of zero angular momentum $j_0(kr)$ in a spherical cavity of radius R. Energy increases with the number of nodes. The heavy curve is the third-lowest $j_1(kr)$ satisfying the same boundary condition.

Problem 2.7 we consider the solutions and normalization when the radius R is very large.

We should note also the behavior in quantum wells of finite depth, where the potential beyond R does not rise to infinity. For simplicity we consider s-states so the second term in Eq. (2.32) is zero. We let $V(r)$ again be zero for $r < R$, but now some positive constant value V_0 for $r > R$. Again the solution for $r < R$ is $\chi = A\sin kr$ and if the energy is less than V_0 the general solution of Eq. (2.32) for $r > R$ is $Ce^{-\mu r} + De^{+\mu r}$ with $\hbar^2\mu^2/(2m) = V_0 - \varepsilon$. Only with $D = 0$ is the solution normalizable at large r so we have the unknowns, A and C and the energy ε to be fixed by matching the slope and the wavefunction at R (since the wavefunction is to be single-valued there and the kinetic energy would be locally infinite with a cusp in the wavefunction). For this simple case, dividing the equations for the two conditions gives

$$\tan kR = -\frac{k}{\mu} . \tag{2.36}$$

This is readily solved numerically. We may for example solve $\hbar^2\mu^2/(2m) = V_0 - \hbar^2k^2/(2m)$ for μ in terms of k and plot the right-hand and left-hand sides of Eq. (2.36) against k as in Fig. 2.11. The intersections indicate the solutions, with the energy determined by the resulting k. The wavefunction corresponding to one such solution is also shown in Fig. 2.11. We see that the electron tunnels into the barrier, thereby "relaxing the wavefunction" and lowering the energy. If we were to increase the height of the barrier (increase V_0) the heavy curve to the left in Fig. 2.11 would rise toward the axis, and the wavenumbers at crossing increase slightly, as does the energy.

For a more complicated $V(r)$ we would need to numerically integrate the Eq. (2.32), which is not so difficult. At small r the behavior of the state is determined by the centrifugal term. If we substitute the form $\chi = Ar^n$ into Eq. (2.32) we obtain

$$-n(n-1)\frac{\hbar^2}{2m}Ar^{n-2} + \frac{\hbar^2 l(l+1)}{2m}Ar^{n-2} + V(r)Ar^n = \varepsilon Ar^n . \tag{2.37}$$

At small enough r all but the two first terms become negligible. Canceling $(\hbar^2/2m)Ar^{n-2}$, we find $n = l + 1$. This is consistent of course with the forms of $j_l(r) = \chi/r$ shown in Fig. 2.10 and with the small-r form which we gave for the $j_l(kr)$.

To do the numerical integration we select the l-value we wish to treat and we may set up a grid of r-values with spacing Δr, perhaps 0.01Å. At the first grid point, $r = \Delta r$, we may use the small-r form $\chi \propto r^{l+1}$ to take χ to

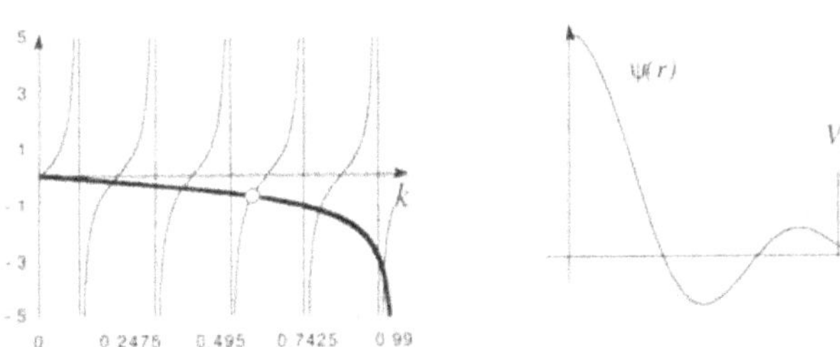

Fig. 2.11. On the left we have plotted the two sides of Eq. (2.36) against k. The intersections of the tankR curves with the heavy curve are the states in the well. To the right is shown the third state, indicated by the circle on the left, as well a schematic representation of the potential.

be $(\Delta r)^{l+1}$ (scaled by any constant we wish). Then at this first grid point $\partial\chi/\partial r \approx (l+1)(\Delta r)^l$ and we guess a value of energy ε so that we may obtain $\partial^2\chi/\partial r^2$ at that grid point directly from Eq. (2.32) using the known $V(r = \Delta r)$. We may then directly calculate the χ, $\chi' = \partial\chi/\partial r$, and $\chi'' = \partial^2\chi/\partial r^2$ at the next grid point from

$$\chi(r+\Delta r) = \chi(r) + \chi'(r)\Delta r + {}^1\!/_2\chi''(r)\Delta r^2,$$

$$\chi'(r+\Delta r) = \chi'(r) + \chi''(r)\Delta r, \tag{2.38}$$

$$\chi''(r+\Delta r) = \frac{2m}{\hbar^2}\left(\frac{\hbar^2 l(l+1)}{2m(r+\Delta r)^2} + \mathrm{V}(r+\Delta r) - \varepsilon\right)\chi(r+\Delta r)\,.$$

The same procedure is repeated, step by step, to large r, giving the solution for the energy ε we guessed. We could instead integrate with some software such as *Mathematica.*

We will ordinarily not have guessed the correct energy, so as we integrate to large r we will find that $\chi(r)$ is growing exponentially, either positively or negatively, as illustrated in Fig. 2.12. If it is in the positive direction, we should increase the energy slightly, which will bend the curve down, and continue until it is diverging in the negative direction, when we begin decreasing the energy until χ again goes positive. It is not difficult to shift up and down until we have a solution. Usually we will know how many nodes are in the solution we want (see Fig. 2.10 or 2.11) and we may need to shift the energy considerably to have the right number of nodes and then adjust up and down to get close to the correct energy. We will do two problems of this type in the coming sections, one for the harmonic oscillator and one for atomic states.

When solving this equation for weak potentials, we may note that for this three-dimensional case a certain strength of a local attractive potential is required to obtain a *bound state*, a state decaying exponentially with r at large r. If the potential is too weak, an integration even at zero energy relative to the potential outside the well will not bend the wavefunction within the well far enough to bring it to zero at large distances. At positive energies relative to the potential outside the solution will always be oscillatory at large enough distances so no localized state can be obtained. This feature arises from our boundary condition at small r. In a one-dimensional case, any net attractive local potential can produce a bound state, if the particle can range in both directions. It will be very shallow (energy near zero) if the potential is quite weak.

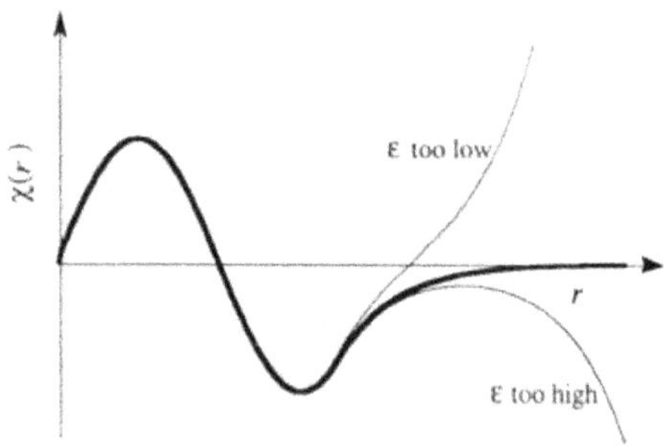

Fig. 2.12. An integration of the radial equation, Eq. (2.32), at an energy ε equal to an eigenvalue, and at slightly higher and slightly lower energies.

2.5 The Harmonic Oscillator

The third important system, in addition to free particles and spherically symmetric systems, is the harmonic oscillator. This is partly because each mode of sound vibrations, discussed in Section 1.8, can be treated as a harmonic oscillator, as can each cavity mode of electromagnetic waves - or light waves. The energy of a simple harmonic oscillator of mass M, with displacement coordinate x, and spring constant κ is $\frac{1}{2}M\dot{x}^2 + \frac{1}{2}\kappa x^2$, which we write in terms of momentum $p = M\dot{x}$ as the Hamiltonian (to be generalized in the following chapter). Then the energy-eigenvalue equation becomes

$$-\frac{\hbar^2}{2M}\frac{\partial^2\psi(x)}{\partial x^2} + \frac{1}{2}\kappa x^2\,\psi(x) = \varepsilon\,\psi(x)\,. \tag{2.39}$$

It is also exactly Eq. (1.21) for one dimension and with a potential given by $^1/_2\kappa x^2$.

We may guess a form of the solution, and confirm that it is correct. The correct guess is the form

$$\psi(x) = A\ \exp(-x^2/(2L^2))\,. \tag{2.40}$$

Substituting this form in Eq. (2.39), and evaluating the derivative, gives

$$-\frac{\hbar^2}{2M}\left(\frac{x^2}{L^4}-\frac{1}{L^2}\right)\psi(x)+\frac{1}{2}\kappa x^2\,\psi(x)=\varepsilon\,\psi(x)\,. \tag{2.41}$$

The factors $\psi(x)$ cancel so this is indeed a solution if $\hbar^2/(ML^4) = \kappa$. Then $\varepsilon = \hbar^2/(2ML^2)$. It is more convenient to write this in terms of the classical vibrational frequency ω (in radians per second), with $\omega^2 = \kappa/M = \hbar^2/(M^2L^4)$. The first step in this equation is simply the classical expression, the second gives $\omega = \hbar/(ML^2)$, and the energy is given by

$$\varepsilon = \frac{1}{2}\hbar\omega. \tag{2.42}$$

This is the *ground state* (lowest-energy state) of the simple harmonic oscillator and the energy is called the *zero-point energy* of the oscillator. We obtain the normalization, $A^2\int_{-\infty,\infty}\exp(-x^2/L^2)dx = A^2L/\sqrt{\pi} = 1$, so

$$A = 1/\sqrt{L\sqrt{\pi}}\,. \tag{2.43}$$

It is also interesting to obtain the mean-squared fluctuation, $\langle x^2\rangle = A^2\int_{-\infty,\infty}\exp(-x^2/L^2)\,x^2dx = L^2/2$. Thus $L/\sqrt{2}$ is the root-mean-square deviation, called the *zero-point fluctuation*, and sometimes written $a_0{}^2 = \langle x^2\rangle$. The state arises from a compromise between kinetic energy, which is lower if the state is spread out and therefore slowly varying, and potential energy, which is lower if the wavefunction is strongly concentrated at the bottom of the well near $x = 0$. In this ground state the expectation value of the kinetic energy and the potential energy are the same, as they are classically, at

$$\frac{1}{2}M\omega^2\frac{L^2}{2} = \frac{1}{2}\kappa\frac{L^2}{2} = \frac{1}{2}\frac{\hbar\omega}{2}. \tag{2.44}$$

Higher-energy solutions are given in terms of Hermite polynomials, H_n, in most quantum texts (e. g., Kroemer (1994), p. 85) as

$$H_n(x/\sqrt{2}L)\exp(-x^2/2L^2). \tag{2.45}$$

The only one of these excited states which we will actually use is the first, and $H_1(Q) = 2Q$. The successively higher Hermite polynomials are successively higher-order polynomials and alternate between even and odd functions. The energies of these harmonic oscillator states are well known,

$$\varepsilon_n = \left(n + \frac{1}{2}\right)\hbar\omega, \quad n = 0, 1, 2, \ldots \tag{2.46}$$

As for many discussions in this text we have solved the simplest problem (the ground state) and quoted the more general results without the mathematical details. In fact, once the solutions are obtained, the techniques for solving the general problem are rarely used. In Problem 2.9 we take an energy appropriate to $n \approx 7$ and integrate the Schroedinger Equation numerically using $\Delta x = 0.1$Å, adjusting the energy to an accuracy of about 1% . We see that there are seven nodes and that the x-value for the largest peak is not so far from the maximum of the classical vibration, obtained from $\varepsilon \approx {}^1/_2\, \kappa x_{max}{}^2$

This ladder of equally spaced energy eigenvalues is characteristic of the harmonic oscillator. It is often said that there are n *vibrational quanta* in the state ψ_n. One consequence of the equally-spaced levels is that if we construct a wave packet of many harmonic-oscillator eigenstates, corresponding to a fixed displacement of the oscillator at the center of the packet, that packet will return to its identical shape every classical period $2\pi/\omega$ since each state will have changed phase by an integral number of 2π's, as seen from Eq. (1.22). [The relative phases of the terms need to be chosen correctly to have the packet remain intact over the entire period; e. g., all in phase when the packet is at its extreme displacement.]

When we describe sound vibrations in a pipe, as in Section 1.8, we can regard the $u \equiv u(0,t)$ of Eq. 1.29 as the displacement of a harmonic oscillator. The total potential energy is proportional to u^2 and the total kinetic energy proportional to $\dot{u}^2$. Thus each mode corresponds to a harmonic oscillator, with a classical frequency given by the $\omega = \sqrt{B/\rho}\, q$ of Eq. (1.31). It is seen that the initial statement that *everything* is both a wave and a particle applies here as does all of the succeeding analysis. The excitation is then quantized with some number of *phonons* , which we think of as particles, in each mode.

This is a considerable conceptual leap, in talking of waves as a function of amplitudes u , rather than as a function of position as we have before.

This leap may have first been made by Debye (1912) in applying quantum theory to sound waves in a solid, as we shall discuss in Section 10.2. We began here with a classical sound wave, pressure as a function of position, and described it with a quantum wavefunction, as a function of amplitude u, which is a harmonic oscillator with quantized excitations. Finally, we called these phonons. This extends to sound waves propagating in three-dimensional space, and each phonon has an energy and momentum, $H = \hbar\omega$, $\mathbf{p} = \hbar\mathbf{k}$. These phonon particles are *indistinguishable* since it has no meaning to "interchange them", a fact which we will make use of when we discuss their statistics in Section 10.2. We can trace the origin of this new particle back to the molecules which made up the air which was vibrating in the pipe. We shall make this generality of the wave-particle-duality statement explicit when we discuss Hamiltonian mechanics next.

The same analysis will apply to light waves, where we may write a quantum wavefunction as a function of the amplitude (vector potential **A** or electric field **E**) and the energy in each mode is quantized, corresponding to some number of photons in each mode. This is a much more familiar particle than a phonon, but no more rigorous nor valid. The photon also is modified by the dielectric properties of the air through which it moves.

Finally, we may easily extend the simple harmonic oscillator to three dimensions, as for an object in a bowl, or an atom which can vibrate in any direction in a well in which it is trapped. If the restoring forces are spherically symmetric, $V(r) = {}^1\!/_2\kappa r^2 = {}^1\!/_2\kappa(x^2 + y^2 + z^2)$ the Hamiltonian becomes separable in the three coordinates and, as in Section 2.2, we can write a product wavefunction $\Psi(\mathbf{r}) = \psi_1(x)\psi_2(y)\psi_3(z)$. Each of the ψ_i are harmonic oscillator states. The energy can be written

$$\varepsilon = (n_x + n_y + n_z + {}^3\!/_2)\hbar\omega. \tag{2.47}$$

The state $\psi_0(x)\psi_0(y)\psi_n(z)$ is vibrating in the z-direction. Since the system has spherically symmetry, it is also possible to write the states $\psi(r)Y_l^m(\theta,\phi)$ if we choose. Thus the ground state can be written

$$\psi_0(x)\psi_0(y)\psi_0(z) = 1/\sqrt{L^3\sqrt{\pi^3}}\,\exp(-r^2/2L^2)\,. \tag{2.48}$$

It is spherically symmetric, $l = 0$. Similarly we can construct p-states $[\psi_1(x)\psi_0(y)\psi_0(z) \pm i\psi_0(x)\psi_1(y)\psi_0(z)]/\sqrt{2} \propto (x \pm iy)\exp(-r^2/2L^2) = r\sin\theta e^{\pm i\phi}\exp(-r^2/2L^2)$, etc..

Chapter 3. Hamiltonian Mechanics

When we generalized the harmonic-oscillator states to sound waves in the preceding section, we saw the extraordinary generality of the wave-particle-duality statement of Section 1.1. The mathematics in which that generality is expressed is Hamiltonian mechanics. We have used the word "Hamiltonian " for the energy, expressed in terms of the coordinates and the momenta. The coordinates could be x, y, and z, or they could be r, θ, and ϕ. Hamiltonian mechanics allows us to write the equations of motion in either set of coordinates, or many other sets. In Section 2.5 we saw that we could even write equations of motion in terms of amplitudes, and these amplitudes might even be electric fields. The real generality of the wave-particle duality is that for any system for which we write the equations of motion in terms of a Hamiltonian, with its coordinates and associated momenta, that system may be represented by a wave as a function of the coordinates, and the momentum may be represented by $\hbar/i$ times the derivative with respect to the coordinate, operating on that wavefunction. Then all of the quantum effects we have discussed are present for that system. This is one rather precise way to state the wave-particle-duality premise, though there are certainly other ways.

Since this mechanics is so central to quantum theory, and since it is not generally included in the early physics courses which are taken by engineers, it is essential to outline the main features here. The history is also of interest (e. g., Thornton (1995)). This dynamics was in some sense developed as independent philosophically from Newtonian mechanics, by seeking a *minimum principle* to describe dynamics. That had its own appeal, but of course the results are mathematically equivalent to ordinary Newtonian mechanics. The first such principle was for optics, given by Hero of Alexander, in the second century BC. He asserted that light followed the shortest distance between two points. This led to the equal angles of incidence and reflection, but failed for diffraction. Fermat in the 17th century postulated that it was the time of flight which was to be minimized, giving the correct law of refraction as well as of reflection. Maupertius in the 18th century first formulated dynamics this way, saying that particles follow a path of minimum *action.* This then led to Lagrangian mechanics later in the 18th century, and Hamiltonian mechanics in the 19th century. Here we bypass that formulation and given the resulting dynamics, showing that it duplicates ordinary mechanics.

3.1 The Lagrangian

As we indicated, Hamiltonian mechanics is based upon Lagrangian mechanics, another method using generalized coordinates. It may be helpful to think of a specific model system as we write quite general statements, as illustration and to confirm that the correct results are obtained. For that purpose we consider a classical problem of a bead strung on a wire as in Fig. 3.1. We also let the wire rotate on an axis vertical in a gravitational field, and the bead may slide without friction along the wire. It would be awkward to work out the dynamics of the bead using force equal to mass times acceleration in Cartesian coordinates. However, it is possible to work it out in terms of a general coordinate such as q giving the distance along the wire, as shown, and Lagrangian or Hamiltonian mechanics. We will not work out this model in detail, which would require specifying the exact shape of the wire, but will illustrate the methods in terms of it.

The first step is to write the kinetic energy T in terms of generalized coordinates, $\{q_j\}$, and their time derivatives $\{\dot{q}\}$, and time t, $T(\{q_j\},\{\dot{q}\},t)$. [It is conventional to use the brackets $\{\}$ to denote a collection of values.] For the system in Fig. 3.1 the kinetic energy includes the kinetic energy of motion perpendicular to the wire, depending upon q, as well as $^1/_2 M\dot{q}^2$. In the simplest case, motion along a straight line, it is of course $^1/_2 M\dot{x}^2$. If the

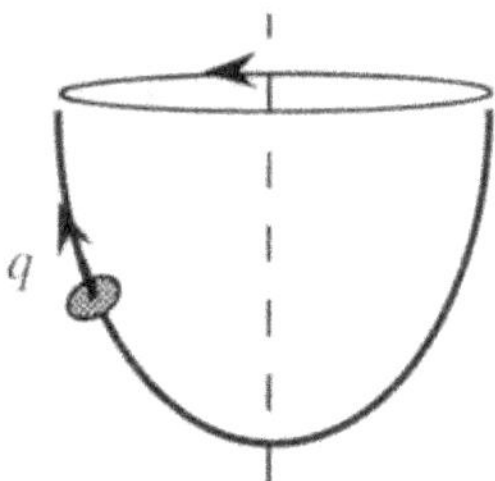

Fig. 3.1. A bead moves without friction along a wire, which rotates around a vertical axis. Its motion could be described in terms of a general coordinate q which is the distance along the wire, measured from the bottom.

forces Q_j along the generalized coordinates are derivable from a potential, $Q_j = -\partial V(\{q_i\},t)/\partial q_j$ then the Lagrangian is given by

$$L(\{q_j\},\{\dot{q}\},t) = T(\{q_j\},\{\dot{q}\},t) - V(\{q_j\},t), \tag{3.1}$$

equal to $\frac{1}{2}M\dot{x}^2$ -V(x) in the simplest case. In the model of Fig. 3.1 the potential is the gravitational potential, but needs to be reexpressed in terms of q . It is also possible to construct a Lagrangian when there is a force applied which is not derivable from a potential, such as the force $-e\mathbf{v}\times\mathbf{H}$ on a charged particle due to a magnetic field, and we shall do that in Section 3.3. We will not make use of it, but Lagrange's Equations of motion are

$$\frac{d}{dt}\frac{\partial L}{\partial \dot{q}_j} - \frac{\partial L}{\partial q_j} = 0. \tag{3.2}$$

For the simplest case, $\partial L/\partial\dot{q} = M\dot{x}$ and Eq. (3.2) becomes $M\ddot{x} + \partial V/\partial x = 0$, the usual equation of motion. A momentum, *conjugate* to each coordinate is defined from the Lagrangian by

$$p_j \equiv \frac{\partial L(\{q_j\},\{\dot{q}\},t)}{\partial \dot{q}_j}, \tag{3.3}$$

equal to $M\dot{x}$ in the simplest case. p_j is called the *canonical momentum* conjugate to the coordinate q_j. All is equivalent to traditional mechanics for the simplest case, but the Lagrangian method could be used for waves in water, or dislocations in crystals, and we shall see that it gives the correct forces when we use it for motion of a charged particle in a magnetic field.

3.2 Hamilton's Equations

The Hamiltonian is given in terms of the Lagrangian by

$$H \equiv \sum_j p_j\dot{q}_j - L \tag{3.4}$$

written as a function of the generalized coordinates, their canonical momenta, and time, $H(\{p_j\},\{q_j\},t)$. We need to use Eq. (3.3) to write each $\dot{q}_j$ in terms of p_j. In the simplest case it is $H(p,x) = p^2/(2M) + V(x)$, equal to the energy as we have used it up to now in our discussions, and we always think of it as the energy. Note that q_j does not necessarily have the units of length, but $q_j p_j$ will always have the units of energy×time, as does $\hbar$.

There are two *Hamilton's Equations* for each generalized coordinate,

$$\dot{q}_j = \frac{\partial H}{\partial p_j}, \qquad \dot{p}_j = -\frac{\partial H}{\partial q_j}. \tag{3.5}$$

For the simplest case they are $\dot{x} = p/M$ and $\dot{p} = M\ddot{x} = -\partial V/\partial x$, respectively. One again relates p to $\dot{x}$ and the other gives the equation of motion. We can do classical mechanics by solving the boxed equations, just as we can with Lagrange's Equations or with mass times acceleration equaling force.

One important classical example may be helpful. To describe a rotating body, we usually specify the rotation with an angular generalized coordinate ϕ. In terms of it the kinetic energy is written $^1/_2 I\dot{\phi}^2$, with I the moment of inertia obtainable (see Fig. 2.8) from the mass density $\rho(\mathbf{r})$ as $I = \int d^3r\, r^2 \sin^2\theta\, \rho(\mathbf{r})$. A generalized force along the angular coordinate is a torque t given by $t = -\partial V(\phi)/\partial\phi$. The Lagrangian is $L = {}^1/_2 I\dot{\phi}^2 - V(\phi)$ and the canonical momentum is the angular momentum, $p = \partial L/\partial\dot{\phi} = I\dot{\phi}$. Then the Hamiltonian is $H(p,\phi) = p^2/(2I) + V(\phi)$. The first of Hamilton's Equations again relates $\dot{\phi}$ and p, and the second says that the rate of change of angular momentum is equal to the torque. All of this is traditional classical mechanics. Note that the coordinate ϕ is dimensionless but the angular momentum has the units of energy×time.

When we go to quantum mechanics, we may proceed just as we did when the coordinates were ordinary spatial position. We write the momentum operator as

$$p_j = \frac{\hbar}{i}\frac{\partial}{\partial q_j} \tag{3.6}$$

as in Eq. (1.11) and the energy operator as

$$H = -\frac{\hbar}{i}\frac{\partial}{\partial t}. \tag{3.7}$$

The classical Hamiltonian is written with the p_j replaced by an operator using Eq. (3.6) and we have the Schroedinger Equation in terms of the generalized coordinates as

$$H\psi(\{q_j\},t) = i\hbar \frac{\partial}{\partial t} \psi(\{q_j\},t) . \tag{3.8}$$

This is what we did for Cartesian coordinates. If we do the same for the angular coordinates, for the case of free rotation, we seek a $\psi(\phi)$ and Schroedinger's Equation becomes, as in Eq. (1.16),

$$-\frac{\hbar}{i}\frac{\partial}{\partial t}\psi(\phi,t) = H\,\psi(\phi,t) = -\frac{\hbar^2}{2I}\frac{\partial^2\psi(\phi,t)}{\partial\phi^2} . \tag{3.9}$$

If we seek energy eigenstates, they are given by $\psi(\phi) = e^{im\phi}/\sqrt{2\pi}$, and if they are to be single-valued and continuous, m must be an integer. These are also eigenstates of angular momentum $(\hbar/i)\partial/\partial\phi$, with eigenvalues $\hbar m$. We would have found this same quantization if we considered a particle of mass M moving on a circle of radius R with no potential energy, equivalent to our free electron in one-dimension with periodic boundary conditions which we treated in Section 2.1, so this may not be surprising. However, we note that we are now applying the wave-particle duality to the rotary motion of a rigid object, which is again a major conceptual leap. We shall return to this when we discuss spins of particles in Section 10.4 and 10.5.

3.3 Including the Vector Potential

When we include magnetic fields in the dynamics of charged particles, it is convenient to use the vector potential, as we did when we wrote out the fields associated with a light wave in Section 1.3. There we noted that the magnetic fields **H** and electric fields **E** could be obtained from the vector potential **A** by Eqs. (1.17),

$$\mathbf{H} = \nabla\times\mathrm{A}(\mathbf{r},t) \quad , \tag{3.10}$$

and

$$\mathbf{E} = -\frac{1}{c}\frac{\partial\mathbf{A}(\mathbf{r},t)}{\partial t} . \tag{3.11}$$

It may be inconvenient for an engineer to work in terms of these units which are almost universally used in quantum physics. Even there one cannot count on formulae being equivalent. Kroemer (1994), for example uses **B** equal to **H**/c in vacuum, and a vector potential equal to our **A** divided by c so that it is **B** which is equal to $\nabla\times\mathrm{A}$ and **E** is equal to $-\partial\mathbf{A}/\partial\mathrm{t}$. We give here the defining equations for our fields, and they can be used in each case to define the properties we predict. $\mathbf{A}(\mathbf{r},t)$ is to satisfy Maxwell's Equations, as in Eq. (1.20), and we will not need them further here. We shall need the energy density (total energy for a volume Ω divided by that volume) associated with uniform fields in vacuum, and it is

$$\frac{E_{\text{field}}}{\Omega} = \frac{E^2}{8\pi} + \frac{H^2}{8\pi} . \tag{3.12}$$

Any effects of dielectric media are treated as additional terms rather than in terms of displacement and induction fields. We write the interactions between an electron of charge $-e$ and the fields as a force $\mathbf{F}$ given by the Lorentz force,

$$\mathbf{F} = (-e)\mathbf{E} + \frac{(-e)\,\dot{\mathbf{r}}}{c} \times \mathbf{H} \tag{3.13}$$

These are enough to define our use of electromagnetic fields. When we actually want to obtain values for forces and accelerations, and magnetic fields are given in gauss, we will need to substitute e in electrostatic units, as indicated at the beginning of Chapter 22 on magnetism.

To include the force, Eq. (3.13), in our dynamics, we need a Lagrangian or a Hamiltonian which will reproduce the dynamics using Lagrange's or Hamilton's Equations. We shall confirm that this is accomplished for a particle of charge $-e$ and mass m by replacing the momentum $\mathbf{p}$ in the Hamiltonian without fields by $\mathbf{p} - (-e/c)\mathbf{A}(\mathbf{r},t)$. There can be additional forces, as arising from an electrostatic potential $\phi(\mathbf{r},t)$ which are included in any potential energy $V(\mathbf{r},t)$ in the starting Hamiltonian. Thus the classical Hamiltonian including the effects of the vector potential is

$$\boxed{H(\mathbf{r},t) = \frac{1}{2m}\left(\mathbf{p} - \frac{-e}{c}\mathbf{A}(\mathbf{r},t)\right)^2 + V(\mathbf{r},t) \quad .} \tag{3.14}$$

Confirming this leads to some important intermediate results. We write the first of Hamilton's Equations (Eq. (3.5)),

$$\dot{\mathbf{r}} = \frac{\partial H}{\partial \mathbf{p}} = \frac{1}{m}\left(\mathbf{p} - \frac{-e}{c}\mathbf{A}(\mathbf{r},t)\right) . \tag{3.15}$$

The meaning of $\partial H/\partial \mathbf{p}$ is $\partial H/\partial p_x \hat{\mathbf{x}} + \partial H/\partial p_y \hat{\mathbf{y}} + \partial H/\partial p_z \hat{\mathbf{z}}$, with $\hat{\mathbf{x}}$, $\hat{\mathbf{y}}$, and $\hat{\mathbf{z}}$ unit vectors in the three directions. This is a shortcut which could be confirmed by writing all vector equations out as three scalar equations. The velocity $\dot{\mathbf{r}}$ retains its usual meaning, but we note that the momentum is no longer $m\dot{\mathbf{r}}$ but has acquired an additional term. It is a complication, but we may proceed. The second of Hamilton's Equations becomes

$$\dot{\mathbf{p}} = -\frac{\partial H}{\partial \mathbf{r}} = -\frac{\partial V}{\partial \mathbf{r}} - \frac{e}{mc}\left(\mathbf{p} - \frac{-e}{c}\mathbf{A}(\mathbf{r},t)\right)\cdot\frac{\partial \mathbf{A}}{\partial \mathbf{r}} . \tag{3.16}$$

[Here the final dot product is with **A**, and $\dot{p}_x$ is obtained from the $\partial/\partial x$.] We want the acceleration, which we obtain by taking the derivative of Eq. (3.15) with respect to time to obtain

$$\ddot{\mathbf{r}} = \frac{\dot{\mathbf{p}}}{m} - \frac{-e}{mc}\frac{d\mathbf{A}}{dt} . \tag{3.17}$$

[Here $d\mathbf{A}/dt = \partial\mathbf{A}/\partial t + \partial\mathbf{A}/\partial\mathbf{r}\cdot\dot{\mathbf{r}}$ and the dot product is with $\partial/\partial\mathbf{r}$. Then $\dot{p}_x$ is obtained from A_x, etc..] We may multiply by m and substitute for $\dot{\mathbf{p}}$ from Eq. (3.16) to obtain

$$m\ddot{\mathbf{r}} = -\frac{\partial V}{\partial \mathbf{r}} - \frac{e\dot{\mathbf{r}}}{c}\cdot\frac{\partial \mathbf{A}}{\partial \mathbf{r}} + \frac{e}{c}\frac{d\mathbf{A}}{dt} . \tag{3.18}$$

We may confirm, by separating the equations into components all the way through, that this is equivalent to

$$m\ddot{\mathbf{r}} = -\frac{\partial V}{\partial \mathbf{r}} - \frac{e\dot{\mathbf{r}}}{c}\times\mathbf{H} - e\mathbf{E} . \tag{3.19}$$

[We use Eqs. (3.10) and (3.11) and, for example, to have **H** in the z-direction we can take $A_y = Hx$. Then an $\dot{\mathbf{r}}$ in the y-direction gives an $m\ddot{\mathbf{r}}$ in the x-direction as it should according to Eq. (3.13).] The final term is the electric field arising from the vector potential, according to Eq. (3.11), in addition to any field arising from other charges, and their resulting potential, which might be included in $\partial V/\partial\mathbf{r}$.

We have confirmed that the Hamiltonian of Eq. (3.14) describes a classical charged particle in an electromagnetic field. In quantum mechanics we follow the procedure of Eqs. (3.6) through (3.8) to obtain the Schroedinger Equation,

$$\boxed{\frac{1}{2m}\left(\mathbf{p} - \frac{-e}{c}\mathbf{A}(\mathbf{r},t)\right)^2\psi(\mathbf{r},t) + V(\mathbf{r},t)\psi(\mathbf{r},t) = i\hbar\frac{\partial\psi(\mathbf{r},t)}{\partial t} ,} \tag{3.20}$$

which we shall have many occasions to use, both for the interaction of electrons with light and for the motion in the presence of a magnetic field. In the case of magnetic fields, we may use units given at the beginning of Chapter 22.

The use of Lagrangians and canonical momentum is illustrated in Problem 3.1 where we consider the quantum mechanics of a ball, translating, spinning, or rolling without slipping.

II. Electronic Structure

Perhaps the greatest accomplishment of quantum theory was providing an understanding of atoms, molecules, and solids in terms of the electrons, which had only been discovered in 1897. Classical physics was not even close to describing their behavior, whereas quantum theory did so quantitatively. It provided an essentially exact description of the hydrogen atom, which includes only a single electron, and which we discuss first. Exact treatment when more than one electron is present is impossible, but very quickly a one-electron approximation was introduced, which was extraordinarily successful, and which provides the basis of our modern understanding of the electronic structure and properties not only of atoms, but of molecules and solids.

Chapter 4. Atoms

The electronic states in atoms, which we think of as spherically symmetric, correspond to the simplest electronic structure. The factorization of the wavefunctions using spherical harmonics, angular-momentum eigenstates which we introduced in Section 2.4, provides the classification of the states and the organization of the subject. The hydrogen atom, with a single electron, is the simplest such atom. We shall then see that an approximate description of the states in other atoms is essentially as simple.

4.1 The Hydrogen Atom

Having just completed a discussion of generalized coordinates and Hamiltonians, we should proceed carefully. The hydrogen atom consists of an electron and a proton so it is describable by the vector position of each particle, six Cartesian coordinates. However, we may transform coordinates to a center-of-mass position $\mathbf{R}$ and the relative coordinate $\mathbf{r}$ from the proton to the electron. The Hamiltonian contains a kinetic energy for the center of mass motion and for the relative motion (with a reduced mass equal to the product of the two masses divided by their sum) and a potential energy of interaction between them, $V(\mathbf{r}) = -e^2/r$. Since there are no terms depending on both $\mathbf{R}$ and $\mathbf{r}$, the wavefunctions for the eigenstates can be factored as $\Psi(\mathbf{R})\psi(\mathbf{r})$, as we saw in Section 2.2. The equation for the center-of-mass wavefunction is just that for a free particle with a mass equal to the mass of the atom moving in free space, and all of the solutions we have obtained for free particles apply. This explains why we can ignore the internal structure of atoms and molecules and apply quantum mechanics to the atom or molecule as a whole. *Everything* is at the same time a particle and a wave. The energies of these states add to the internal energies of the atom, which is the part in which we are interested. $\psi(\mathbf{r})$ is really in terms of a relative coordinate, but because the mass difference is so large, the reduced mass is almost exactly equal to the electron mass, and the nucleus moves very little in comparison to the electron. Thus it may be simplest to think of the nucleus as a classical positive charge at a fixed **R** and the electron orbiting in the potential $-e^2/r$ relative to that position. It is easier to think about it that way and the only error we make is in using a mass slightly too large, which we could correct if we chose. This treating of nuclei as classical charges when doing electronic structure is sometimes called the *Born-Oppenheimer Approximation*, and we make it here.

With spherical symmetry the wavefunction in the relative coordinate can be written

$$\psi(\mathbf{r}) = (\chi(r)/r)\, Y_l^m(\theta,\phi) \tag{4.1}$$

with $\chi(r)$ determined from a radial Schroedinger Equation for the function, Eq. (2.32), which becomes

$$-\frac{\hbar^2}{2m}\frac{\partial^2\chi}{\partial r^2} + \frac{\hbar^2 l(l+1)}{2mr^2}\chi - \frac{e^2}{r}\chi = \varepsilon\chi. \tag{4.2}$$

We shall, as with the harmonic oscillator, treat the simplest case, and write the results for the rest. For $l = 0$ we may try $\chi(r) = re^{-\mu r}$ (corresponding to $\psi(r) \propto e^{-\mu r}$). Eq. (4.2) becomes

$$-\frac{\hbar^2\mu^2 re^{-\mu r}}{2m} + \frac{\hbar^2\mu e^{-\mu r}}{m} - e^2e^{-\mu r} = \varepsilon re^{-\mu r}. \tag{4.3}$$

This will be a solution if the second and third terms cancel, $\hbar^2\mu/m = e^2$, which fixes μ as $e^2m/\hbar^2 = 14.4/7.62$ Å^{-1} = 1/(0.529 Å). Then the energy is given by

$$\varepsilon = -\frac{\hbar^2\mu^2}{2m} = \frac{-e^4m}{2\hbar^2}. \tag{4.4}$$

This is the Rydberg, $14.4^2/(2\times7.62) = 13.6$ eV, using Eq. (1.10). This is the ground state, with normalized wavefunction given by

$$\psi(r) = \sqrt{\frac{\mu^3}{\pi}}\, e^{-\mu r}\ . \tag{4.5}$$

[Note that this is $R(r)$ times $Y_0{}^0(\theta,\phi) = 1/\sqrt{4\pi}$.]

The higher-energy s-states, obtained from Eq. (4.2) with again $l = 0$, are successive polynomials in r times $e^{-\mu r}$, closely analogous to the higher states of the harmonic oscillator. The energies are

$$\varepsilon_n = \frac{-e^4 m}{2\hbar^2}\frac{1}{n^2} \tag{4.6}$$

for polynomials of order $n - 1$. They are obtained analytically in most quantum texts, can be easily obtained as we did the $n = 1$ ground state for small values of n, or can be obtained numerically by the method we use in Problem 4.3.

The index n is the *principal quantum number,* $n = 1, 2, ...$ States of $l \neq 0$ are obtained from Eq. (4.2), including the second (centrifugal) term. At small r they approach r^l , rather than the constant for s-states. Their energies are found also to be given by Eq. (4.6), but for p-states there are solutions with energy ε_n only for $n = 2, 3, ...$ In general the lowest eigenvalue corresponds to $n = l + 1$, and there are $2l + 1$ eigenvalues for each, corresponding to different orientations of the angular momentum The corresponding eigenvalues are sketched in Fig. 4.1.

These results obtain only for a potential equal to $V(r) = -Ze^2/r$, with an increased charge Z scaling the wavefunction and the energy. However, we shall see that all atoms can be approximately represented in terms of a more complicated $V(r)$. Making that change shifts each of the energies shown in Fig. 4.1, the extra attractive potential near the nucleus lowering the s-state energies relative to the p-state energies of the same n in particular. Each level can be followed as the nuclear charge is increased from element to element and successive levels are occupied, providing the organization of the electronic structure of all of the atoms, as we shall see.

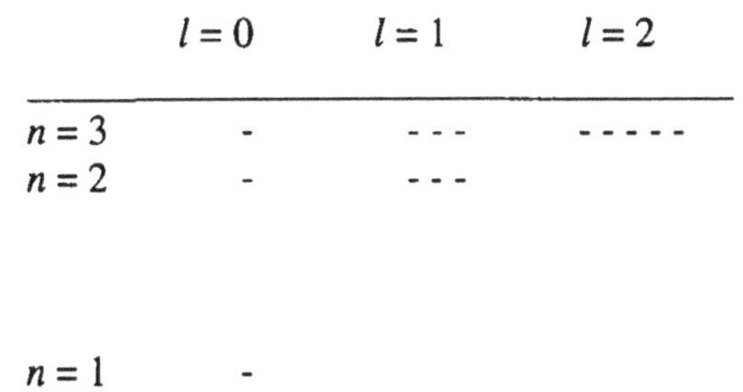

	$l = 0$	$l = 1$	$l = 2$
$n = 3$	-	- - -	- - - - -
$n = 2$	-	- - -	
$n = 1$	-		

Fig. 4.1. The atomic term values, or energy eigenvalues or levels, for the hydrogen atom. Levels of the same n make up *shells*, sometimes subdivided to s-shells and p-shells.

4.2 Many-Electron Atoms

We may move on to the second atom in the Periodic Table, helium with two electrons per atom. It is then necessary to describe the system with a wavefunction depending upon the coordinates of the two electrons, $\Psi(\mathbf{r}_1, \mathbf{r}_2)$ and the Hamiltonian operator becomes

$$H = -\frac{\hbar^2\nabla_1{}^2}{2m} - \frac{\hbar^2\nabla_2{}^2}{2m} - \frac{Z_N e^2}{r_1} - \frac{Z_N e^2}{r_2} + \frac{e^2}{|\mathbf{r}_1 - \mathbf{r}_2|}, \tag{4.7}$$

with $Z_N = 2$ the nuclear charge. It is the final term which causes the problem. Without it the Hamiltonian would be of the form $H(\mathbf{r}_1) + H(\mathbf{r}_2)$ and the wavefunction could be factored as $\psi(\mathbf{r}_1,\mathbf{r}_2) = \psi(\mathbf{r}_1)\psi(\mathbf{r}_2)$, but with that term the six-dimensional problem is only tractable numerically, and becomes quite impossible as we go to heavier elements, more electrons, and therefore more coordinates.

In the early days of quantum mechanics it was recognized that only such separable problems can be solved, so it was asked how good an approximation can it be to treat it as a separable problem. That question can quite unambiguously be answered with a *variational* calculation, a general and powerful method for quantum mechanics.

The general idea of the variational calculation is that any Hamiltonian H, such as Eq. (4.7), has a set of eigenstates Ψ_j with energies ε_j, which we may not know. However, any approximate state can be expanded in the complete orthogonal set, $\{\Psi_j\}$, as $\Psi = \Sigma_j\, u_j\, \Psi_j$. Then for the particular case of two particles, for that approximate state $\Psi(\mathbf{r}_1,\mathbf{r}_2)$, the expectation value of the energy is

$$\frac{\int d^3r_1 d^3r_2 \Psi(\mathbf{r}_1,\mathbf{r}_2)^* H \Psi(\mathbf{r}_1,\mathbf{r}_2)}{\int d^3r_1 d^3r_2 \Psi(\mathbf{r}_1,\mathbf{r}_2)^* \Psi(\mathbf{r}_1,\mathbf{r}_2)} = \frac{\Sigma_j u_j{}^* u_j \varepsilon_j}{\Sigma_j u_j{}^* u_j}. \tag{4.8}$$

The form on the right is obtained by noting that the terms with $\int d^3r_1 d^3r_2 \Psi_i(\mathbf{r}_1,\mathbf{r}_2)^* H \Psi_j(\mathbf{r}_1,\mathbf{r}_2) = \varepsilon_j \int d^3r_1 d^3r_2 \Psi_i(\mathbf{r}_1,\mathbf{r}_2)^* \Psi_j(\mathbf{r}_1,\mathbf{r}_2))$ for i and j different give zero, by the orthogonality of different eigenstates. The result will always be higher than the ground state energy, the lowest ε_j, since every other contribution to the final form is higher. Thus the best possible approximate solution of any given form for the ground state energy can be obtained by minimizing the left side of Eq. (4.8) for the approximate form. The result generalizes immediately to any number of electrons. We shall in fact see that this variational calculation ordinarily gives not only an estimate of the ground-state energy, but the next-lowest state orthogonal to the lowest, etc., giving approximate estimates of the entire range of states. Thus the variational method provides a complete, though approximate, theory of the electronic structure of many-electron systems. A simple variational calculation for the hydrogen ground state is carried out in Problem 4.1.

For the two-electron problem, we make the *one-electron approximation* for the state, $\Psi(\mathbf{r}_1,\mathbf{r}_2) = \psi(\mathbf{r}_1)\psi(\mathbf{r}_2)$, which would be correct if it were possible to separate the Hamiltonian as $H(\mathbf{r}_1) + H(\mathbf{r}_2)$. In some sense this variational approximation for the wavefunction embodies the physical idea which we expressed at the end of the last section, that the hydrogen-like one-electron states retain meaning throughout the periodic table. When the physical concept is sound, the variational calculation tends to be successful. In this case, it provides the basis for our understanding of atoms, molecules, and solids.

We are, then, to minimize

$$E = \frac{\int \psi_2{}^*(\mathbf{r}_2)\psi_1{}^*(\mathbf{r}_1)H\psi_1(\mathbf{r}_1)\psi_2(\mathbf{r}_2)d^3r_1d^3r_2}{\int \psi_2{}^*(\mathbf{r}_2)\psi_1{}^*(\mathbf{r}_1)\psi_1(\mathbf{r}_1)\psi_2(\mathbf{r}_2)d^3r_1d^3r_2} \tag{4.9}$$

with respect to ψ_1 and ψ_2, using the Hamiltonian H from Eq. (4.7). This is done by adding $\delta\psi_1{}^*(\mathbf{r}_1)$ to $\psi_1{}^*(\mathbf{r}_1)$ and asking that the result be stationary with any arbitrary $\delta\psi_1{}^*(\mathbf{r}_1)$. [This is simplest if we take $\psi_1{}^*(\mathbf{r}_1)$ and $\psi_1(\mathbf{r}_1)$ to be independent of each other, and it can be confirmed by writing out real and imaginary parts that this yields the correct result.] This leads to the *Hartree Equations*, which are much like the energy eigenvalue equation,

$$-\frac{\hbar^2}{2m}\nabla^2\psi_1(\mathbf{r}) - \frac{Ze^2}{r}\,\psi_1(\mathbf{r}) + V_{ee}(\mathbf{r})\,\psi_1(\mathbf{r}) = \varepsilon_1\,\psi_1(\mathbf{r}) \tag{4.10}$$

and the corresponding equation for $\psi_2(\mathbf{r})$ is obtained the same way. Here

$$V_{ee}(\mathbf{r}_1) = \int \psi_2{}^*(\mathbf{r}_2)\frac{e^2}{|\mathbf{r}_1-\mathbf{r}_2|}\,\psi_2(\mathbf{r}_2)d^3r_2 \tag{4.11}$$

is an average potential from the other electron, which arises in the one-electron approximation, but is not meaningful otherwise. A real electron sees another point-like electron moving in the system. However, this approximate description enables us to solve the one-electron equation, Eq. (4.10). We may need to iterate the result, obtaining the probability density $\psi_1(\mathbf{r})^*\psi_1(\mathbf{r})$ from a solution of Eq. (4.10), using it to obtain a new $V_{ee}(\mathbf{r}_2)$ from the counterpart of Eq. (4.11), which is used in the $\psi_2(\mathbf{r})$ counterpart of Eq. (4.10) to obtain $\psi_2(\mathbf{r})$ and therefore $\psi_2(\mathbf{r})^*\psi_2(\mathbf{r})$, etc..

The $\psi_i(\mathbf{r}_i)$ obtained in this way would be used to evaluate the energy E from Eq. (4.9). We might hope that we could obtain that total energy as the sum, over occupied states, of the *one-electron energies* ε_j as

$$E \approx \sum_j \varepsilon_j , \tag{4.12}$$

as was true when we factored the three-dimensional states in Eq. (2.5). This we shall in fact assume, but there are a number of complications which we discuss, though briefly, because they are so fundamental to our understanding of real systems. However, for almost all of this text, the one-electron eigenvalue equation, Eq. (4.10), and the interpretation, Eq. (4.12), of the eigenvalues will be all that we need.

The first complication is that Eq. (4.10) includes in ε_1 the Coulomb interaction between the two electrons, and that energy is included again in the evaluation of ε_2. It is necessary really to subtract that interaction counted twice, and that will also be true when we treat collections of atoms. For these collections we must also add the Coulomb repulsion between the nuclei. However, it turns out that as long as the atoms remain approximately neutral, the *change* in energy as the atoms are moved is quite well given by the change in the sum of eigenvalues, as given in Eq. (4.12). Thus we will ordinarily be able to make this tremendously simplifying approximation, Eq. (4.12). We will need to be careful and make corrections when the approximation of neutral atoms is not good.

There is a second complication in that the real many-electron wavefunction for electrons, $\Psi(\mathbf{r}_1,\mathbf{r}_2)$, is antisymmetric with respect to the interchange of the two electrons,

$$\Psi(\mathbf{r}_2,\mathbf{r}_1) = -\Psi(\mathbf{r}_1,\mathbf{r}_2), \tag{4.13}$$

as we shall see in Section 10.5. We can incorporate this antisymmetry in our variational calculation by using an approximate wavefunction

$$\Psi(\mathbf{r}_1,\mathbf{r}_2) = \frac{\psi_1(\mathbf{r}_1)\psi_2(\mathbf{r}_2) - \psi_2(\mathbf{r}_1)\psi_1(\mathbf{r}_2)}{\sqrt{2}}. \tag{4.14}$$

An immediate consequence of this form is the *Pauli Exclusion Principle* that two electrons cannot occupy the same state, since then the wavefunction becomes zero. We have used this before, but this is the origin. Eq. (4.14) is called the *Hartree-Fock Approximation*, and it leads to an additional term in the $V_{ee}(\mathbf{r})$ of Eq. (4.10) which is called the *exchange interaction* and adds an exchange energy which we shall need to mention at various points. This exchange energy arises from the fact that the two-electron wavefunction, Eq. (4.14), clearly approaches zero as $\mathbf{r}_1$ is close to $\mathbf{r}_2$ and as a result the expectation value of the Coulomb repulsion $e^2/|\mathbf{r}_1 - \mathbf{r}_2|$ is reduced. We shall see in Section 10.5 that this reduction applies only to electrons of the same spin, since the states for electrons of opposite spin are already antisymmetric due to the opposite spins, and the spatial wavefunction of Eq. (4.14) has the minus replaced by a plus. We shall also in Section 10.5 extend these antisymmetric wavefunctions to systems involving many electrons.

An important consequence of this exchange energy is *Hund's Rule* which states that when only some orbitals of the same energy (as in Fig. 4.1) are occupied by electrons, the energy will be lower if their spins are the same as each other. Then the corresponding magnetic moments line up to produce magnetic properties for the atom. The same effect in metals such as iron produces ferromagnetism in those metals as we shall discuss in Chapter 22.

Another important feature of the exchange energy in Hartree-Fock theory is the fact that we can add to the electron-electron interaction, Eq. (4.11), the interaction of each electron *with its own electric charge distribution*, if we also add the exchange interaction of each electron with itself; the two self-interactions cancel exactly in Hartree-Fock theory and have the advantage that then the Hamiltonian entering the Eqs. (4.10) for different electrons is exactly the same, while Eqs. (4.10) and (4.11) in general produce different potentials for different electron states. Working with a single potential is a considerable simplification, partly because the different one-electron states are then automatically orthogonal to each other.

Modern calculations for systems with many atoms seldom use Hartree-Fock, but use what is called *Density-Functional Theory* and an approximation to it called the *Local-Density Approximation*. In this approximation the exchange interaction, as well as all the corrections to making a one-electron approximation, $\Psi(\mathbf{r}_1,\mathbf{r}_2) \approx \psi(\mathbf{r}_1)\psi(\mathbf{r}_2)$, in the first place, are incorporated in an effective potential $V_{ee}(\mathbf{r})$ which is assumed to depend only upon the electron density $\rho(\mathbf{r})$ at the point $\mathbf{r}$. The largest contribution to this potential is the exchange interaction, which is usually evaluated for a free-electron gas at the same density. It is found to be

$$E_{ex} = -\frac{3e^2k_F}{4\pi} \text{ per electron,} \tag{4.15}$$

where k_F is of course the Fermi wavenumber for a free-electron gas at that density. It is a numerical constant times e^2k_F as one can see from the form of the integral. [A discussion of Local-Density Theory, with references, is given for example by Hafner (1987), 315ff.]

As we indicated at the outset, for our purposes the result is that for helium, and for all other heavier atoms, *there exists an effective potential which appears in the Schroedinger Equation and the eigenvalue equation, Eq. (4.10), and using it we may proceed in a one-electron approximation, treating the total energy as the sum of the one-electron eigenvalues.*

This effective potential for the atoms is taken to be spherically symmetric, so that the electronic states can by factored into radial functions and spherical harmonics, and classified according to the angular-momentum quantum numbers l, and m, and numbered by the principal quantum numbers n according to the scheme given for hydrogen in Section 4.1. Such a potential exists for each element, and the resulting eigenstates, or *energy levels*, are filled with the appropriate number of electrons for that element. We fill the 1s levels (with terminology given for spherical systems in Section 2.4, 1s- meaning $n = 1, l = 0$) first and then the s- and p-levels for $n = 2$, then successively the $n = 3, n = 4, \ldots$ levels corresponding to successive rows in the periodic table. For each row the levels of the particular n which are being filled are called *valence* states, as distinct from those of lower n which are called *core* states. In the midst of these series, we fill d-levels ($l = 2$) through the transition-metal series, and f-levels ($l = 3$) when they are low enough in energy, but we shall do little with these systems in this text. Extensive treatment is given in Harrison (1999) in much the same spirit as we use here for s- and p-levels. The levels which determine the chemical and physical properties of the elements are the valence levels, in the shells which are partially occupied in the atom, the highest occupied and lowest unoccupied energy levels. Core levels, much lower in energy, are of no consequence since they are so closely tied to the nucleus that they do not change as the atoms are rearranged; higher levels are empty and do not affect the total energy, Eq. (4.12).

The corresponding valence electronic energy levels for the atoms, obtained by Mann (1967) in the Hartree-Fock Approximation, are listed in Table 4.1. They may be thought of as the removal energy for the corresponding electron from the isolated atom, taking that electron to large distances at rest, as would be anticipated from Eq. (4.12). This is only approximate; the experimental removal energies for sodium through argon are [ionization potentials from the CRC Handbook (Weast (1975)] in eV, 5.14, 7.64, (both corresponding to s-states), 5.98, 8.15, 10.48, 10.36, 13.01 and 15.75 (all for p-states). These give a fair assessment of the degree of validity of these numbers. Removal of a second electron will require several electron volts of additional energy since it comes from a positively-charged atom. The needed correction is a consequence of the approximations in the treatment of electron-electron interactions discussed above.

It is usual to represent the states occupied in the atom (the *configuration*

Table 4.1. Hartree-Fock term values for valence levels (Mann (1967)). The first entry is ε_s, the second is ε_p (values in parentheses are highest core level; values with * are extrapolated). All are in eV. Transition metals would appear to the left of Cu, Ag, and Au; f-shell metals would appear to the right of Ba and Ra.

I	II	III	IV	V	VI	VII	VIII	IA	IIA
							He	**Li**	
$n = 1$							-24.98	-5.34 $n = 2$	
	Be	**B**	**C**	**N**	**O**	**F**	**Ne**	**Na**	
$n = 2$	-8.42	-13.46	-19.38	-26.22	-34.02	-42.79	-52.53	-4.96 $n = 3$	
	-5.81*	-8.43	-11.07	-13.84	-16.77	-19.87	-23.14	(-41.31)	
	Mg	**Al**	**Si**	**P**	**S**	**Cl**	**Ar**	**K**	**Ca**
$n = 3$	-6.89	-10.71	-14.79	-19.22	-24.02	-29.20	-34.76	-4.01	-5.32
	-3.79*	-5.71	-7.59	-9.54	-11.60	-13.78	-16.08	(-25.97)	(-36.48)
Cu	**Zn**	**Ga**	**Ge**	**As**	**Se**	**Br**	**Kr**	**Rb**	**Sr**
-6.49	-7.96	-11.55	-15.16	-18.92	-22.86	-27.01	-31.37	-3.75	-4.86
-3.31*	-3.98*	-5.67	-7.33	-8.98	-10.68	-12.44	-14.26	(-22.04)	(-29.88)
Ag	**Cd**	**In**	**Sn**	**Sb**	**Te**	**I**	**Xe**	**Cs**	**Ba**
-5.99	-7.21	-10.14	-13.04	-16.03	-19.12	-22.34	-25.70	-3.37	-4.29
-3.29*	-3.89*	-5.37	-6.76	-8.14	-9.54	-10.97	-12.44	(-18.60)	(24.60)
Au	**Hg**	**Tl**	**Pb**	**Bi**	**Po**	**At**	**Rn**	**Fr**	**Ra**
-6.01	-7.10	-9.83	-12.49	-15.19	-17.97	-20.83	-23.78	-3.21	-4.05
-3.31*	-3.83*	-5.24	-6.53	-7.79	-9.05	-10.34	-11.65	(-17.10)	(-22.31)

of the atom) by giving the quantum number n of the level, followed by its orbital momentum, s, p, d, or f, and an exponent indicating the number of electrons. Thus the ground state of boron is $1s^22s^22p$. The cores are sometimes written in parentheses $(1s^2)2s^22p$, or omitted.

The systematics of the values are quite simple and are worth noting. The first element in the IA column is lithium and its 2s-state ε_s value would be that of the hydrogen 2s-state, $e^4m/(8\hbar^2) = -3.4$ eV, except for the effects of the extra attractive potential from the additional two protons in the nucleus and the extra two core electrons close to the nucleus. These drop the energy to -5.34 eV for Li, and the corresponding shift gets smaller as we move down in the periodic table. -3.4 eV is roughly right for the IA series, particularly if we remember that they are somewhat deeper at the top of the table.

As we increase the nuclear charge in each row, going from Li to Be to B to C, etc., it is customary to write the number of valence electrons (rather than the total nuclear charge Z_N) as $Z = 1, 2, 3, 4...$, so Z is now the Roman numeral at the top of the columns in Table 4.1. The additional attractive potential continues to lower the s-state energy and we see that ε_s increases almost linearly with Z,

$$\varepsilon_s \approx -\frac{e^4 m}{8\hbar^2} Z = -3.4Z \text{ eV}, \tag{4.16}$$

This is not a prediction, but a plausible empirical trend. This works particularly well with the lower rows, such as Na, Mg, Al, etc. where Eq. (4.16) gives -3.4, -6.8. -10.2, -13.6, -17.0, -20.4, -23.8, and -27.2 for columns I through VIII. A final rule of thumb is that for the entire table,

$$\varepsilon_p \approx \varepsilon_s/2. \tag{4.17}$$

We shall confirm that this is expected from pseudopotential calculations in Problem 4.1b. The two rules are so simple that we can always remember approximately the energy levels for all of these elements. They will prove very useful when we begin studying molecules and solids.

4.3 Pseudopotentials

The similarity of the energy levels in successive rows, quantified by these approximate rules for the atomic term values, is the ultimate basis of the *periodic table of the elements*, which classifies the elements by the *valence* I through VIII at the top of Table 4.1. In one way the similarity is surprising since in each successive row there is additional structure, corresponding to an additional node, in the wavefunctions. It is a fact that this additional structure in the wavefunction, associated with the atomic core, does not greatly affect the energy of the states, nor therefore most chemical and physical properties of the atom. This fact is made explicit in the concept of a *pseudopotential* which replaces the true intricate wavefunction by a simple *pseudowavefunction*.

There are many ways to formulate such pseudopotentials (for discussion see Harrison (1966)), but one of the simplest, and the one we use here, is the Ashcroft (1966) *empty-core pseudopotential*. In this approximation the potential for the free atom is replaced by the true potential outside some "core radius" and zero inside, with the core radius adjusted so that the lowest-energy s-state for that modified potential has energy equal to the valence s-state. We may illustrate this using the hydrogen 2s-state, shown in Fig. 4.2, along with the hydrogen potential. A pseudopotential $w(r)$, also shown, may be constructed such that the 1s-state is raised in energy up the original 2s-state energy, $-{}^1/_8 me^4/\hbar^2$. This nodeless pseudowavefunction is also shown in Fig. 4.2. In the same way, the core radius could be adjusted such that the pseudowavefunction had energy equal to any of the valence s-state energies for the alkali metals Li, Na, K, etc., listed in Table 4.1.

The resulting pseudopotential can be used to describe the valence states of the corresponding metal, but has eliminated the core states. The pseudopotential will be much the same for all elements in one column of the periodic table, since they have similar valence s-state energies, though the real wavefunctions are quite different in the core region. In that way it makes explicit the periodicity of the elements. More importantly, this pseudopotential is sufficiently weak that it becomes understandable that the electrons in metals are so much like free-electrons, as discussed in Section 2.2, in spite of the fact that the potentials are so large that they introduce several nodes in the true wavefunction near each nucleus.

More generally a pseudopotential can be constructed for elements from the Z 'th column in the periodic table as

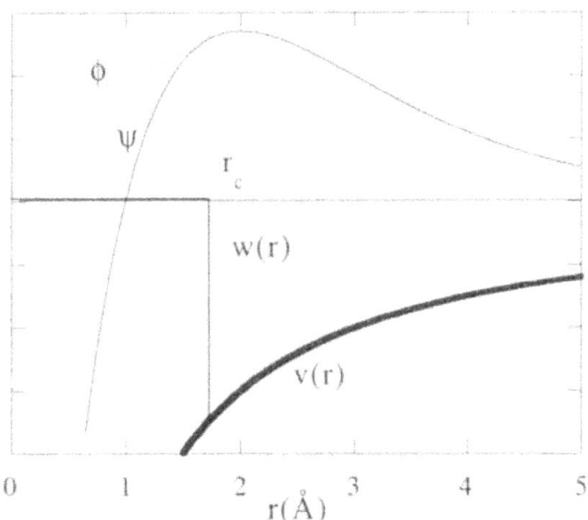

Fig. 4.2. $v(r)$ is the potential $-e^2/r$ for hydrogen, and ψ is the corresponding 2s-state, with one node. The empty-core pseudopotential $w(r)$ has the core radius r_c adjusted such that its 1s-state, the pseudowavefunction ϕ, has the same energy as the hydrogen 2s-state.

$$w(r) = \begin{cases} 0 & \text{for } r < r_c \\ -Ze^2/r & \text{for } r > r_c, \end{cases} \tag{4.18}$$

with generally an additional potential for $r > r_c$ to approximate the potential arising from the charge of the other valence electrons (perhaps even approximated by a uniform charge density). Since the core radius turns out to lie generally outside the region where the electron density of the cores is appreciable, they do not affect the potential. In Problem 4.3 we obtain the core radii for lithium and sodium, using the s-state energies from Table 4.1. This can be done with a numerical integration of the radial Schroedinger Equation just as we obtained the harmonic-oscillator function in Problem 2.9. We also use this same pseudopotential for sodium to calculate the lowest p-state energy, to be compared with the $\varepsilon_p \approx \varepsilon_s/2$ of Eq. (4.17). We find that generally the empty-core pseudopotentials obtained in this way give the valence p-states in accord with Table 4.1 roughly on the scale of accuracy of the agreement between that table and the experimental ionization potentials.

When we put these atoms together to form solids, the representation of the atomic potentials by weak pseudopotentials will turn out to be an

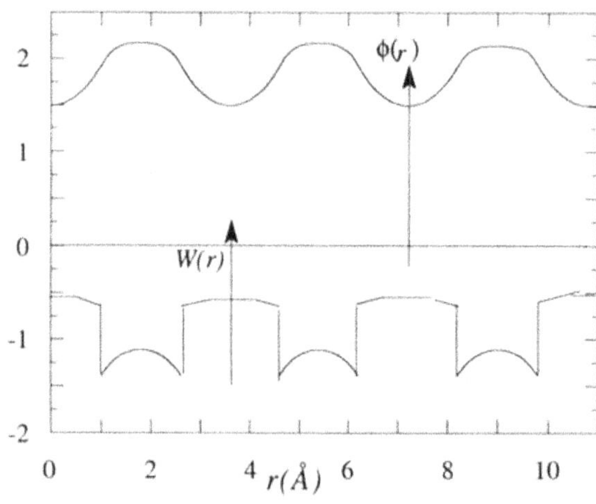

Fig. 4.3. The pseudopotentials for sodium atoms, obtained in Problem 4.3 and similar to that shown in Fig. 4.2, are added for atoms at the sodium spacing of 3.66 Å to give the $W(r)$ shown. The lowest electronic pseudowavefunction in the metal is given approximately by a sum of the corresponding atomic pseudowavefunctions, shown as $\phi(r)$.

extraordinary simplification. We shall see in Chapter 13 that it is a good approximation to simply add the pseudopotentials for all of the atoms, as illustrated in Fig. 4.3. Then the pseudowavefunction for the electronic state in the metal is approximately given by the sum of atomic pseudowavefunctions, just as the full wavefunction is written as a sum of full atomic states in Chapter 6. This pseudowavefunction is shown also in Fig. 4.3. We may note in passing that the net effect of the pseudopotential is a repulsion, reducing the pseudowavefunction near the atomic site. The sum of pseudowavefunctions in a metal is nearly enough constant (in strong contrast to a sum of full atomic states) that we can approximate it as a constant, and correct for the effect of the weak pseudopotential, making many calculations for the solid elementary. For treating molecules in Chapter 5, we return to a description in terms of full atomic states.

4.4 Nuclear Structure

For most purposes in this text, the only properties of the nucleus which will be needed are its mass and its charge. However, the nucleus does have structure on its own, which is heavily influenced by the quantum-mechanical effects we have discussed. Further, we shall see in Section 10.4 that the nuclear structure will actually determine how molecules can tumble, and the zero spin we shall find for helium is responsible for its superfluidity which we discuss in Section 10.3. Here we give only the most basic aspects of the application of quantum theory to nuclear structure. More details are available from many sources, such as *Nuclear Structure Theory* by Irvine (1972).

The nucleus is composed of *nucleons*, the positively charged *protons* and the *neutrons*, without charge. Both have a spin angular momentum of $^1/_2\hbar$, as do electrons, and very nearly the same mass as each other, about 2000 times that of the electron. The small difference is important because the neutron, being slightly heavier and having therefore greater rest energy, can beta-decay (with a half-life of several minutes) into a proton by emitting an electron (beta-ray, or β-ray) and a neutrino, as we shall discuss in Section 9.5. This need not occur in the nucleus where the Coulomb energy from the other protons raises the energy of the proton. They both have magnetic moments but, not surprisingly, they are different. They are held together in the nucleus by strong, short-range forces, which are very much the same between any pair, protons with protons, protons with neutrons, and neutrons with neutrons. We shall describe the origin of these forces, π-mesons, in Section 17.4.

The experimental properties of the ground state of the nucleus are very reminiscent of that of a drop of ordinary liquid, on a very much smaller scale; the diameter of the nucleus as determined by scattering experiments is several times 10^{-13} cm. The corresponding view is called the *liquid-drop model*, discussed in the earliest days by Bohr. These drop-like properties are that the nuclei are approximately spherical and the volume, as measured by scattering experiments, is approximately proportional to the number of nucleons making up the nucleus. Further, the binding energy of the nucleus - the energy required to separate it into individual nucleons - is also approximately proportional to the number of nucleons. Since the diameter of the nucleons is small on the scale of the nucleus, the nucleus might best be thought of as a drop of liquid metal, in which the distance between nucleons, thought of as ion cores, is large compared to the core diameter.

Much more detailed properties of the nucleus can be obtained by taking the same one-particle view which we developed for electrons in Section 4.2, called the *shell model* for the nucleus. This approximation can be justified by the same variational calculation which we used for electrons. Then we say that each nucleon moves in the average potential arising from the interaction with all of the other nucleons. This potential should be spherically symmetric for the spherical nucleus and the liquid-drop model would suggest a square-well potential such as we discussed in Section 2.4. Then the one-particle states for a proton will again have the angular dependence of the spherical harmonics and the radial wavefunctions, $j_l(kr)$, and energies can be calculated just as we calculated the states for electrons. Further, the protons have half-integral spin and will obey the Pauli Principle, filling the lowest-energy states just as they were filled with electrons in atoms and metals. Since the neutrons have almost the same strong internuclear interaction, the neutron states will have very similar energies, with the shifts due to lack of charge being quite small. The neutrons also

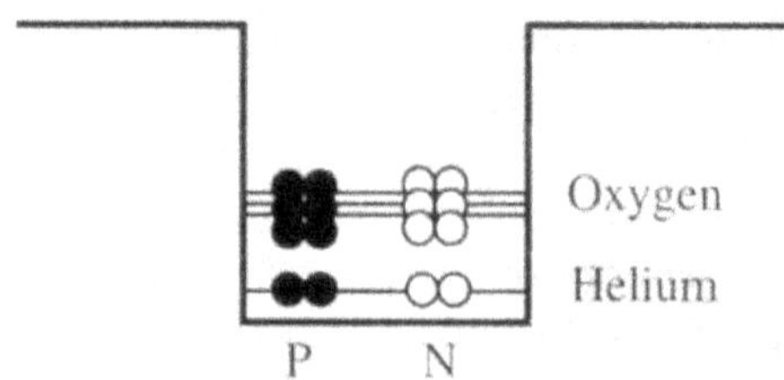

Fig. 4.4. The one-nucleon potential in the shell model is a spherical well of approximately constant depth. The 1s-state is lowest, and can be occupied by two protons and two neutrons, giving the helium nucleus. The next level is a p-state, which can accommodate up to six additional protons and six additional neutrons. For successive additional protons this gives Li, Be, B, C, N, and O. The full shell is O^{16}, oxygen-16.

obey the Pauli principle, and fill the neutron states independently of the filling of the states by protons since the neutrons and protons are not identical to each other and can both occupy the same states. This is just as electrons - or protons - of different spin can occupy the same orbital.

This shell model immediately suggests some of the most important properties of nuclei. First, as in atoms, there will be shells of levels of increasing energy. If there are two protons in the nucleus, helium, they will occupy the lowest s-state, as in Fig. 4.4. We see in Problem 4.2 that the next highest state in the spherical cavity is a p-state, so in lithium, with three protons, the third will go into that p-state at considerably higher energy. This gives a correction to the liquid-drop model, indicating that the total binding of the nucleons is not *exactly* proportional to the number of nucleons, but those with a newly filled closed shell will be extra stable. This happens again with a total of eight protons (oxygen) where the p-shell is filled (Fig. 4.4) and in the fluorine nucleus the additional proton must go into the 2s-state.

At the same time that we are filling the proton states, it will be favorable to fill the neutron states, of very nearly the same energy. If in lithium (three protons) we had not put two neutrons into the neutron 1s-state, this third proton would decay by emitting a positron (the antiparticle version of beta-decay) to transmute the nucleus to helium with an additional neutron. There must always be approximately the same number of protons and neutrons for these light nuclei. In particular, the helium nucleus with two protons and two neutrons filling the 1s-nuclear states is especially stable, as is the oxygen nucleus with eight protons and eight neutrons, both shown in Fig. 4.4. When a nuclear shell of protons is partly filled, the same shell of neutrons can be partly filled with a different number of neutrons without producing the instability mentioned for lithium, allowing different *isotopes* of the same element, nuclei with the same number of protons, but different numbers of neutrons.

As we move to increasingly large numbers of protons and neutrons in the nucleus, the depth of the square well binding the nucleons remains approximately constant because the nucleon-nucleon interaction is of so short a range that each nucleon sees only a few neighbors at one time. Thus the well expands in volume and the one-particle states become closer together. As in adding atoms to a metal the Fermi energy remains about the same, as does the cohesive energy per nucleon, and the volume increases, all as suggested by the liquid-drop model. However, as there are more and more protons, the Coulomb interaction between them raises the proton energy more and more above the neutron energy and it becomes favorable to have *more* neutrons than protons, up to 50% more for the heavier elements. Otherwise the protons would emit positrons to produce more neutrons.

This simplest shell model also describes excited states of nuclei, analogous to excited electronic states of atoms. Also as in atoms a nucleus in an excited state can emit photons and drop to the ground state. These processes can be calculated just as we calculate them for electrons in atoms, molecules, and solids. In the case of nuclei, which are so strongly bound, the energy differences are huge and the photons have energies of the order of millions of electron volts, *gamma rays*, rather than a few electron volts for electronic transitions in atoms.

The shell model provides an understanding of the magnetic moments of nuclei. In particular, in the helium nucleus with the proton 1s-state filled with both spins, and of the neutron 1s-state filled, the nucleus has no net spin and no magnetic moment. The same is true of the nucleus of oxygen with its 2p-shells completely filled. We shall discuss some of the consequences of these zero spins in Chapter 10. The magnetic moments of *other* nuclei allow nuclear magnetic resonance (NMR) when magnetic fields are applied and microwave radiation is used to cause transitions between different orientations of the nuclear magnetic moment.

The shell model also provides the basis for the theory of fission and fusion of nuclei. It is of course an approximate theory as is our theory of electronic states in atoms, but again a very successful one. For the case of fission of a heavy nucleus, such as uranium with a ratio of neutrons to protons of 1.6, into two lighter nuclei, with smaller ratios for the stable isotopes, it is not surprising that extra neutrons are emitted. These neutrons causing fission of other uranium nuclei is of course the origin of the chain reactions in nuclear reactors and bombs. Much more detailed theory is necessary to describe such processes well. One of the most important refinements of the theory is the addition of spin-orbit coupling, which we shall describe for electronic systems in Section 22.5.

There is also structure to the nucleons, each being constructed of three *quarks*, held together by *gluons*. Indeed the quarks may be without mass, so the nucleon mass arises from the binding of the quarks together. The corresponding Standard Model of fundamental particles is beyond the scope of this text, and of this author. Isolated quarks have not been observed, and indeed they may be unobservable in principle. Just as the ends of a string cannot be isolated, pulling quarks apart may require enough energy to produce the new quarks needed to form new nucleons. This may be the most suitable point to stop the discussion at the fundamental-particle end. In this realm nature has given us an extraordinary variety of systems, but again quantum mechanics governs the behavior of those systems.

Chapter 5. Molecules

When we bring atoms together to form molecules, we may use the same one-electron approximation which we used for atoms, and the potential, or pseudopotential, is approximately a superposition of atomic potentials, $V(\mathbf{r}) = \Sigma_j \, v(\mathbf{r} - \mathbf{r}_j)$, summed over the positions $\mathbf{r}_j$ of the nuclei. However, we have lost the spherical symmetry, the corresponding factorization of the wavefunction, and reduction to a simple radial Schroedinger Equation. An alternative approximation has proven very successful, the representation of the states as *Linear Combinations of Atomic Orbitals*, the LCAO method. It allows meaningful molecular states even including only the valence atomic states, those listed in Table 4.1, in that representation. Further, it is applicable to solids as well as to molecules. Strictly speaking, we shall not use the LCAO method here, but the concept of the LCAO method, and obtain some of the parameters needed for the calculation from other sources. Such an approach is generally called *Tight-Binding Theory*. We apply this theory first to a molecule composed of two lithium atoms, which is a simple prototype of molecules in general, and then move on to other molecules with new features. The hydrogen molecule, H_2, might be simpler, but with no cores it is scarcely any kind of prototype.

5.1 The Li_2 Molecule

For the lithium atom the valence s-state was obtained from

$$-\frac{\hbar^2}{2m}\nabla^2\psi(\mathbf{r} - \mathbf{r}_j) + v(\mathbf{r} - \mathbf{r}_j)\psi(\mathbf{r} - \mathbf{r}_j) = \varepsilon_{2s}\psi(\mathbf{r} - \mathbf{r}_j) \qquad (5.1)$$

For the molecule we add the potentials centered at two positions, $\mathbf{r}_1$ and $\mathbf{r}_2$, giving again a one-electron equation, Eq. (5.1), but with $v(\mathbf{r} - \mathbf{r}_j)$ replaced by $V(\mathbf{r}) = v(\mathbf{r} - \mathbf{r}_1) + v(\mathbf{r} - \mathbf{r}_2)$, shown at the bottom in Fig. 5.1. This Hamiltonian is symmetric around the nuclear midpoint ($\mathbf{r} \rightarrow -\mathbf{r}$ relative to this center) and it follows that the eigenstates can be taken as symmetric or antisymmetric relative to this point. [This follows because if $\psi(\mathbf{r})$ is an eigenstate of H, then $\psi(-\mathbf{r})$ is an eigenstate of the same energy, as are $\psi(\mathbf{r}) \pm \psi(-\mathbf{r})$ each of which is either even or odd or zero.] Thus if we are to approximate the eigenstates of the molecule as linear combinations of the atomic s-states, we shall approximate them by

$$\psi(\mathbf{r}) = \frac{\psi_{2s}(\mathbf{r} - \mathbf{r}_1) \pm \psi_{2s}(\mathbf{r} - \mathbf{r}_2)}{\sqrt{2}} \quad . \tag{5.2}$$

The combination with the plus has lower energy, is the *bonding state*, and is plotted in Fig. 5.1. The combination with the minus is the *antibonding state*, has one more node and therefore has higher energy. We might expect these tight-binding states to be more accurate when the two atoms are further apart, but they turn out to be meaningful at the observed spacing of the molecule, 2.67Å.

We may estimate the energy of either state as the expectation value of

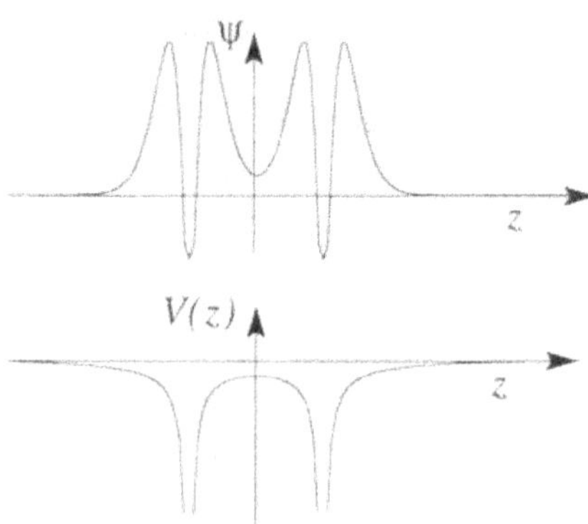

Fig. 5.1. Below is a plot of the potential for a Li_2 molecule, plotted along the z-axis through both nuclei, and measured from the midpoint. Above is the even (bonding) combination of lithium 2s-states.

the Hamiltonian with respect to this approximate state, $\langle H\rangle = \int\psi^*(\mathbf{r})H\psi(\mathbf{r})\, d^3r \,/ \int\psi^*(\mathbf{r})\psi(\mathbf{r})\, d^3r$ as in Eqs. (1.13) and (1.14). This is a good point to introduce the *Dirac notation* for the states and expectation values which we showed in Fig. 2.9 and which we shall use throughout the text. Each state $\psi(\mathbf{r})$ is represented by a *ket*, written $|\psi\rangle$, usually with the symbol ψ replaced by a symbol distinguishing the state. We used such symbols for states in spherical potentials for Fig. 2.8. Here the atomic state $\psi_{2s}(\mathbf{r} - \mathbf{r}_1)$ might be written $|1\rangle$. The complex conjugate of this state is written as a *bra*, $\langle 1|$. When the bra and the ket face each other (a bracket), we are to integrate over all coordinates involving these states,

$$\langle 2|H|1\rangle \equiv \int\psi_{2s}(\mathbf{r} - \mathbf{r}_2)^* H\psi_{2s}(\mathbf{r} - \mathbf{r}_1)\, d^3r. \tag{5.3}$$

The numbers $H_{ij} = \langle i|H|j\rangle$ are called *matrix elements*, making up a *matrix*, with as many rows and columns as we have states.

In terms of this notation, we have written our bonding state as $(|1\rangle + |2\rangle)/\sqrt{2}$. Note that if the states $|1\rangle$ and $|2\rangle$ are normalized, $\langle 1|1\rangle = \langle 2|2\rangle = 1$, the bonding state is approximately normalized, $(\langle 1|1\rangle + \langle 1|2\rangle + \langle 2|1\rangle + \langle 2|2\rangle)/2 \approx 1$ if the *overlap* $\langle 1|2\rangle = \langle 2|1\rangle$ is small. In tight-binding theory we will take these overlaps, representing the nonorthogonality of the two atomic states, to be zero. We shall partly correct for this approximation by adjusting other parameters which enter our calculations, and partly by introducing shortly their real effect in holding the atoms apart. If we do neglect these overlaps, the energy of the bonding state of Eq. (5.2) becomes

$$\varepsilon_B = \langle H\rangle = \frac{\int\psi^*(\mathbf{r})H\psi(\mathbf{r})\, d^3r}{\int\psi^*(\mathbf{r})\psi(\mathbf{r})\, d^3r} = \frac{\langle 1|H|1\rangle + \langle 2|H|2\rangle + 2\langle 1|H|2\rangle}{2}. \tag{5.4}$$

We have taken $\langle 2|H|1\rangle = \langle 1|H|2\rangle$, which is true if the two states are real functions. This is obvious for the integration over the potential, and proven by two partial integrations for the kinetic energy operator. The mathematical statement is that $\langle 2|H|1\rangle^* = \langle 1|H|2\rangle$, being called the *Hermitian* property of the Hamiltonian matrix, and it will apply to all operators we consider. We further note that $\langle 1|H|1\rangle$ is the energy of the first atomic state, which we write ε_s, though it could be shifted from the free-atom value by the potential from the neighboring atom. Finally, we write the matrix element $\langle 1|H|2\rangle$ between two neighboring atomic s-states as $V_{ss\sigma}$, a notation we shall use in the remainder of the text. The σ subscript is redundant in this case, in representing the component of angular momentum around the internuclear axis as $m = 0$, always zero for s-states. When there is one unit of angular momentum around this axis we shall use the subscript π, analogous to the "p" in p-states, etc. This set of standard approximations, which has proven very useful, leads us from Eq. (5.4) to

$$\varepsilon_B = \varepsilon_s + V_{ss\sigma}. \tag{5.5}$$

The same approximations for the antibonding state, the minus in Eq. (5.2), leads to $\varepsilon_A = \varepsilon_s - V_{ss\sigma}$.

We shall find in Section 6.2 that in solids there is a simple approximate formula for this $V_{ss\sigma}$ for two atoms separated by the distance d , which should also apply approximately in molecules. It is

$$V_{ss\sigma} = -\frac{\pi^2}{8}\frac{\hbar^2}{md^2}. \tag{5.6}$$

It is negative as we might expect since $<1|H|2>$ is

$$<1|(-\frac{\hbar^2\nabla^2}{2m}+v(\mathbf{r}-\mathbf{r}_1)+v(\mathbf{r}-\mathbf{r}_2))|2> = <1|\,\varepsilon_s + v(\mathbf{r}-\mathbf{r}_1)|2> \approx <1|v(\mathbf{r}-\mathbf{r}_1)|2>. \tag{5.7}$$

In the first step we noted that $|2>$ is an eigenstate of the Hamiltonian with only the potential $v(\mathbf{r}-\mathbf{r}_2)$, and in the final step we again neglected $<1|2>$. In the final form, we note that $|1>$ and $|2>$ are of the same sign where they overlap and the potential is negative, leading to a negative $<1|H|2>$. Further, $V_{ss\sigma}$ decreases at large spacing, as we should expect.

If we are willing to use this approximate expression, Eq. (5.6), and the approximations which led us to Eq. (5.5), we obtain the electronic structure of this Li_2 molecule, in the same sense we obtained the electronic structure for the isolated atoms, but the predictions for the molecule are much richer. We saw in Eq. (4.12) that the total energy may be thought of as the sum of the energies of occupied states in the system. At least any change in energy of the system as the atoms are rearranged can be estimated as the change in that energy. Each lithium atom began with one electron in a state with energy ε_s, and with two electrons in core states with energies which have negligible change as the atoms are rearranged. Thus we now have the ability to calculate the total energy of the molecule, which allows us to calculate the molecular binding energy and the vibrational frequency, as well as the electric polarizability and the optical spectra. Some of these are carried out in Problems 5.1 and 5.2

5.2 The Variational Method

For Li_2 we could use the symmetry of the molecule to write down the molecular electronic states directly, but when that symmetry is not present, we use the variational method, which we introduced in Section 4.2. We again wish to write the molecular state as a linear combination of atomic orbitals $|j\rangle$,

$$|\psi\rangle = \Sigma_j u_j |j\rangle , \tag{5.8}$$

but now we do not know the coefficients. If we did, we would evaluate the energy as in Eq. (5.4) as

$$\varepsilon \equiv \frac{\langle\psi|H|\psi\rangle}{\langle\psi|\psi\rangle} = \frac{\Sigma_{ji} u_j{}^* u_i \langle j|H|i\rangle}{\Sigma_{ji} u_j{}^* u_i \langle j|i\rangle} . \tag{5.9}$$

The idea of the variational method is that the energy obtained from Eq. (5.9) will always be higher than the real *ground-state* (lowest) energy eigenvalue, as we showed in Section 4.2. The best estimate of the ground state which we can make is the lowest expectation value we can obtain using that form. This was illustrated in Problem 4.1 by approximating the hydrogen atomic state by a form $A\, e^{-\alpha r^2}$ and adjusting α to obtain the lowest energy state. The form is not very close to the real hydrogen state which we described in Section 4.1, but by doing the variation we obtain a reasonable estimate of the ground state.

Returning to the approximate expansion in atomic states, the best estimate we can obtain for the ground state is the lowest possible energy from Eq. (5.9). Thus we should vary the u_i in Eq. (5.9) to obtain the lowest energy to get the best estimate of the ground state energy. When we do this, we will find not only our best estimate of the ground state, but the lowest possible energy of a state orthogonal to our ground-state estimate, which is our best estimate of the second-lowest state. Similarly, we obtain estimates for as many eigenstates as we have terms in our expansion, Eq. (5.8).

We can easily obtain the minimum by setting the derivative of Eq. (5.9) with respect to each u_i equal to zero, but we shall obtain the same result using *Lagrange multipliers* (e. g., Mathews and Walker (1964), p. 313), a method which will be very useful at other points in the book, particularly for statistical physics in Chapter 10. We develop it first for a case with only two terms in the expansion, Eq. (5.8), and take the coefficients u_i to be real, which is not an important limitation. In tight-binding theory we neglect the nonorthogonality $\langle 1|2\rangle$ as we did for Li_2, and normalize the states, $g(u_1,u_2)$

$\equiv u_1^2 + u_2^2 - 1 = 0$. Then the denominator in Eq. (5.9) is one, and we must vary the u_1 and u_2 (the variations are written du_1 and du_2) subject to retaining the normalization condition $\partial g/\partial u_1 du_1 + \partial g/\partial u_2\, du_2 = 0$. The condition for a minimum ε from Eq. (5.9) is then $\partial\varepsilon/\partial u_1 du_1 + \partial\varepsilon/\partial u_2\, du_2 = 0$. We take the two terms to opposite sides of both equations and divide, to see that

$$\frac{\partial\varepsilon/\partial u_1}{\partial g/\partial u_1} = \frac{\partial\varepsilon/\partial u_2}{\partial g/\partial u_2} \equiv \lambda \tag{5.10}$$

with the common ratio λ called the Lagrange multiplier. These two equations with λ are,

$$\begin{aligned} \partial\varepsilon/\partial u_1 - \lambda\partial g/\partial u_1 &= 0\,, \\ \partial\varepsilon/\partial u_2 - \lambda\partial g/\partial u_2 &= 0\,, \end{aligned} \tag{5.11}$$

exactly the conditions we obtain if we minimize the energy $<\psi|H|\psi> - \lambda(<\psi|\psi> - 1)$ as if u_1 and u_2 were independent variables without any condition. The two equations are solved together to obtain the state. For this case, the two equations, using $<\psi|H|\psi> = \Sigma_{ji} u_j{}^* u_i <j|H|i>$, are

$$\begin{aligned} <1|H|1>u_1 + <1|H|2>u_2 &= \lambda u_1, \\ <2|H|1>u_1 + <2|H|2>u_2 &= \lambda u_2. \end{aligned} \tag{5.12}$$

(We have divided out a factor of two and noted that $<1|H|2> = <2|H|1>$ for this case and written the form which is the proper generalization to other systems.) If we multiply the first equation by u_1 and the second by u_2 and add the two equations, we see that λ is our estimate of the energy, ε in Eq. (5.9), and so we may replace λ by ε in Eq. (5.12).

For the more general case where we expand the state in a large number N of terms, as in Eq. (5.8), we write $<j|H|i> = H_{ji}$ and these variational equations generalize to the matrix equation

$$\Sigma_i\, H_{ji} u_i = \varepsilon u_j, \tag{5.13}$$

with N equations, numbered by j. Solving these N linear algebraic equations yields N eigenvalues ε with their N orthogonal eigenstates $\{u_i\}$. The lowest eigenvalue ε is our estimate of the ground state, the next is our estimate of the first excited state (the lowest energy state of the form of Eq. (5.8) and orthogonal to the approximate ground state), etc.

This is an extraordinarily useful and powerful method. No matter how difficult the problem, if our physical intuition suggests the nature of the states we may write an approximate state Eq. (5.8) corresponding to that intuition and obtain algebraically the best solutions consistent with that intuition. One reason it is as accurate as it is, is that because of the variational condition, if we have made some small error $\delta\psi$ in the state, the error in the energy eigenvalue will only be of order $\delta\psi^2$, and even smaller.

We used a variational method to obtain the Hartree Equations for many-electron systems, which should give the best product wavefunction, and the Hartree-Fock Equations which give the best one-electron approximation with antisymmetric states. Similarly, the Bardeen-Cooper-Schrieffer (1957) ground-state wavefunction gives the best "paired" electron state (allowing for uncertain electron numbers, as we shall see). Here we use the variational calculation to obtain the best approximate one-electron states based upon linear combinations of atomic orbitals.

5.3 Molecular Orbitals

The first generalization we make is for an expansion again in only two states, but with different energies. In a molecule NaLi it is based on the s-states from each atom, having energies from Table 4.1 of ε_1 = -4.96 eV for sodium and ε_2 = -5.34 eV for lithium. Making the same approximations as for Li_2, the two variational Eqs. (5.12) become

$$\begin{aligned} \varepsilon_1 u_1 + V_{ss\sigma} u_2 &= \varepsilon u_1, \\ V_{ss\sigma} u_1 + \varepsilon_2 u_2 &= \varepsilon\, u_2. \end{aligned} \tag{5.14}$$

They may be solved together by eliminating the u_1 and u_2 to obtain

$$\varepsilon = \frac{\varepsilon_1 + \varepsilon_2}{2} \pm \sqrt{\left(\frac{\varepsilon_1 - \varepsilon_2}{2}\right)^2 + V_{ss\sigma}^2}\ . \tag{5.15}$$

We see that this leads to the correct result, $\varepsilon = \varepsilon_s \pm V_{ss\sigma}$, for the case in which $\varepsilon_1 = \varepsilon_2 = \varepsilon_s$, and substituting either of these values into Eqs. (5.14) yields $u_2 = \mp u_1$, with magnitude equal to $1/\sqrt{2}$ for normalization. Eq. (5.14) also gives the correct results when $V_{ss\sigma} = 0$; they are $\varepsilon = \varepsilon_1$ and ε_2.

This problem of two coupled levels arises so frequently that it is convenient to write it in general form. For two levels, coupled by a *covalent energy* of magnitude V_2, and differing in energy by twice a *polar energy* $V_3 = (\varepsilon_2 - \varepsilon_1)/2$, the resulting state energies are as in Eq. (5.15),

$$\varepsilon = \bar{\varepsilon} \pm \sqrt{V_2^2 + V_3^2}\ , \qquad (5.16)$$

with of course $\bar{\varepsilon} = (\varepsilon_2 + \varepsilon_1)/2$. The energy may be substituted back into the variational equations to obtain the coefficients u_i. It may be readily verified that the results may be written in terms of the *polarity*

$$\alpha_p = \frac{V_3}{\sqrt{V_2^2 + V_3^2}} \qquad (5.17)$$

as

$$u_1 = \sqrt{\frac{1 + \alpha_p}{2}}\ , \quad u_2 = \sqrt{\frac{1 - \alpha_p}{2}} \qquad (5.18)$$

for the lower, or *bonding*, state and

$$u_1 = \sqrt{\frac{1 - \alpha_p}{2}}\ , \quad u_2 = -\sqrt{\frac{1 + \alpha_p}{2}} \qquad (5.19)$$

for the upper, or *antibonding*, state. We may easily confirm that these lead to the correct results for the case $V_3 = 0$ and for the case $V_2 = 0$, and that the bonding and antibonding states are orthogonal to each other and normalized.

An interesting application of such equations is the calculation of the polarizability of the Li_2 molecule, carried out in Problem 5.2. V_3 is zero, but if we apply an electric field $\mathbf{E}$ along the internuclear axis $\mathbf{d}$, an energy difference for the two states arises equal to $\varepsilon_2 - \varepsilon_1 = -(-e)\mathbf{E}\cdot\mathbf{d}$. This then gives rise to a dipole $\mathbf{p} = -e\mathbf{d}(u_2^2 - u_1^2)/2$ for each of two electrons. In Problem 5.2 we obtain the polarizability α defined by $\mathbf{p} = \alpha\mathbf{E}$, neglecting higher-order terms in $\mathbf{E}$.

It is instructive to look briefly at the effects of the nonorthogonality $\langle i|j\rangle$ of the two atomic states, which we have thus far neglected. We can in fact include them in the variational energy of Eq. (5.9), which we do for the case of two coupled states with $V_3 = 0$. We write $\varepsilon_1 = \langle 1|H|1\rangle = \langle 2|H|2\rangle$ and note $\langle 2|1\rangle = \langle 1|2\rangle$ if both are real and correspondingly $\langle 2|H|1\rangle =$

$<1|H|2>$. Then for this case, we have seen that because of the symmetry states can be written with $u_2 = \pm u_1$. Eq. (5.9) becomes

$$\varepsilon \equiv \frac{<\psi|H|\psi>}{<\psi|\psi>} = \frac{\varepsilon_1 \pm <1|H|2>}{1 \pm <1|2>} \tag{5.20}$$

Half the energy difference between the antibonding state ($u_2 = - u_1$ if the atomic wavefunctions are the same on the two sites) and the bonding state, which we identify with the covalent energy, is

$$\begin{aligned} V_2 &= \frac{\varepsilon_1 + <1|H|2>}{2(1+<1|2>)} - \frac{\varepsilon_1 - <1|H|2>}{2(1-<1|2>)} \\ &= \frac{\varepsilon_1<1|2> - <1|H|2>}{1 - <1|2>^2} = - \frac{<1|v(\mathbf{r}-\mathbf{r}_1)|2>}{1 - <1|2>^2}. \end{aligned} \tag{5.21}$$

In the final step we used the first equality in Eq. (5.7).

We may also obtain the average of the bonding and antibonding states from Eq. (5.20), substituting the V_2 from the second form in Eq. (5.21), to find that it is given by exactly $\varepsilon_1 + <1|2>V_2$. This is illustrated in Fig. 5.2 where we confirm that if we neglect the nonorthogonality $<1|2>$, the antibonding and bonding states are split equally up and down, and that the effect of the nonorthogonality is simply to shift both levels upward. In Li_2, with two electrons in the bonding state this nonorthogonality adds an *overlap repulsion* to the system, an increase in energy as the atoms are brought together, of

$$V_0(d) = 2<1|2>V_2(d) \propto \frac{1}{d^4}. \tag{5.22}$$

To obtain the final form we note that Hoffmann (1963) speculated that $<1|H|2>$ could be related approximately to $<1|2>$ by keeping the first term in the middle form in Eq. (5.7), rather than the second term as we did.

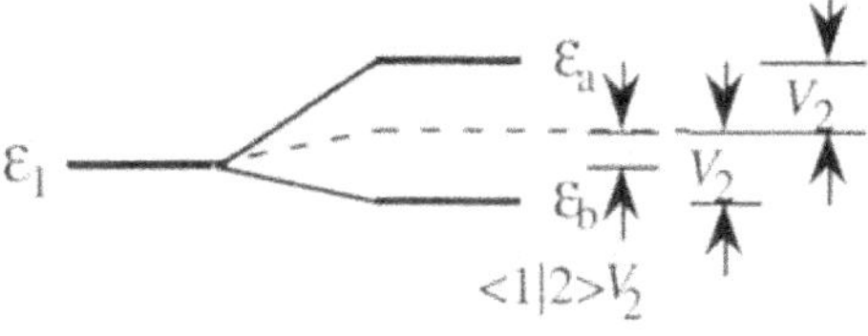

Fig. 5.2. When two levels at energy ε_1 are coupled, they split to a bonding energy level at ε_b and an antibonding level at ε_a, differing in energy by twice the covalent energy V_2. The average energy shifts up by the product of this covalent energy and the nonorthogonality $<1|2>$.

Making it symmetric and inserting a scale factor K , he wrote the central assumption of *Extended Hückel Theory* ,

$$<1|H|2> \approx K\frac{\varepsilon_1 + \varepsilon_2}{2}<1|2>. \tag{5.23}$$

This has turned out to be sufficiently successful that we might use it to suggest that both terms in the numerator of the middle form in Eq. (5.21) are proportional to $<1|2>$ so that if $<1|2>$ is not too large in the denominator, both factors in Eq. (5.22) may have similar dependence upon d as in Eq. (5.6). It then follows that the repulsion $V_0(d)$ is approximately proportional to $1/d^4$ as we wrote in Eq. (5.22).

This overlap repulsion is essential to the understanding of the molecule. Without it the energy from the two electrons in the bond (with energy given by Eq. (5.16) with in this case $V_3 = 0$ and with the minus sign) would continue to drop as d decreases and V_2 increases as $1/d^2$. However, the overlap repulsion, varying as A/d^4 will always win at low enough d , and we may adjust the coefficient A so that the minimum comes at the observed spacing. This is all we shall need to calculate the energy gained in the formation of the molecule, given the spacing of that molecule, as we see in Problem 5.1, where the repulsion cancels half the gain from bond formation.

When the two atomic levels have different energy, so V_3 is not zero, a similar analysis (Harrison (1980) Appendix B) shows that we must modify V_3 to $^1/_2(\varepsilon_2 - \varepsilon_1)/\sqrt{1 - <1|2>^2}$ and then the splitting is again correctly given by Eq. (5.16) and again the overlap repulsion is given by Eq. (5.22) and the V_2 of Eq. (5.21).

The extension of this approach to the effects of nonorthogonality when there are more than two coupled levels is considerably more complicated (Van Schilfgaarde and Harrison (1986)). It will be adequate here to continue treating the atomic states as orthogonal, and to approximate the effects of nonorthogonality by an overlap repulsion as given in Eq. (5.22). Then the problem is reduced to the solution of a set of algebraic equations, Eqs. (5.13), one for each orbital included in the expansion. The energy eigenvalues are obtained by solving the secular equation, setting the secular determinant equal to zero.

$$\mathrm{Det}(H_{ij} - \varepsilon\delta_{ij}) = 0. \tag{5.24}$$

We obtain N solutions ε_k if there are N rows, and N columns, of the Hamiltonian matrix. For each eigenvalue ε_k we substitute back in Eq. (5.13) to obtain the corresponding eigenvector $|k> = (u_1, u_2, ...)_k$. We note here a generality of the definition of states $|k>$ in Dirac notation. We have thought of such states as functions of position $\psi_k(\mathbf{r})$, but now can equivalently think of them as vectors in N-dimensional space with components u_1, u_2....

5.4 Perturbation Theory

We imagine a starting Hamiltonian H^0 which is sufficiently simple that we can calculate its energy eigenstates $|j\rangle$ and their energies ε_j^0 . This might be the Hamiltonian for an isolated atom. Then we add a small correction, a perturbation, H^1 which might be a small change in potential. If the perturbation is small enough that we may neglect any change in the eigenstates themselves, there may still be a change in the energy. We evaluate that energy as $\langle j|H^0 + H^1|j\rangle = \varepsilon_j^0 + \langle j|H^1|j\rangle$, and the second term is called the *first-order term in perturbation theory* for the energy of the state. Any other correction to the energy must come from changes in the state itself.

We may actually obtain those corrections directly for the case of two coupled levels, for which we obtained the exact energy in Eq. (5.15), by regarding the coupling $V_{ss\sigma}$ as the perturbation and expanding Eq. (5.15) in $V_{ss\sigma}$. This yields

$$\varepsilon = \frac{\varepsilon_1 + \varepsilon_2}{2} \pm \sqrt{\left(\frac{\varepsilon_1 - \varepsilon_2}{2}\right)^2 + V_{ss\sigma}^2}$$

$$= \frac{\varepsilon_1 + \varepsilon_2}{2} \pm \frac{\varepsilon_1 - \varepsilon_2}{2}\left(1 + \frac{V_{ss\sigma}^2}{2[(\varepsilon_1 - \varepsilon_2)/2]^2} + \ldots\right) = \varepsilon_1 + \frac{V_{ss\sigma}^2}{\varepsilon_1 - \varepsilon_2} + \ldots. \tag{5.25}$$

for the plus sign, and the same final form for the minus sign with 1 and 2 interchanged. The shift given by $V_{ss\sigma}^2/(\varepsilon_1 - \varepsilon_2)$ is called the *second-order term in perturbation theory* for the energy of the state. There are higher-order terms, but if the perturbation is small compared to the difference in the energies of the coupled states they will be smaller, and they are usually not included. This correction has arisen entirely from the changes in the state.

We may also obtain the change in the state itself by making a similar expansion of the state by expanding the coefficients given in Eq. (5.18) for small $V_{ss\sigma}$. We then find that the state $|1\rangle$ is modified as

$$|1\rangle \rightarrow |1\rangle + \frac{V_{ss\sigma}}{\varepsilon_1 - \varepsilon_2}|2\rangle + \ldots \tag{5.26}$$

The second term is called the first-order correction to the state.

These results may be directly generalized to many coupled states by systematically expanding the Eqs. (5.13) in a perturbation H^1 which is small. It leads to an energy

$$\varepsilon_j = \varepsilon_j^0 + \langle j|H^1|j\rangle + \sum_i \frac{\langle j|H^1|i\rangle\langle i|H^1|j\rangle}{\varepsilon_j^0 - \varepsilon_i^0} + \ldots \tag{5.27}$$

and a state given by

$$|j\rangle \rightarrow |j\rangle + \sum_i \frac{|i\rangle\langle i|H^1|j\rangle}{\varepsilon_j^0 - \varepsilon_i^0} + \ldots \tag{5.28}$$

These are useful only if the matrix elements $\langle i|H^1|j\rangle$ are small compared to the energy differences $\varepsilon_j^0 - \varepsilon_i^0$ so that the expansions converge and the discarded terms are smaller. If the energy differences are not small, we must return to the square-root form in Eq. (5.25) for those two levels, sometimes called "degenerate perturbation theory". Frequently then the coupling with the remaining states can be included using the perturbation theory of (5.27).

We note from the final term in Eq. (5.27) that if the state $|j\rangle$ is lower in energy than the state $|i\rangle$ the denominator will be negative and then since the numerator $\langle j|H^1|i\rangle\langle i|H^1|j\rangle = |\langle j|H^1|i\rangle|^2$ is always positive the energy of the state $|j\rangle$ will be further lowered. Similarly the energy of the higher state will be raised further. This *repulsion* of the levels is illustrated in Fig. 5.3. This same effect occurs in classical physics. If two harmonic oscillators of different frequency are coupled by a term bilinear in the two displacements, $\kappa u_1 u_2$, the lower-frequency mode will be lowered further in frequency and the frequency of the higher-frequency mode will be raised.

We may similarly interpret the perturbation theory correction to the state. We see from Eq. (5.28) that if the state $|j\rangle$ is lower in energy than the state $|i\rangle$, the denominator will again be negative. If the coupling is negative, as is $V_{ss\sigma}$, then the second state will be added in a bonding relationship, with

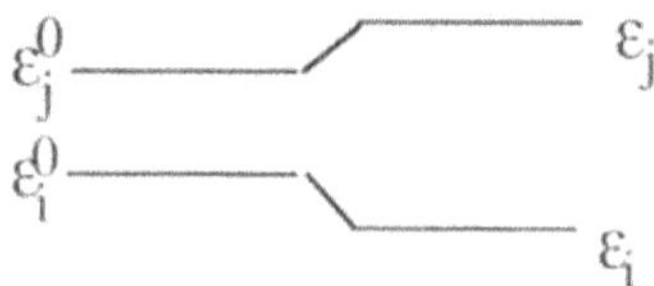

Fig. 5.3. Two coupled levels shift their energies away from each other, according to Eq. (5.27), an effect called the repulsion between levels.

no node between the atoms, consistent with a lowering in the energy. If the second state is of *lower* energy, it will be added in an antibonding relationship, consistent with a raising of the energy.

The perturbation theory corresponding to Eqs. (5.27) and (5.28) has proven to be extremely fruitful and will be used extensively here. In Chapter 7 we shall generalize it to include perturbations which depend upon time. In Problem 5.3 we apply it to calculate the polarizability of the Li_2 molecule by noting that an electric field couples the occupied bonding levels to the antibonding levels, thereby lowering their energy in proportion to the square of the electric field according to Eq. (5.27). The corresponding change in energy of the molecule is equated to the change of energy $-{}^1/_2\alpha E^2$ of a system of polarizability α to obtain that polarizability, an alternative method to that used in Problem 5.2. This same approach is also used in Problem 5.3 to find the polarizability of a quantum-well state in a semiconductor.

5.5 N_2, CO, and CO_2

We turn next to molecular orbitals in a series of other molecules, each of which introduces important new features. The nitrogen atom has a configuration (defined in Section 4.2) of $(1s^2)2s^23p^3$ so that clearly p-states will be involved in the molecular orbitals. We are still making the approximation that the orbitals can be written as a linear combination of atomic orbitals, $|MO\rangle \approx \Sigma_j\, u_j\, |j\rangle$, but now the sum can contain eight terms, the 2s-orbital and the three 2p-orbitals on each atom. We again neglect any effect of the molecular formation on the 1s-core electronic states. In principle, this requires the solution of eight simultaneous equations, Eqs. (5.13), but the high symmetry of the molecule greatly simplifies the

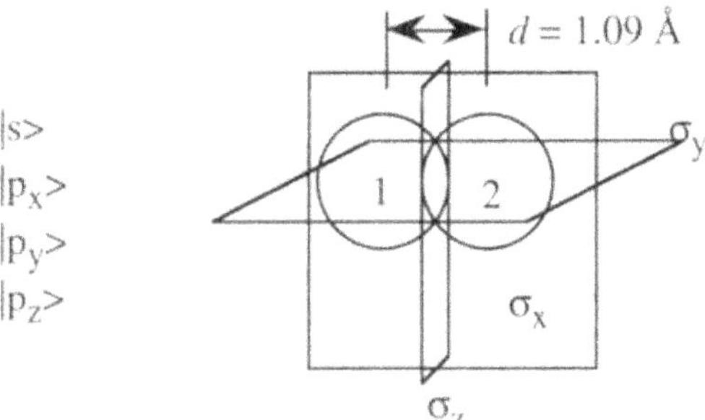

Fig. 5.4. The N_2 molecule shown to the right has reflection symmetry σ_x and σ_y in two perpendicular planes through the two nuclei and σ_z in the plane bisecting the internuclear vector. All p-states on both atoms are taken to be odd under either the σ_x, the σ_y, or the σ_z reflection in a plane containing the atomic nucleus.

calculation. The molecule, and therefore the Hamiltonian, has three independent reflection planes, illustrated in Fig. 5.4, and every molecular orbital can be chosen to be even or odd under each of the reflections.

We consider first states which are odd under the reflection σ_y shown. In fact only the two $|p_y\rangle$ states are odd under that reflection so any molecular orbital odd under that reflection must contain only those two atomic states, as illustrated in Fig. 5.5. The molecular orbitals will be even under the reflection σ_x; there are no states based upon these atomic orbitals which are odd under both σ_x and σ_y, though atomic d-states would allow such symmetry. This becomes just like the Li_2 molecule, with bonding and antibonding states (even and odd, respectively, under σ_z) of energies equal to the atomic-state energy plus or minus the matrix element between them, which in this case is called $V_{pp\pi}$. The two subscripts "p" are because it is a coupling between p-states, and the π represents one unit of angular momentum, $m = \pm 1$, around the z-axis.(Section 2.4) If we construct such atomic states with one unit of angular momentum around the molecular axis, $(|p_x\rangle \pm i|p_y\rangle)/\sqrt{2}$, their matrix element is called $V_{pp\pi}$ and it is equal to the matrix element between two states $|p_y\rangle$. We shall see in Section 6.2 that its magnitude is approximately the same as $V_{ss\sigma}$, given in Eq. (5.6). Thus the energy of the resulting two "π-states" is

$$\varepsilon_\pi = \varepsilon_p \pm V_{pp\pi}. \tag{5.29}$$

Obviously the energy of the π-states based upon atomic orbitals $|p_x\rangle$ is the same. We have used symmetry to obtain four of the eight states immediately.

We proceed to the remaining four, which are even under σ_x and σ_y. They will be linear combinations of the states $|p_z\rangle$ and $|s\rangle$. We may in fact take even and odd combinations of the s-states which, exactly as for Li_2, have energy $\langle\psi|H|\psi\rangle = \varepsilon_s \pm V_{ss\sigma}$, though we shall see that they are coupled to the corresponding states $(|p_{z1}\rangle \pm |p_{z2}\rangle)/\sqrt{2}$, which have energy $\varepsilon_p \pm V_{pp\sigma}$ with $V_{pp\sigma}$ given by the same form as in Eq. (5.6) but, as we shall see in Section 6.2, $-\pi^2/8$ replaced by $+3\pi^2/8$. The subscript σ again refers to zero

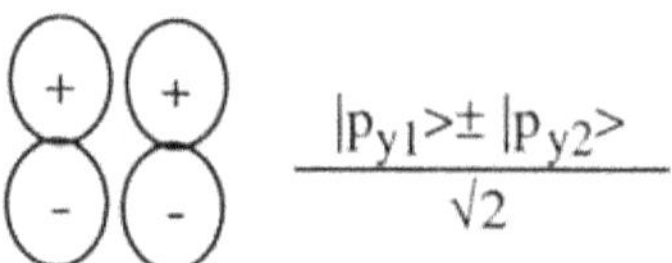

Fig. 5.5. Only the $|p_y\rangle$ states are odd under σ_y and can therefore enter a molecular orbital which is odd under that reflection. These molecular orbitals will be even under σ_x and can be even or odd under σ_z.

angular momentum around the internuclear axis, and in this case it is chosen by convention to correspond to both orbitals having the same orientation, which we represent graphically as ⊖⊕ ⊖⊕. Where the two orbitals overlap they are of opposite sign so, with a negative potential, we have a positive matrix element $V_{pp\sigma}$ by the same argument which gave a negative $V_{ss\sigma}$ and a negative $V_{pp\pi}$.

Now, if we seek a molecular orbital which is even under the reflection σ_z , it must be a combination of the bonding s-state, and the symmetric combination of p-states,⊖⊕ ⊕⊖, as illustrated in Fig. 5.6. This is called the bonding p_σ-state, and the matrix element between the two p-orbitals is the negative $-V_{pp\sigma}$. We may easily evaluate the coupling between these two bonding states, $(1/\sqrt{2})(\langle s_1| + \langle s_2|)|H|(|p_{z1}\rangle - |p_{z2}\rangle)/\sqrt{2} = -V_{sp\sigma}$, where $V_{sp\sigma}$ is a coupling between an s-state and a p-state, with orientation Ⓢ ⊖⊕. $V_{sp\sigma}$ will be taken in Section 6.2 to be given by Eq. (5.6) with $-\pi^2/8$ replaced by $+\pi/2$. It is positive since the wavefunction of the s-state is taken positive in the region of overlap while the p-state is negative. The coupling between orbitals with the opposite orientation, ⊖⊕ Ⓢ, is negative. The energy of these two coupled bonding states is obtained as in Eq.(5.15) as

$$\varepsilon = \frac{\varepsilon_s + V_{ss\sigma} + \varepsilon_p - V_{pp\sigma}}{2} \pm \sqrt{\left(\frac{\varepsilon_s + V_{ss\sigma} - \varepsilon_p + V_{pp\sigma}}{2}\right)^2 + V_{sp\sigma}^2} \ . \quad (5.30)$$

We may construct a state from the antibonding combination of s-states and of p-states in the same way, and obtain exactly the same expression but with the sign changed in front of $V_{ss\sigma}$ and $V_{pp\sigma}$. This gives the other four states, called σ-states, in addition to the π-states.

It can be helpful to illustrate this calculation with an energy diagram, as in Fig. 5.7. On the left are the starting levels, with the introduction of couplings as we move to the right. The 1s-levels lie far below. We have

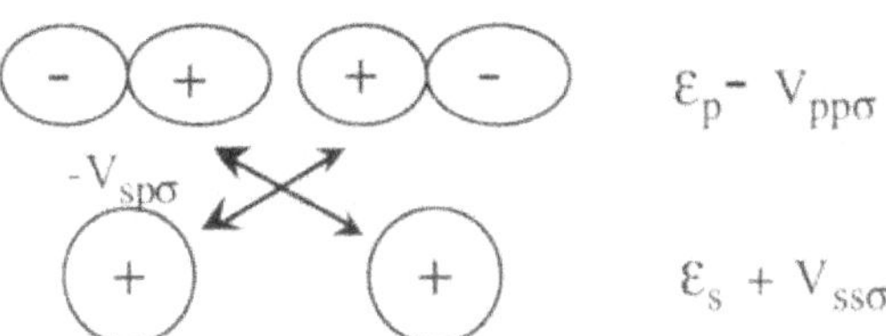

Fig. 5.6. Construction of the bonding σ-states in N_2 from s-states and σ-oriented p-states (no angular momentum around the internuclear axis). The bonding state is even under all three reflections of Fig. 5.4.

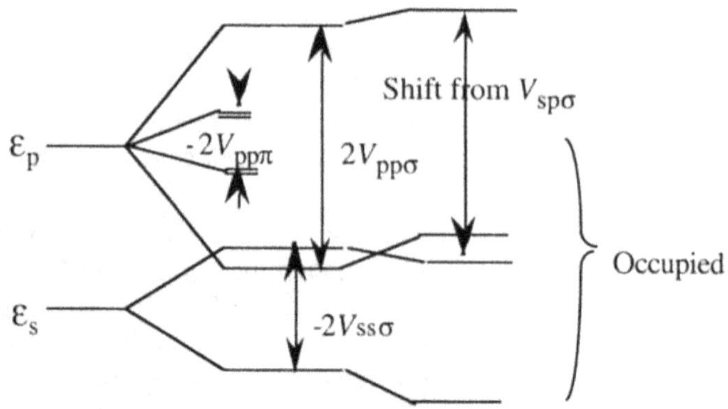

Fig. 5.7. A schematic energy-level diagram illustrating the formation of molecular levels in nitrogen. First the π-bonds and antibonds are formed, then the σ-bonding and σ-antibonding s- and p-levels, which are then coupled to form the final set. The ten valence electrons per molecule fill the lowest five levels.

five valence electrons from each nitrogen atom which fill the lowest five levels, each with a spin-up and a spin-down electron, as indicated.

We may evaluate each of these energies, using the s- and p-state energies from Table 4.1 and evaluating the couplings for the formulae given with $d =$ 1.09Å. These are listed in Table 5.1, along with values obtained from a much earlier, but much more complete calculation. They are listed with standard notation as σ or π, and indicating g for even (gerade in German) if the state is even under inversion through the midpoint between the atoms, and u (ungerade) for odd. (This is different from *reflection* symmetry for the π-states).

The comparison is informative. The values for the deeper levels are of the correct general order although the calculation done here was almost trivial. It is usual that the higher-energy states, which in fact are in the range of other atomic states which were not included in the calculation, are very poorly given. It is often not important since the highest occupied levels (called HOMO's, Highest Occupied Molecular Orbitals, by chemists), and the lowest empty levels (called LUMO's, Lowest Unoccupied Molecular Orbitals), are most important. The effects of nonorthogonality, which we include only as a separate overlap repulsion, are not included and our values, deeper than the full calculation, are partly explained by that. All of these discrepancies seem usually to be much worse in the first row of the periodic table, and would be less for the heavier elements. In Problem 5.4 we calculate the change in the sum of one-electron energies in forming the molecule, and divide by two (approximate correction for overlap repulsion) to estimate the cohesion. Usually we overestimate couplings $V_{ss\sigma}$, etc.,

Table 5.1. Energies, in eV, of the molecular orbitals for N_2 calculated in the text, compared with values from Ransil (1960).

	Simple Theory	Ransil
$3\sigma_u$	13.1	30.0
$1\pi_g$	-5.9	8.2
$2\sigma_u$	-21.5	-19.4
$1\pi_u$	-21.8	-14.8
$3\sigma_g$	-25.6	-15.1
$2\sigma_g$	-46.1	-38.6

and bond energies by a factor of two for first-row systems, but here it is a considerably greater over-estimate in comparison to the observed 9.7 eV per molecule, and we may expect to do better on other systems than N_2 and the CO treated in Problems 5.4 and 5.5.

There is a short cut, which is of no importance here but becomes essential in other systems. It is the use of *hybrid* orbitals, combinations of two orbitals of different energy on the same atom. We are assuming that the σ-molecular orbital can be expanded in terms of the four nitrogen orbitals, and we could equally as well use sp-hybrids,

$$|h\rangle = \frac{|s\rangle \pm |p_z\rangle}{\sqrt{2}}, \tag{5.31}$$

on each atom. These four hybrids, each with energy expectation value of $(\varepsilon_s + \varepsilon_p)/2$, are equivalent for expansion, but they allow an approximation. We note that the s-state and the p-state wavefunctions are both positive to the right of the atom (Ⓢ + ⊖⊕) and add, while they cancel on the left. In that sense the orbital with the plus in ± "leans" to the right and that with the minus leans to the left. We might expect the coupling between the inward-pointing hybrids $(V_{ss\sigma} - 2V_{sp\sigma} - V_{pp\sigma})/2 = -4.04\hbar^2/(md^2)$ to be much larger than that between outward-point hybrids $(V_{ss\sigma} + 2V_{sp\sigma} - V_{pp\sigma})/2 = -0.90\ \hbar^2/(md^2)$, and it clearly is. We may in fact neglect any coupling with an outward-pointing hybrid, leaving them at their hybrid energy and include only the coupling, with magnitude which we called the *covalent energy* V_2, between hybrids directed into the bond. If we do that, we obtain two hybrid energies (*nonbonding states*) at $(\varepsilon_s + \varepsilon_p)/2 = -20.0$ eV and a bond and antibond at -45.9 eV and +5.9 eV. This is not so far from the corresponding -21.5, -25.6, -46.1, and +13.1 which we obtained for the first column in Table 5.1. We see in Problem 5.4 that it of course also predicts a very similar cohesion to that of the full calculation.

It may not be worth making the approximation when the symmetry is so high that we can obtain a solution by the solution of quadratic equations, Eq. (5.30). However, if we go to carbon monoxide, which has the same number of electrons per molecule but no reflection symmetry, we must solve four simultaneous equations to obtain the σ-orbital energies. However, in Problem 5.5 we form sp-hybrids on the oxygen and carbon, and can obtain the bond and antibond energies from the solution of a quadratic equation. This enables a simple estimate of the cohesion, as for N_2. The use of hybrids is of even greater advantage in crystalline diamond or silicon, where it allows the approximation of treating the crystal as made up of independent bonds.

We may consider one more molecule, CO_2, which forms with the three nuclei in a straight line, equally spaced, with carbon in the center and with a carbon-oxygen distance of $d = 1.16$ Å, as shown in Fig. 5.8. We now have twelve orbitals, but all three reflection symmetries of Fig. 5.4. We proceed part way with the analysis, which is straightforward, though a little intricate. Beginning with states which are odd under the reflection σ_y, there is one p_y-state on each atom . If we seek a molecular orbital also odd under the reflection σ_z it cannot include the p_y-state on the carbon atom (which we number as 2) in the center and will simply be $(|p_{y1}\rangle - |p_{y3}\rangle/\sqrt{2}$, with molecular-orbital energy $\varepsilon_p(O)$ since it contains no state coupled to these two atomic orbitals. The states even under σ_z will contain an even combination $(|p_{y1}\rangle + |p_{y3}\rangle/\sqrt{2}$ coupled to the carbon p-state $|p_{y2}\rangle$ by $\sqrt{2}V_{pp\pi}$ and have energy

$$\varepsilon_\pi = \frac{\varepsilon_p(O) + \varepsilon_p(C)}{2} \pm \sqrt{\left(\frac{\varepsilon_p(O) - \varepsilon_p(C)}{2}\right)^2 + 2V_{pp\pi}^2}\,. \qquad (15.32)$$

This, and the $|p_x\rangle$ counterparts, are *multicenter bonds*, involving orbitals from three atoms. The energy could be directly evaluated from the orbital energies from Table 4.1 and $V_{pp\pi}$ evaluated in terms of the oxygen-carbon spacing d.

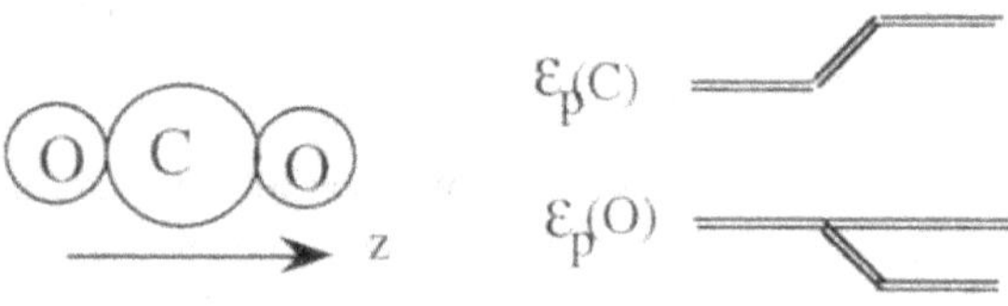

Fig. 5.8. The carbon dioxide molecule is linear, as shown to the left. The three $|p_y\rangle$ orbitals form three π-states, as shown to the right, each degenerate with the π-states based upon the $|p_x\rangle$ orbitals as shown.

The remaining six molecular orbitals are even under both σ_x and σ_y and are based upon an s- and a p_z-state on each atom. The molecular orbital even under σ_z contains only the carbon s-state, the even combination of oxygen s-states $(|s_1> + |s_3>)/\sqrt{2}$ and the even combination of oxygen p_z-states. This requires the solution of a cubic equation. Similarly the molecular orbital odd under σ_z contains only the carbon p_z-state and the odd combinations of oxygen s- and p-states, again requiring solution of a cubic equation. In fact the oxygen s-states are so deep in energy that it would not be a bad approximation to neglect their coupling to the carbon states and take them as two states at energy $\varepsilon_s(O)$, leaving only quadratic equations and solutions of the form of Eq. (15.32), but with $V_{pp\pi}$ replaced by $V_{pp\sigma}$ for the odd molecular orbital and with $V_{pp\pi}$ replaced by $V_{sp\sigma}$ and $\varepsilon_p(C)$ replaced by $\varepsilon_s(C)$ for the even molecular orbital. These would again represent three-center bonds. A further approximation, which would be less accurate and of no advantage, would be to make sp-hybrids on the carbon atom as we did for CO, pointing to right and to left. Even if we neglect the coupling of the right-pointing hybrid to orbitals on the left oxygen, we have three coupled orbitals and solution of a cubic equation is required. We now have a formulation in terms of two-center bonds, but nothing is gained from the further approximation.

In this case, and in the case of more complicated molecules, the tight-binding approximation should have similar validity to that for the diatomic molecules, and all of the parameters are obtainable in the same way. It is also possible to estimate all of the same properties for these other molecules which we estimated for Li_2 in Problems 5.1 and 5.2.

Chapter 6. Crystals

The same tight-binding expansion in atomic states which we used for molecules is applicable also to crystalline solids. However, in solids we may also use an alternative description in terms of free-electron states because of the fact, which we saw in Section 4.3, that the effect of the atomic potentials can be represented by that of a weak pseudopotential. In fact, when we combine the two approaches, in analogy with the complementarity principle, it will provide us the parameters, such as $V_{ss\sigma} = -(\pi^2/8)\hbar^2/(md^2)$, needed to carry out the calculations. We begin with a one-dimensional chain of atoms.

6.1 The Linear Chain

The simplest generalization to many-atom systems is a chain of lithium atoms, illustrated in Fig. 6.1. It is convenient to use periodic boundary conditions, as in Section 1.7, which corresponds to bending the chain into a ring so that the last atom is coupled to the first. As for Li_2 we approximate the orbital for the entire chain as a sum of N s-states, one on each atom, $|\psi\rangle = \Sigma_j u_j|s_j\rangle$, and the variational equations, Eq. (5.13), become

$$u_j\varepsilon_s + V_{ss\sigma}(u_{j+1} + u_{j-1}) = u_j\varepsilon\ , \tag{6.1}$$

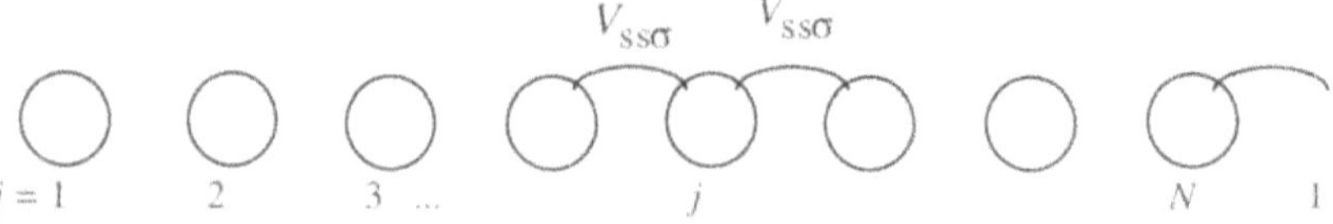

Fig. 6.1. A row of N lithium atoms, each with a valence 2s-state of energy ε_s, coupled by $V_{ss\sigma}$ to the s-state on each of its neighboring atoms. With periodic boundary conditions, the Nth is also coupled to the first.

since H_{ij} is only nonzero for $i = j$ at ε_s, and for nearest neighbors at $V_{ss\sigma}$. There are N such equations. We guess a solution $u_j = e^{ikdj}/\sqrt{N}$, with the factor $1/\sqrt{N}$ for normalization. ($\Sigma_j e^{ikdj}|s_j\rangle/\sqrt{N}$ is called a *Bloch sum.*) Substituting this form into Eq. (6.1), a factor $e^{ikdj}/\sqrt{N}$ cancels out, leaving

$$\varepsilon_k^s \equiv \varepsilon = \varepsilon_s + V_{ss\sigma}(e^{ikd} + e^{-ikd}) = \varepsilon_s + 2V_{ss\sigma}\cos kd\,. \tag{6.2}$$

We have added labels on the energy ε. Since the dependence upon j canceled out Eqs. (6.1) is satisfied for all j within the chain, but we must choose kdN equal to an integral multiple of 2π so that it is satisfied for j = 1 and $j = N$. This is the same condition on k as with periodic boundary conditions for free electrons in Section 1.7.

We have found the best tight-binding estimate for the energy eigenstates in the chain. The results are illustrated in Fig. 6.2 for a chain of N = 8 atoms. Then the $8kd$ must equal an integral multiple of 2π, and points indicate the wavenumbers at which this is true. Two points which differ in n by eight, or a multiple of eight, give coefficients $u_j = e^{ikdj}/\sqrt{N}$ which are identical (differing by factors of $e^{i2\pi}$), so they are the same state, and of

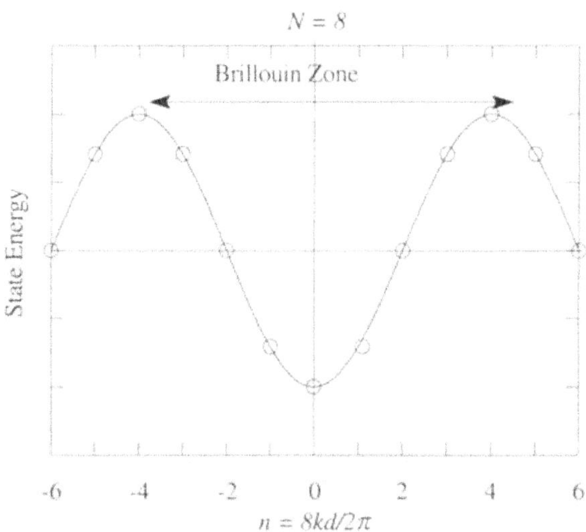

Fig. 6.2. The energies of tight-binding states for a row of eight lithium atoms are indicated by points. The n values -3, -2, ...3,4 represent the full set of eight states, and all other integers repeat these states.

course have the same energy. We limit the values $-N/2 < n \leq N/2$, or $-\pi/d < k \leq \pi/d$, which is called the *Brillouin Zone*, the set of smallest wavenumbers for each state. This same limitation to a Brillouin Zone will apply to three-dimensional systems.

The π-states in benzene, which we discussed in Section 2.1, are exactly a case of a chain with periodic boundary conditions, in this case with $N = 6$. The benzene molecule is flat, as illustrated in Fig. 6.3, and the molecular orbitals which are odd under reflection in the plane will consist entirely of $|p_z\rangle$ states if the z-axis is normal to the plane. Each is coupled to its neighbor and the π-bands are obtained exactly as in Eq. 6.2 as

$$\varepsilon_k^{\pi} = \varepsilon_p + 2V_{pp\pi}\cos kd\,. \tag{6.3}$$

In this case, with $N = 6$, the allowed states are at $k = 2n\pi/6d$, with $n =$ -2, -1, 0, 1, 2, and 3 in the Brillouin Zone. There are six electrons available to occupy these states. Because these energies are given relative to the free-atom states, we may obtain the change in energy of these six electrons as they doubly occupy the lowest levels in the formation of the benzene. These give the π-bonding contribution to the cohesion of benzene, as calculated in Problem 6.1. This calculation of cohesion was not possible for the free-electron description without the ε_s reference. However, the free-electron description is also meaningful, and the levels in the Brillouin Zone in Fig. 6.2 are qualitatively similar to a free-electron parabola. We note from Eq. (6.3) that for this to be true $V_{pp\pi}$ must be negative so that it is minimum at $k = 0$. Further, if we match up the band width of $-4V_{pp\pi}$ from the Eq. (6.3) to the free-electron width for wavenumbers at the Brillouin Zone boundary, $-\hbar^2(\pi/d)^2/(2m)$ we obtain the $V_{pp\pi} = -(\pi^2/8)\hbar^2/(md^2)$ which we used in the preceding chapter. We shall explore this comparison in detail for a three-dimensional structure in the next section.

We chose periodic boundary conditions so that we could obtain solutions $u_j = e^{ikdj}/\sqrt{N}$, and this eliminated the effects of ends to a chain. We can, however, treat finite chains by noting that we could also satisfy the Eqs. (6.1) within the chain using solutions $u_j = e^{ikdj} - e^{-ikdj} = 2i\sin(kdj)$, or $\sqrt{2/N}\sin(kdj)$ normalized. Then if we choose boundary conditions such that u_j is

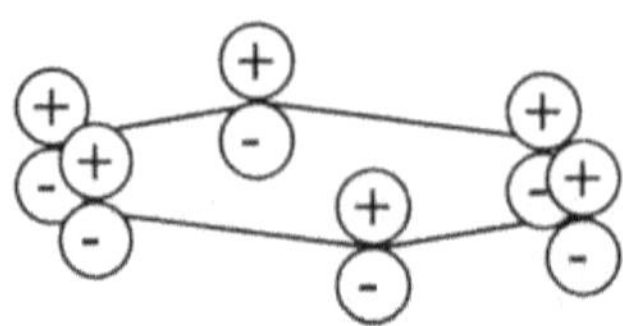

Fig. 6.3. A benzene molecule, viewed from the side, with six π-orbitals oriented perpendicular to the plane giving rise to six π-states.

zero for $j = 0$ and $j = N+1$, as illustrated in Fig. 6.4, we will satisfy the equations for $j = 1$ to N even though there are no terms in the state for $j = 0$ and $j = N+1$. We have obtained a solution for a finite chain. From a free-electron point of view the allowed k's are for vanishing boundary condition a distance d from the last atom, rather than $d/2$ from the last atom as we assumed in Problem 2.2 when we broke the benzene chain. In Problem 6.1 we redo the breaking of the benzene chain for this more appropriate, tight-binding, view. We noted also the change in boundary conditions when we treated metal surfaces in Chapter 2.

The generalization to an s-band in a three-dimensional system is immediate. For a simple-cubic lattice, in particular, each atom has a nearest neighbor at $\pm d$ in the x-direction, and the same in the y- and z-directions. The states are again written as a linear combination of s-states on atoms at position $\mathbf{r}_j$ with coefficients $u_j = e^{i\mathbf{k}\cdot\mathbf{r}_j}/\sqrt{N}$, and with nearest-neighbor coupling the bands become

$$\varepsilon_{\mathbf{k}}{}^{s} = \varepsilon_s + 2V_{ss\sigma}(\cos k_x d + \cos k_y d + \cos k_z d) \tag{6.4}$$

We imagine a large crystal in the shape of a rectangular parallelepiped, N_x atoms along the x-direction, N_y and N_z along the other two, since it is the simplest case and most results are insensitive to the boundary conditions, as we have seen. With periodic boundary conditions, the restrictions on $\mathbf{k}$ are the same as for free-electrons in a rectangular box of dimensions, $L_x = N_x d$, $L_y = N_y d$, and $L_z = N_z d$ which we treated in Section 2.2 and the density of states in wavenumber space is the same. If we wished to study surfaces we could use vanishing boundary conditions but it is clear that we must apply these conditions a distance d from the last atoms as in Fig. 6.4, so the effective dimensions are $L_x = (N_x+1)d$, etc. This also allows us to construct explicit wavefunctions valid even at the corner atom in the crystal, and such wavefunctions can be used to study electron tunneling from a metallic tip, as we shall see in Section 8.3.

The generalization to more realistic structures than simple cubic is straightforward, and discussed in Section 13.2. Of more immediate

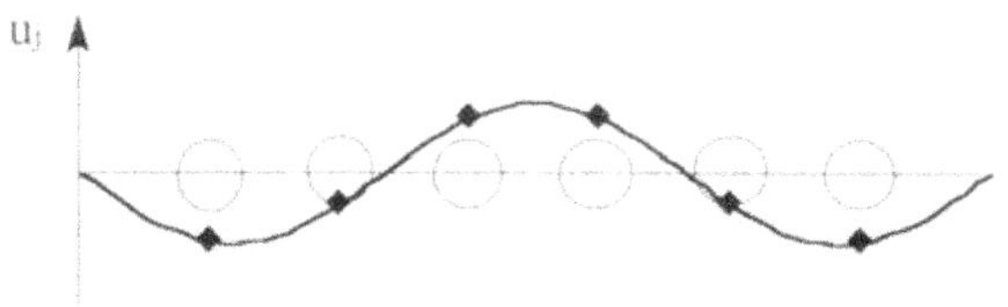

Fig. 6.4. Solving the tight-binding equations for a finite chain of $N = 6$ atoms, by requiring the coefficients to go to zero at $j = 0$ and 7.

importance is the incorporation of p-states, as we did in going to the nitrogen molecule and we turn to that next.

6.2 Free-Electron Bands and Tight-Binding Parameters

We see that the boundary conditions determine the wavenumbers at which states occur, but the bands themselves, $\varepsilon_k{}^s = \varepsilon_s + 2V_{ss\sigma}\cos kd$ in one dimension, depend only upon the crystal structure and the electronic states on the atoms which make up the crystal. We have replotted our tight-binding s-band for a chain of atoms as the heavy line in Fig. 6.5. We may argue that because of the weakness of pseudopotentials a free-electron description is also meaningful and we have also drawn in a free-electron parabola with the $k = 0$ energy adjusted to the s-band minimum, and $V_{ss\sigma}$ chosen equal to $-(\pi^2/8)\,\hbar^2/(md^2)$ such that the two bands match also at the edges of the Zone. When we include $V_{sp\sigma}$ we shall see that this is the appropriate point for matching.

The free-electron bands extend to wavenumbers beyond the Brillouin Zone, as shown in the figure. By the convention used in tight-binding chains we would translate these bands back, by some multiple of $2\pi/d$, so that they were plotted in the Brillouin Zone, also shown in Fig. 6.5. They must have some correspondence to tight-binding bands from other atomic orbitals for the constituent atoms.

In particular, we might imagine tight-binding bands arising from p-states oriented along the chain axis. They are calculated just as were the s-bands leading to $\varepsilon_k{}^p = \varepsilon_p + 2V_{pp\sigma}\cos kd$. They would in fact match the (translated) free-electron bands at the Zone center and Zone edges if $V_{pp\sigma} = +(3\pi^2/8)\hbar^2/(md^2)$ and ε_p is suitably chosen, as illustrated above in Fig. 6.5. Indeed, this is a sensible relation to make since p-states oriented along the chain, each have a node at the atomic site. Thus the state at $k = 0$ has one node at each nuclear site and one midway between each atom, two per distance d. This is to be compared with a free-electron state, $\sin kd$ for $k = 2\pi/d$ (the wavenumber of the state translated to $k = 0$) which also has two nodes for each distance d. The identification requires a positive $V_{pp\sigma}$ but that is just what is expected for these p-states, as we indicated in Section 5.5, because the wavefunctions have opposite signs in the region where they overlap.

One thing has been left out, the coupling between neighboring s- and p-states $V_{sp\sigma}$ which coupled, for example, the bonding s-state and the bonding p-state in the N_2 molecule. It is not difficult to see that for tight-binding s- and p-band states at $k = 0$ there is no coupling between the tight-binding band states. This follows because an s-state on one atom is coupled to a p-state to the right by $V_{sp\sigma}$ and to a p-state to the left by $-V_{sp\sigma}$ (because the

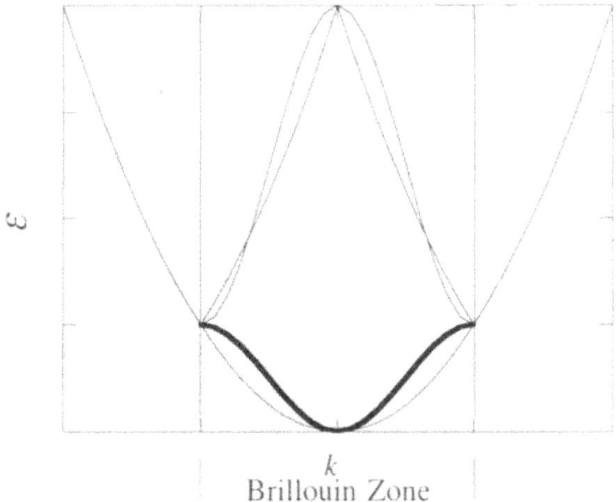

Fig. 6.5. The free-electron and tight-binding bands for a chain of atoms. The heavy line is an s-band, $\varepsilon_k{}^s$, for nearest-neighbor coupling, with parameters adjusted so that it fits the free-electron band at the center and edges of the Brillouin Zone. The free-electron band which continues beyond the Zone is redrawn within the zone, by the convention used in crystals, and parameters for a p-band, $\varepsilon_k{}^p$, are adjusted to fit these translated free-electron bands.

positive lobe of the p-state is nearest) and the two terms cancel. Similarly at the Brillouin Zone edge the two coupling terms cancel. However, at other wavenumbers, the two are only partly out of phase and the cancellation is not complete, but given by $2i\, V_{sp\sigma}\sin kd$. The net effect is that the s-bands and p-bands of Fig. 6.5 (or the simple-cubic counterpart) are combined as

$$\varepsilon_k = \frac{\varepsilon_k{}^s + \varepsilon_k{}^p}{2} \pm \sqrt{\left(\frac{\varepsilon_k{}^p - \varepsilon_k{}^s}{2}\right)^2 + 4V_{sp\sigma}{}^2\sin^2 kd}\ . \tag{6.5}$$

We see that the sp-coupling does not shift the bands where we matched them to obtain $V_{ss\sigma}$ and $V_{pp\sigma}$, but it causes the bands at intermediate wavenumbers to move away from each other and become more free-electron-like. Everything has worked out to make this matching of free-electron and tight-binding bands appropriate and matching at the Zone edge as in Fig. 6.5 was the appropriate choice.

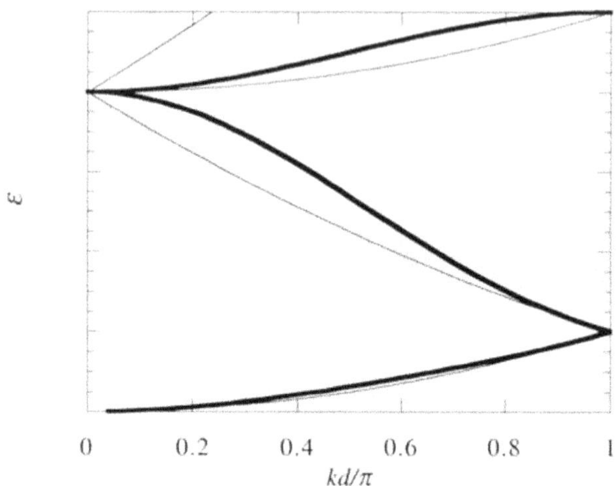

Fig. 6.6. Energy bands in a simple-cubic crystal, for **k** in a [100] direction to the Brillouin Zone face. Light lines are free-electron states, reduced to the Brillouin Zone. Heavy lines are tight-binding sp-bands with parameters fit to match the free-electron bands. The top two complete bands shown are π-like and are doubly degenerate, two bands at the same energy.

There are various ways to adjust $V_{sp\sigma}$ to fit the free-electron bands, but probably the best is to note that the magnitude of the coupling between s-and p-bands, $2V_{sp\sigma}\sin kd$, grows linearly with change in wavenumber from the Brillouin-Zone edge. Thus it can be adjusted to make the bands linear in k at the Zone edge, with slope equal to that for the free-electron bands, rather than horizontal, as the tight-binding bands are seen to be in Fig. 6.5. This is simple to do. Writing the energy at the Zone boundary $\varepsilon_{ZB} = (\pi^2/2)\,\hbar^2/(md^2)$, and keeping terms only linear in the δk measured from the edge, Eq. (6.2) becomes $\varepsilon_k = \varepsilon_{ZB} \pm 2V_{sp\sigma}\, d\, \delta k$. Setting this equal to $\varepsilon_{ZB} + \partial\varepsilon_k/\partial k \;\; \delta k = \hbar^2(\pi/d)\delta k\, /m$ gives $V_{sp\sigma} = (\pi/2)\,\hbar^2/(md^2)$, chosen positive because with the orientation of orbitals chosen by convention, Ⓢ ⊖⊕, the orbitals are of opposite sign where they overlap. The resulting tight-binding sp-bands are given by the lowest two heavy-lines in Fig. 6.6. An alternative matching would be to adjust $V_{sp\sigma}$ such that the curvature of the band $\partial^2\varepsilon_k/\partial k^2$ at the bottom of the band has the free-electron value of $\hbar^2/m$, as done in Problem 6.2c, giving a value nearer 1.9 $\hbar^2/(md^2)$ than the 1.57 $\hbar^2/(md^2)$ we find here. It can be seen from Fig. 6.6 that the larger value is needed to lower the s-like tight-binding band further to the free-electron band. One might also suggest the geometric mean of $V_{ss\sigma}$ and $V_{pp\sigma}$, equal to $\sqrt{3}\,(\pi^2/8)\,\hbar^2/(md^2) = 2.14\,\hbar^2/(md^2)$.

Before collecting these results together, we should consider a three-dimensional, simple-cubic system. We gave the s-band for such a structure in Eq. (6.4), and we see that it yields the same band width from $\mathbf{k} = 0$ and $\mathbf{k}$ equal to π/d in an x-direction, so the predicted $V_{ss\sigma}$ is unchanged from one dimension, and the same applies to the p_σ-bands, p-states oriented along $\mathbf{k}$, so in fact even our fit of $V_{sp\sigma}$ remains appropriate. However, the free-electron bands have an important new feature in three dimensions. We are translating all free-electron states back to wavenumbers within the Brillouin Zone, $-\pi/d < k_x \le \pi/d$, $-\pi/d < k_y \le \pi/d$, and $-\pi/d < k_z \le \pi/d$. The Brillouin Zone has become a cube in three-dimensional wavenumber space, centered at $\mathbf{k} = 0$. For our bands along the x-direction, from $\mathbf{k} = 0$, we have translated those at $\mathbf{k} = (k_x - 2\pi/d)\hat{\mathbf{x}}$ to $0 \le k_x\hat{\mathbf{x}} \le \pi\hat{\mathbf{x}}/d$, with $\hat{\mathbf{x}}$ a unit vector in the x-direction. We must also translate the states for $\mathbf{k} = k_x\hat{\mathbf{x}} - (2\pi/d)\hat{\mathbf{y}}$, with $\hat{\mathbf{y}}$ a unit vector in the y-direction, back from the side to the same line. These states have energy $\hbar^2[(2\pi/d)^2 + k_x^2]/2m$ and are shown as the higher light line in Fig. 6.6, emerging horizontally from the vertical axis, and rising parallel to the lowest free-electron band. There are in fact four such bands of the same energy, translated to this line by $\pm(2\pi/d)\mathbf{y}_1$ and $\pm(2\pi/d)$ times a unit vector $\mathbf{z}_1$ in the z-direction. At $\mathbf{k} = 0$ one combination of these states is the state we identified with the p_σ-bands, but with p-states oriented along the y-direction. As we increase k_x, these become exactly π-like bands, p_y-states, with the phase of the coefficients changing along the x-axis. Adding the coupling with all six neighbors we obtain bands, $\varepsilon_k{}^p = \varepsilon_p + 2V_{pp\sigma} + 2V_{pp\pi}(1 + \cos k_x d)$. Adjusting these to fit the free-electron bands $\hbar^2[(2\pi/d)^2 + k_x^2]/2m$ at the Zone center and edge, we find that $V_{pp\pi} = -(\pi^2/8)\hbar^2/(md^2)$, as we found for the π-bands in benzene. Combining all of these results we have the "universal" parameters for sp-bonded systems,

$$V_{ss\sigma} = -\frac{\pi^2}{8}\frac{\hbar^2}{md^2}, \qquad V_{sp\sigma} = +\frac{\pi}{2}\frac{\hbar^2}{md^2},$$
$$V_{pp\sigma} = +\frac{3\pi^2}{8}\frac{\hbar^2}{md^2}, \qquad V_{pp\pi} = -\frac{\pi^2}{8}\frac{\hbar^2}{md^2}. \tag{6.6}$$

Another combination of these four free-electron bands can be identified with the p_z π-bands along the k_x axis. The other two of these free-electron bands would need to be identified with atomic d-states and the highest free-electron band, rising linearly from the vertical axis, would need to be identified with at atomic s-state of higher quantum number than that identified with the lowest free-electron band. None of these additional bands will be of interest here.

We might note one other remarkable feature of the three-dimensional bands. We saw that the s- and p-band states were not coupled at the center of the Brillouin Zone nor at the face of the Brillouin Zone, $(\pi/d, 0, 0)$, because of the symmetry of the states. This is of course true at the center of the other five faces of the Brillouin Zone. This justified our use of those points for determining $V_{ss\sigma}$ and $V_{pp\sigma}$. It is not difficult to show that, in addition to the face center with energy $(\pi^2/2)\,\hbar^2/md^2$, there are two other sets of points within the Brillouin Zone for which the s- and p-band states are uncoupled. They are at the center of the cube edges, with energy $2(\pi^2/2)\,\hbar^2/md^2$, and at the cube corner, with energy $3(\pi^2/2)\,\hbar^2/md^2$. However, we see from Eq. (6.4) that in going from the cube center successively to these three points, the tight-binding energy rises by $-4V_{ss\sigma}$ at each step so that *we would have obtained exactly the same formula for* $V_{ss\sigma}$ had we used any two of these four points. Furthermore, the states based upon p-states oriented along **k** are found to have energies identical to the s-band energies at these points with the parameters of Eq. (6.6) and therefore again, any pair of the four points could have been used to obtain the same values. It appears also that the π-like states at these four points are also consistent with the free-electron bands, a remarkable consistency.

The evaluations for these points with wavenumbers not parallel to a cube axis required a construction of matrix elements between p-states which we have not needed before, but will need in the next section and it should be explained at this point. A p-state oriented along the x-axis, described in Section 2.4, can be written $R_1(r)\hat{\mathbf{x}}\cdot\mathbf{r}$, with $R_1(r)$ a function only of radial distance and $\mathbf{x}_1$ a unit vector in the x-direction. Then clearly a p-state oriented along a direction $\hat{\mathbf{x}}\cos\phi + \hat{\mathbf{y}}\sin\phi$, with $\hat{\mathbf{y}}$ a unit vector in the y-direction, can be written $R_1(r)(\hat{\mathbf{x}}\cos\phi + \hat{\mathbf{y}}\sin\phi)\cdot\mathbf{r} = \cos\phi\, R_1(r)\hat{\mathbf{x}}\cdot\mathbf{r} + \sin\phi\, R_1(r)\hat{\mathbf{y}}\cdot\mathbf{r}$ so that the p-states can be divided into components just as vectors are. If we have a p-state with axis and angle ϕ from the internuclear axis to an s-state, we simply divide it into a σ-oriented component with a factor $\cos\phi$, and coupling $\pm V_{sp\sigma}$ and a π-oriented component which is not coupled to the s-state. This is illustrated in Fig. 6.7. Similarly, for two coupled p-states we may divide both into σ- and π-components (two orientations of π-components) to obtain the matrix elements for the two p-states, also illustrated in Fig. 6.7. For d-states the decomposition is more intricate and is given in the Slater-Koster (1954) tables.

These fits to the free-electron bands also lead to values for $\varepsilon_p - \varepsilon_s$ which are proportional to $\hbar^2/(md^2)$, as seen in Problem 6.2. These values are in some accord with experiment, but vary much more with spacing than real

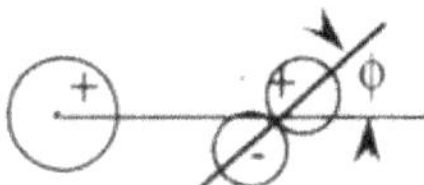
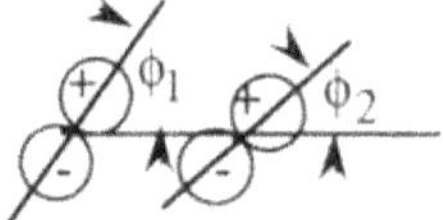

Fig. 6.7. p-states can be decomposed like vectors into σ-oriented and π-oriented components with coefficients $\cos\phi$ and $\sin\phi$, respectively. Only the σ-component is coupled to an s-state, so the coupling between the orbitals shown to the left is $\cos\phi V_{sp\sigma}$. Between those to the right, which are coplanar, the coupling is $\cos\phi_1\cos\phi_2 V_{pp\sigma} + \sin\phi_1\sin\phi_2 V_{pp\pi}$ If one was rotated out of the plane by an angle ϕ_p, there would be an additional factor of $\cos\phi_p$ in both terms.

systems. It is of some interest that such a relation can be used in reverse to predict equilibrium spacings for solids in terms of the free-atom term values, and is in semiquantitative accord with experiment for elements. However, it would not allow us to distinguish the different atoms in compounds, for which we use the free-atom term values as we did for molecules in the preceding chapter. Similarly, we use the free-atom term values for the elements rather than free-electron fits to the term values.

Eq. (6.6) for the s- and p-state couplings was derived for a simple-cubic structure, for which the calculation is particularly simple. The results are quite similar, but not identical, if the same calculation is carried out for a different structure. For example, in the tetrahedral structure of diamond and silicon, the fit yields $V_{ss\sigma} = (9\pi^2/64)\,\hbar^2/(md^2)$, only 1.5% larger than the simple-cubic value. If we are specifically treating such tetrahedral solids, we can as well use these tetrahedral results, or in fact coefficients fit to the known band structures of the semiconductors themselves. That is what was done in Harrison (1980) and (1999), using an average from silicon and germanium. It is just as easy to use such a coefficient, taken as universal, as to use $-\pi^2/8$, but for the purposes of this text we take the values from Eq. (6.6).

It seems also reasonable to use these couplings for molecules since they represent the coupling between atomic orbitals and these change little in the formation of molecules or solids. That is in fact exactly what we did in Chapter 5 in treating N_2, CO, and CO_2. If on the other hand, we go to structures with more neighbors, such as the face-centered-cubic structure of copper and aluminum with twelve nearest neighbors to each atom, our coefficients lead to bands which are much too broad, and a fit of free-electron bands to those structures yields smaller - and more appropriate - values to be used there. The difference appears to come from our neglect of the nonorthogonality, and its absorption into an overlap repulsion. Including nonorthogonalities systematically is much more complicated, as discussed in Section 5.3, but it avoids this difficulty of tight-binding theory in carrying over parameters to much different circumstances.

Finally, we should note that the variation of each coupling as $1/d^2$ can only apply near the observed spacings, which are much the same in different structures. Indeed it is only near the observed spacings that the bands *are* free-electron-like. At much larger spacings the couplings drop exponentially as the wavefunctions do and at smaller spacings the nonorthogonalities grow and modify the couplings. However, for a very wide range of calculations we are only interested in spacings near equilibrium and the theory is very powerful. It even has remarkable physical predictions, such as that we can estimate the change in some properties of silicon with pressure by interpolating, as a function of spacing, between that property for unstrained silicon and for diamond.

6.3 Metallic, Ionic, and Covalent Solids

We give a very brief account of the electronic structure of solids, described much more fully, by a factor of fifty, in Harrison (1999). The goal in *Applied Quantum Mechanics* is to provide the tools which make such an analysis possible, not to carry it out.

It is remarkable that both the free-electron and tight-binding limits are meaningful for describing solids, but for particular solids they are not equally convenient. When the energy bands are partly occupied, the defining property of metals since then a small electric field shifts electrons between states to allow current to flow, the free-electron limit is ordinarily much simpler. When for example we sum the energy over occupied states, it is very simple for a free-electron gas, as we saw in Section 2.2. When the energy bands are each either completely full or completely empty, characteristic of insulators since then a small field cannot cause redistribution of the electrons, a tight-binding view is ordinarily simpler.

For this distinction it is useful to look again at the Periodic Table of the elements, the central part of which is shown in Fig. 6.8. All elements outside of Columns IV and VIII are metals, and those shown are considered free-electron metals, with weak pseudopotentials and as many valence electrons per atom as the column number, as assumed in Problem 2.3. To the left of Cu, Ag, and Au are the transition metals, with partly-filled d-shells (atomic states with $l = 2$). The electron states based upon these d-states are best treated in tight-binding theory. [Extensive analysis of transition metals, and all other types of solids in these terms is given in Harrison (1999).] Similarly, to the right of Ba are the rare earths and to the right of Ra are the actinides, all with partly-filled f-shells ($l = 3$). The f-states are also best treated in tight-binding theory.

Elements in Column VIII have eight valence electrons, filling the s-states and all of the p-states. [Helium has no valence p-states, but the

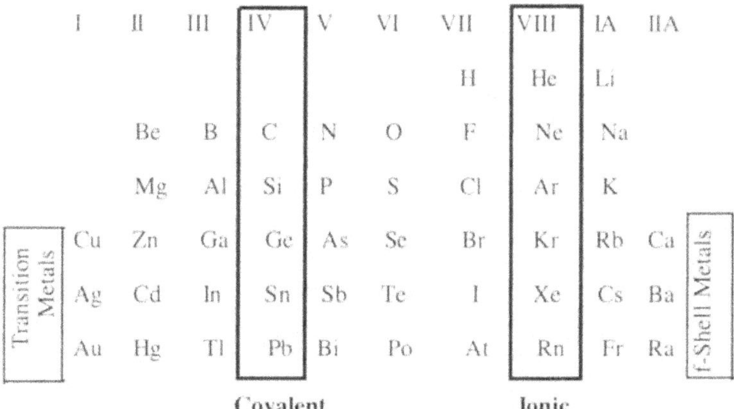

Fig. 6.8. The periodic table of the elements, arranged to show the electronic structure of solids. The nonmetals in the center are bounded by Column-IV elements which form covalent semiconductors based upon two-electron bonds, and by Column-VIII elements which form insulators based upon full electronic shells. Listed elements beyond these two columns are simple metals. Those to the left form covalent compounds with the nonmetals; those to the right form ionic insulators with the nonmetals. To the left of Column I are the transition (d-shell) metals, to the right of Column IIA are the rare earths and actinides (f-shell metals).

situation is similar.] The states above these are empty and far removed in energy. In this situation the atoms are chemically inactive. They cannot accept electrons from other atoms, nor can their electrons be easily removed. When these atoms interact with each other, any bonding energy is canceled by an antibonding energy. Thus they form inert gases. Even as solids, held together by van-der-Waals forces which we discuss in Section 12.2, electrons cannot move from atom to atom and the crystals are insulating, and transparent to visible light, which does not have photons of sufficient energy to excite the electrons.

Elements in Column IV form semiconductors, which are insulating for a different reason. Though they have only two of the six p-levels per atom occupied, they form completely filled and completely empty bands. This is most easily understood in terms of the hybrid states we introduced in Chapter 5. Instead of forming two orthogonal hybrids from a p-state and the s-state, as we did for N_2, we form four orthogonal hybrids on each atom, each oriented in the direction of one corner of a tetrahedron with the nucleus at the center. Then if the atoms are arranged such that each has four neighbors in these same directions, the diamond structure, we may form independent bonds with each neighbor, using the two hybrids directed into the bond, just as we formed the σ-bond in nitrogen. Each atom contributes one electron to each bond, filling all bonding states and leaving all antibonding states empty. With the formation of four hybrids, there are no nonbonding states left over.

The bond states are coupled to neighboring bond states, just as the lithium s-states were coupled to their neighbors in Fig. 6.1, and broaden into fully occupied bands called *valence bands* for a semiconductor. The antibonding states are coupled to their neighbors forming empty bands, called *conduction bands*. In this case the gap between valence and conduction bands is only an electron volt or so and some electrons are thermally excited into the conduction band, as we shall see in Chapter 10, making them weakly conducting, semiconductors.

One construction of four hybrid states on an atom is given by

$$\begin{aligned} |h_1> &= (|s> + |p_x> + |p_y> + |p_z>)/2, \\ |h_2> &= (|s> + |p_x> - |p_y> - |p_z>)/2, \\ |h_3> &= (|s> - |p_x> + |p_y> - |p_z>)/2, \\ |h_4> &= (|s> - |p_x> - |p_y> + |p_z>)/2. \end{aligned} \tag{6.7}$$

The three p-states in the first can be added, as in Fig. 6.7, and seen to be a p-state $\sqrt{3}\,|p>$ oriented along a cube diagonal in the cube defined by the x-, y-, and z-axes. It is normalized, with a probability 3/4 on the p-state and 1/4 on the s-state, called an *sp^3-hybrid,* in contrast to the sp-hybrids in nitrogen. The other three are along other cube diagonals and are orthogonal hybrids, $<h_1|h_2> = (<s|s> + <p_x|p_x> - <p_yp_y> - <p_y|p_y>)/4 = 0$. From these four orthogonal atomic states, $|s>$, $|p_x>$, $|p_y>$, and $|p_z>$, only four orthogonal orbitals can be constructed. In a similar way three orthogonal hybrids (sp^2-hybrids) can be constructed from the $|s>$, $|p_x>$, and $|p_y>$ states, oriented in the direction of the three neighboring atoms in a graphite xy-plane, or the two neighboring carbon atoms and a neighboring hydrogen atom in benzene. Again fully occupied valence bands are formed, and the π-states are formed from the remaining $|p_z>$ orbitals as we saw in Section 6.1. In the case of tetrahedral bonds, Eq. (6.7), the coupling between the two hybrids directed into one bond is $-V_2$, with V_2 the covalent energy given by

$$V_2 = (-V_{ss\sigma} + 2\sqrt{3}\,V_{sp\sigma} + 3V_{pp\sigma})/4 = 4.44\,\hbar^2/(md^2), \tag{6.8}$$

where in the last step we used Eq. (6.6). An analysis with couplings based upon the known energy bands of silicon and germanium (Harrison (1999)) gives a slightly smaller value of 3.22 $\hbar^2/(md^2)$. This covalent energy characterizes the strength of the covalent bond which separates the occupied and empty states in these covalent semiconductors. The other nonmetals between Columns IV and VIII in Figure 6.8 find other ways to make covalent bonds. In Column V three bonds are formed with neighbors with a doubly occupied nonbonding state. In Column VI (Se and below) two bonds are formed with two doubly-occupied nonbonding states per atom, and in Column VII, one bond is formed so that they are molecular gases.

Many compounds can be understood from the same point of view, using "theoretical alchemy". For example, when compounds are formed between a nonmetallic element and a metal to the left, it is ordinarily in a tetrahedral structure for which we may again construct an sp^3-hybrid on each atom, but the hybrids have different hybrid energies, $\langle h_1|H|h_1\rangle = (\varepsilon_s + 3\varepsilon_p)/4$. Half the difference in the two hybrid energies is called the *polar energy*, as described in Section 5.3, and doubly-occupied *polar* bonds are formed, leading again to semiconducting behavior, and properties calculated in terms of the parameters we have given (Harrison (1999)). Such a series of compounds can be made from a single row of the periodic table by starting with the Column-IV element, Ge, and transferring one proton from alternate nuclei to those between, leaving the first as a Ga nucleus and transmuting the second to As. (See Fig. 6.8.) The bonds shift slightly toward the As, but are qualitatively the same. A second transfer produces Zn and Se nuclei, and the third transfer produces Cu and Br nuclei. This is a series of isoelectronic compounds of increasing polarity, Ge, GaAs, ZnSe, and CuBr, with similar electronic structure. The "covalent" in Fig. 6.8 indicates the formation of such compounds with metals to the left.

Similarly if compounds are formed with equal numbers of metal atoms from the right (columns IA, IIA), they can be understood beginning with inert-gas atoms placed in a crystalline array and transferring protons to make a series such as Ar, KCl, CaS, understood as was the inert gas atom, as made up of closed-shell *ions* (charged atoms), in this case full-shell configurations but charged, rather than neutral atoms. In these cases the atoms which *receive* protons become metals, while in the covalent solids the atoms which lost protons became metals. In the case based upon inert-gas atoms the tetrahedral arrangement needed for covalent-bond formation is not useful and these *ionic* solids form in more closely-packed structures, such as the rock-salt structure which would arise from beginning with simple-cubic inert-gas atoms. The "ionic" in Fig. 6.8 indicates the formation of such compounds with metals to the right.

The ionic compounds turn out in some ways to be the simplest to understand. Proceeding not from theoretical alchemy, but from for example potassium and chlorine atoms, we could imagine bringing them together to their final structure, and transferring one electron per atom pair, to gain $\varepsilon_s(K) - \varepsilon_p(Cl)$ equal to 9.77 eV from Table 4.1, not far from the observed cohesion of KCl of 6.9 eV per atom pair (Harrison (1999), p. 337). Further, the lowest-energy excitation for the system might be expected to be this same 9.77 eV, not far from the observed band gap of 8.4 eV (ibid. p. 333). We shall see part of the reason that such a simple analysis works in Section 20.1, but a further discussion is beyond the scope of this text, as is a discussion of the extraordinary variety of other compounds which can be understood using these same covalent and ionic concepts (Harrison (1999)).

Before leaving the electronic structure of crystals, we should note an important difference which arises when we consider systems with d-electrons or f-electrons. The difference is the strong localization of these states around the nucleus, which greatly reduces the coupling between such states on neighboring atoms. Because of this they form bands very narrow in comparison to the free-electron bands arising from s- and p-states.

The reason for this can be seen already classically. The d-states have two units of angular momentum, as we saw in section 2.4, but in transition elements they have energies near the s-state energies (since we are successively filling d-states as we move across the series with a single electron, or two electrons, in the s-state). If we then think of this d-state orbit as a classical circular orbit, as illustrated in Fig. 6.9, we then imagine an orbit with no angular momentum, an s-state, at the same energy. It is oscillating through the nucleus, moving always radially if it has no angular momentum. However, when it is at the radius of the circular d-state orbit it has the same kinetic energy, directed now inward or outward, as in Fig. 6.9. Thus it will move far out from the radius of the d-state orbit, corresponding to a much larger orbit. The argument for localization of f-states, with three units of angular momentum, is even stronger.

This effect not only makes the orbit for d-states much more localized, but it causes the coupling between d-states on neighboring atoms to drop much more rapidly with increasing distance d. It is in fact found to drop

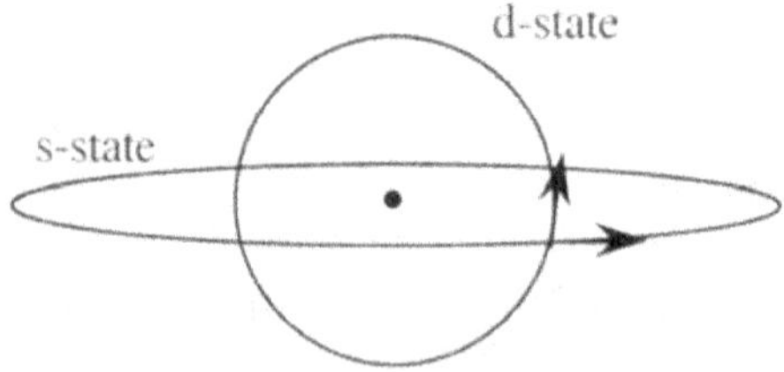

Fig. 6.9. An electron in a d-state represented as a circular classical orbit, has all of its kinetic energy directed tangentially. An electron in an s-state at the same energy, represented as a classical orbit passing very close to the nucleus, has the same kinetic energy when at the same radius, but since it is directed outward (or inward) it carries the electron much further from the nucleus. This explains the high localization of d-states in transition-metal atoms.

approximately as $1/d^5$. The coupling between neighboring d-states with no angular momentum around the internuclear axis is given, for example, by

$$V_{dd\sigma} = -\frac{45}{\pi}\frac{\hbar^2 r_d^3}{md^5} \tag{6.9}$$

with r_d a "d-state radius" characteristic of the element in question, tabulated in Harrison (1999) (e. g., 1.03 Å for Ti, 0.74 Å for Fe, 0.69 Å for Cu). The coupling $V_{dd\pi}$ is given by the same expression, with the leading factor replaced by $30/\pi$, and $V_{dd\delta}$ by the same expression with the leading factor replaced by $-15/(2\pi)$. We will not have occasion to use these expressions. It is interesting to note that the element-dependent parameter r_d^3 was necessary to obtain the units of energy, with a $1/d^5$ dependence. For f-states the coupling drops more rapidly, as r_f^5/d^7 with the f-state radius providing the correct units. For the s- and p-states with coupling varying approximately as $1/d^2$ the corresponding orbital radius would enter to the zero power, suggesting correctly that the formulae for the coupling are independent of element, as given by Eq. (6.6). We note, however, that in all cases these formulae for couplings are only approximations.

III. Time Dependence

Electronic structure is primarily concerned with the lowest-energy state of a system, the ground state. This encompasses many properties, but often we are interested in higher-energy states, and transitions between different states. Up the this point it might appear that the goal of quantum mechanics was to find solutions, or approximate solutions, to the Schroedinger Equation. For the rest of the text we move on to other problems. The parallel in classical mechanics would be first addressing the calculation of normal modes of vibration, and then moving on to trajectories, collisions, and the myriad of other problems an engineer or scientist must deal with. In quantum mechanics the basic premise is still the wave-particle duality, with $\hbar$ providing the relation between the two, but we now seek the answer to different questions.

Chapter 7. Transitions

7.1 A Pair of Coupled States

We have seen already in Eq. (1.22) that the time dependence of an energy eigenstate of energy ε_j is given by a factor $e^{-i\varepsilon_j t/\hbar}$. If a state is expanded in states of more than one energy, each term varies with its own factor. This leads to complicated time dependence, just as when a classical violin string vibrates in several modes. In the case of two coupled electronic states, as in the polar molecular orbital described in Section 5.3, we may see that it gives an oscillation of the electron between the two atoms. We write the state as a sum of the bond state, with coefficient u_b at $t = 0$ and time dependence $e^{-i\varepsilon_b t/\hbar}$, and the antibonding state, with coefficient u_a and time dependence $e^{-i\varepsilon_a t/\hbar}$.

$$|\psi> = u_b\left(\sqrt{\frac{1+\alpha_p}{2}}\,|1> + \sqrt{\frac{1-\alpha_p}{2}}\,|2>\right) e^{-i\varepsilon_b t/\hbar}$$

$$+\, u_a\left(\sqrt{\frac{1-\alpha_p}{2}}\,|1> - \sqrt{\frac{1+\alpha_p}{2}}\,|2>\right) e^{-i\varepsilon_a t/\hbar}. \tag{7.1}$$

We may confirm that the electron will be on atom 1 with probability one, at $t = 0$, if the expansion coefficients are chosen as $u_b = \sqrt{(1+\alpha_p)/2}$ and $u_a = \sqrt{(1-\alpha_p)/2}$. The terms in Eq. (7.1) may then be rearranged to obtain

$$|\psi\rangle = [(\cos\omega_0 t + i\alpha_p \sin\omega_0 t)|1\rangle + i\sqrt{1-\alpha_p^2}\sin\omega_0 t\ |2\rangle]e^{-i(\varepsilon_a + \varepsilon_b)t/(2\hbar)} \quad (7.2)$$

with $\omega_0 = (\varepsilon_a - \varepsilon_b)/(2\hbar)$. The probability density on atom 2 is

$$P_2 = (1-\alpha_p^2)\sin^2\omega_0 t = \frac{1-\alpha_p^2}{2}(1-\cos 2\omega_0 t) \quad (7.3)$$

and on atom 1 is one minus this. We see that a fraction $1-\alpha_p^2$ of the charge oscillates back and forth with a frequency given by the bonding-antibonding splitting divided by Planck's constant, $2\sqrt{V_2^2 + V_3^2}/\hbar$. For two coupled atomic states of the same energy ($V_3 = 0$), the transfer is complete each period. If the energies of the atomic states are different, and the coupling V_2 between the two atomic states is very small, the amount of the transfer is proportional to the square of that coupling and the frequency is determined by the energy difference as $\omega = 2V_3/\hbar = (\varepsilon_2 - \varepsilon_1)/\hbar$.

This latter case will be of most interest since we shall see that the corresponding oscillating dipole will radiate energy with a frequency $(\varepsilon_2 - \varepsilon_1)/\hbar$, corresponding to $\hbar\omega = \varepsilon_2 - \varepsilon_1$, and at a rate proportional to the square of the electron-light coupling. In the process, the electron drops (or makes a transition) to the lower-energy state. However, in the model we are discussing, without light, no transition occurs - the system simply oscillates. We shall see that a real transition can occur only when there is a range of energies for the final states, a range of frequencies of light in the case of emission of light.

The requirement for a range of frequencies is already there in classical physics. A pair of coupled oscillators, of different frequencies, will transfer vibrational energy back and forth between the two oscillators. However, if an oscillator is coupled to a system with many frequencies, as in the oscillator coupled to a taught wire, illustrated in Fig. 7.1, the vibrational energy of the oscillator will be dissipated into the modes of the wire. Similarly an electron in a quantum-well state can tunnel into a continuum of electronic states beyond the barrier, also illustrated in Fig. 7.1.

7.2 Fermi's Golden Rule

In order to understand quantum transitions, then, we consider a Hamiltonian which has a state $|0\rangle$ at energy ε_0 coupled by matrix elements H_{0j} to states $|j\rangle$ with a range of energies ε_j. This could be the system in Fig. 7.1b. We expand the wavefunction in all of these states, including the time dependence $e^{-i\omega_j t}$ with $\omega_j = \varepsilon_j/\hbar$, but allowing additional time dependence of the coefficients which will arise from the coupling,

Fig. 7.1. Part a. A classical oscillator, coupled to a system with many modes of similar frequency, will dissipate its vibrational energy into those modes. Similarly, Part b, an electron in a quantum well can tunnel into a range of continuum states, treated in Problem 7.2.

$$|\psi\rangle = u_0(t)e^{-i\omega_0 t}|0\rangle + \Sigma_j\, u_j(t)e^{-i\omega_j t}|j\rangle. \qquad (7.4)$$

We shall set u_0 equal to one, and all other coefficients to zero, at time $t = 0$, and calculate the rate that the coefficients u_j grow with time for small t.

We substitute Eq. (7.4) into the Schroedinger Equation, $i\hbar\partial|\psi\rangle/\partial t = H|\psi\rangle$. We obtain $u_j(t)$ by multiplying on the left by a particular $\langle j|$, noting $\langle j|i\rangle = \delta_{ij}$, to obtain

$$i\hbar\frac{\partial u_j(t)}{\partial t}e^{-i\omega_j t} + \hbar\omega_j u_j(t)e^{-i\omega_j t} = u_j(t)\varepsilon_j e^{-i\omega_j t} + H_{j0}u_0 e^{-i\omega_0 t} + \ldots \qquad (7.5)$$

The additional terms would be from any coupling H_{ji} between the state $|j\rangle$ and others, $|i\rangle$, of that collection of states. We shall see that the u_i will be of first order in the coupling so these terms would be second order, written $O(H_{ji}{}^2)$, and can be dropped at small t. The terms in $\hbar\omega_j$ and ε_j cancel and at small times we can take $u_0(t)$ equal to $u_0(0) = 1$. We multiply the remaining term on each side by $e^{i\omega_j t}$ to obtain

$$i\hbar\frac{\partial u_j(t)}{\partial t} = H_{j0}e^{-i(\omega_0-\omega_j)t} + O(H_{j0}{}^2). \qquad (7.6)$$

The final term indicates the second-order terms which we have neglected. We may integrate directly from t=0 to get

$$u_j(t) = \frac{H_{j0}}{\hbar(\omega_0 - \omega_j)}\left(e^{-i(\omega_0-\omega_j)t} - 1\right), \qquad (7.7)$$

or

$$u_j{}^*(t)u_j(t) = \frac{H_{0j}H_{j0}}{\hbar^2}\frac{4\sin^2[(\omega_0 - \omega_j)t/2]}{(\omega_0 - \omega_j)^2}. \qquad (7.8)$$

The right-hand fraction is a function of the energy difference $E = \hbar(\omega_0 - \omega_j)$ between the two coupled states which is very interesting. It is peaked at $E = 0$, more so as time proceeds, and has an integral over E which is $2\pi\hbar t$. In other words, as time proceeds its properties approach the mathematical properties of $2\pi\hbar t$ times an energy delta function, $\delta(E) = \delta(\varepsilon_0 - \varepsilon_j)$. This *Dirac delta function* $\delta(E)$ is defined to be nonzero only near $E = 0$, and to have $\int\delta(E)dE = 1$ (to be distinguished from the *Kronecker delta function* defined in Eq. (1.24).) This is illustrated by the plot of the final factor in Eq. (7.8), divided by $2\pi\hbar t$ in Fig. 7.2. Thus the total probability of a transition having occurred increases linearly in time, and is more and more concentrated in the energy-conserving states as time proceeds. If we sum $u_j{}^*(t)u_j(t)$ over all states $|j\rangle$ we obtain the total probability that a transition has occurred, and if we divide the result by t we obtain the transition rate, which we might write $1/\tau = (\partial/\partial t)\Sigma_j u_j{}^*(t)u_j(t)$ or $\Sigma_j u_j{}^*(t)u_j(t)/t$. We obtain it by replacing the final factor in Eq. (7.8) by $\delta(\varepsilon_j - \varepsilon_0)/(2\pi\hbar)$ and summing over j to obtain *Fermi's Golden Rule*, or the *Golden Rule of Quantum Mechanics*. Sometimes it is also referred to as time-dependent perturbation theory since it carried terms only to second order in the perturbation which coupled the starting states. Thus the transition rate is

$$\frac{1}{\tau} \equiv \frac{\partial}{\partial t}\sum_j u_j{}^*(t)u_j(t)\Big|_{t=0} = \frac{2\pi}{\hbar}\sum_j H_{0j}H_{j0}\,\delta(\varepsilon_0 - \varepsilon_j)\,. \tag{7.9}$$

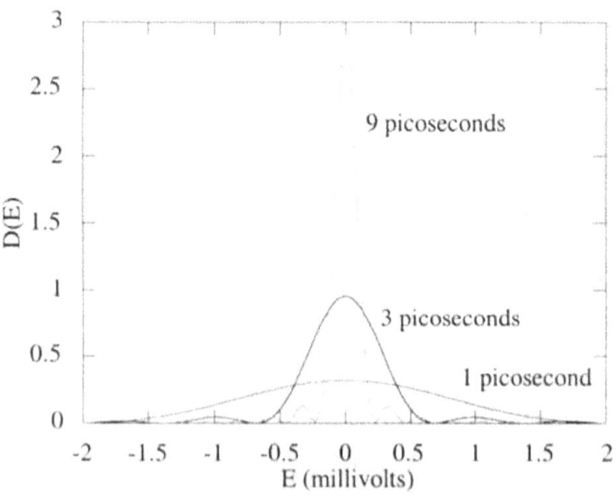

Fig. 7.2 A plot of $D(E) = 2\hbar\sin^2[Et/2\hbar]/(\pi t E^2)$, for which $\int dE\, D(E) = 1$, is given for different times.

As its name implies, it is one of the most important equations in quantum theory. The sum over states coupled by H_{0j} will always be converted to an integral on energy over a density of states, and the integral over the delta function will give a one, with H_{0j} evaluated at the energy for which the argument is zero. Since $\int d\varepsilon\ \delta(\varepsilon - \varepsilon_0)$ is dimensionless, $\delta(\varepsilon - \varepsilon_0)$ has the units of reciprocal energy. With H_{0j} having units of energy, and $\hbar$ of course energy-times-time, the right side of Eq. (7.9) has units of one over time as it must. The meaning of the τ is that the probability of occupation of the initial state is dropping as $e^{-t/\tau}$. We shall see in Chapter 19 (Eqs. (19.7) and (19.15)) that an equivalent form can be derived for the absorption of energy in a classical system such as that illustrated in Fig. 7.1a.

7.3 Scattering in One and Three Dimensions

It is important to illustrate the use of the Golden Rule immediately, and we return to the simple case of a the one-dimensional chain of lithium atoms discussed in Section 6.1 and sketched in Fig. 7.3. We gave the equations from which the states could be determined in Eq. (6.1),

$$u_j\varepsilon_s + V_{ss\sigma}(u_{j+1} + u_{j-1}) = u_j\varepsilon\,. \tag{7.10}$$

The eigenstates were given as $u_j = e^{ikdj}/\sqrt{N}$. If we change one lithium atom to a sodium atom, with an s-state differing in energy by $\delta\varepsilon_s = 0.38$ eV (Table 4.1), the expectation value for the energy of each such state is shifted by only $\delta\varepsilon_s/N$, but there is also a coupling introduced between any two states of wavenumbers k and k' of $H_{k'k} = \delta\varepsilon_s e^{i(k-k')di}/N$ if that impurity is at the position $j = i$. This perturbation can produce a transition of an electron initially in the state k to a state k' moving in the opposite direction with the same energy. We calculate the rate directly using Eq. (7.9). It is

$$\frac{1}{\tau} = \frac{2\pi}{\hbar}\sum_{k'}\frac{\delta\varepsilon_s{}^2}{N^2}\delta(\varepsilon_{k'} - \varepsilon_k) = \frac{2\pi}{\hbar}\frac{Nd}{2\pi}\int dk'\frac{\delta\varepsilon_s{}^2}{N^2}\delta(\varepsilon_{k'} - \varepsilon_k)\,. \tag{7.11}$$

In the first step we noted that the two matrix elements were the complex conjugate of each other so the phase factors canceled. In the second step we

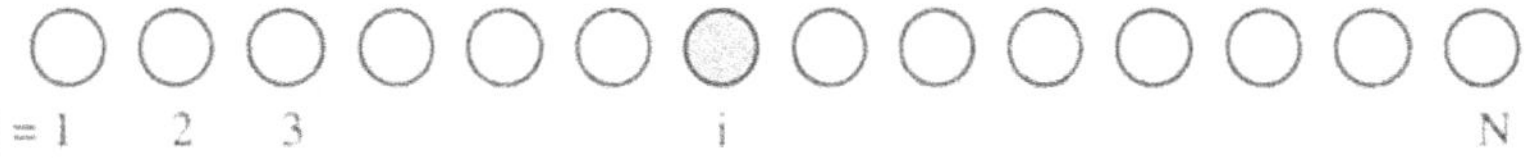

Fig. 7.3. The one-dimensional chain of Section 6.1, with the atom at the position $j = i$ replaced by an impurity.

noted that the spacing between successive wavenumbers allowed by periodic boundary conditions is $2\pi/L = 2\pi/Nd$ so that the number of states in the interval dk' is $Nd\ dk'/(2\pi)$ and the integral as written, times the factor $Nd/2\pi$, is equal to a sum over those states. We substitute for dk' using the differential relation $d\varepsilon_{k'} = -2dV_{ss\sigma}\sin k'd\ dk'$ (from $\varepsilon_k = \varepsilon_s + 2V_{ss\sigma}\cos kd$). Then the integral can be performed, giving a contribution from the delta function where $\varepsilon_{k'} = \varepsilon_k$ of

$$\frac{1}{\tau} = \frac{\delta\varepsilon_s{}^2}{2|V_{ss\sigma}|\sin kd\ N\hbar} = \frac{d\ \delta\varepsilon_s{}^2}{v\ N\hbar^2} . \tag{7.12}$$

In the final form we used the electron speed $v = (1/\hbar)d\varepsilon_k/dk$, and that form is correct also if we use the form $\varepsilon_k = \hbar^2k^2/(2m)$ throughout, rather than the tight-binding form, since the corresponding factors came from changing variables from k' to $\varepsilon_{k'}$ for the integration.

We may check that the units are correct and that the rate goes to zero as $\delta\varepsilon_s$ goes to zero, as it should. We may also note that the scattering rate is proportional to $1/N$, which it should be since the probability per unit time of hitting the impurity is inversely proportional to the length. We might have expected the rate to increase with velocity for the same reason, but in this case we seem to have found that the chances of reflecting upon impact are proportional to $1/v^2$. We shall confirm that this result is correct at the beginning of the next chapter when we evaluate the reflection from an impurity in a chain in a more direct and accurate way. We might have thought of calculating scattering to states moving in the same direction as the initial state, forward scattering, but then the delta function takes us to the same state and it would not be regarded as a scattering event. Note that we have calculated the density of states in wavenumber space for a single spin, The reason is that the perturbation does not couple the initial electron state to states of different spin. In cases with spin-orbit coupling such spin-flip scattering can be allowed, and is treated in parallel to the calculation here.

The calculation is similar for an impurity in a three-dimensional crystal. For a tight-binding s-band the coupling between two states is $H_{\mathbf{k}'\mathbf{k}} = \delta\varepsilon_s e^{i(\mathbf{k}-\mathbf{k}')\cdot\mathbf{r}_i}/N$ if the impurity is placed at the site $\mathbf{r}_i$. The first form in Eq. (7.11) is essentially the same and, since the product of matrix elements is independent of the direction of $\mathbf{k}'$, the sum over $\mathbf{k}'$ can be written as $(\Omega/(2\pi)^3)\int 4\pi k'^2dk'$ if we take the bands to be spherically symmetric as for free electrons. Here $\Omega/(2\pi)^3$ is the density of states in wavenumber space if the volume of the system is Ω and $4\pi k'^2dk'$ is the volume in wavenumber space of a shell of radius k' and thickness dk' . In the more usual case the matrix elements will also depend upon $\mathbf{k}' - \mathbf{k}$, and therefore on the angle θ

between **k** and **k'**. We would then perform the integral $(\Omega/(2\pi)^3)\int 2\pi k'^2 dk' \sin\theta \, d\theta$.

We may again convert the integral over k' to an integral over $\varepsilon_{k'}$ and use the delta function to obtain

$$\frac{1}{\tau} = \frac{\delta\varepsilon_s{}^2 k^2 \Omega}{\pi\hbar^2 v N^2} = \frac{\delta\varepsilon_s{}^2 m^2 v \, \Omega}{\pi\hbar^4 N^2} \; . \tag{7.13}$$

In the last form we used the free-electron $v = \hbar k/m$. Again the various factors are understandable and in this case the proportionality to v is as we might have guessed. In Problem 7.1 we redo this calculation for an impurity in a two-dimensional plane of atoms.

We may return to the emission from a local state illustrated in Fig. 7.1 b. There will be tunneling matrix elements between the localized state $|0\rangle$ and states $|k\rangle$ outside of the well, and ordinarily they will vary smoothly with k, rather than having magnitude independent of k as assumed for our scattering calculations above. Then we can again replace the sum over final states by an integral over k in order to obtain the rate of decay. Such a rate is calculated in Problem 7.2 for transitions from the lowest s-state in a spherical bowl. The matrix elements between an s-state and states of higher angular momentum vanish, so the sum is only over the s-states outside of the well.

Chapter 8. Tunneling

8.1 Transmission in a 1-D Chain

Perhaps the simplest problem involving quantum-mechanical tunneling is the one-dimensional chain which we treated in Section 7.3. If the shift in the impurity level $\delta\varepsilon_s$ takes that level above the energy of the electron state we consider, then the transmission of that electron through the impurity can be considered tunneling; the electron is transmitted though it has insufficient energy to be on the intervening atom. It is a good case to treat because it is so simple and because we can compare our result with that obtained with the Golden Rule.

The Eq. (6.1) for obtaining the (variational) states in a one-dimensional chain, which we used in Section 7.3 to calculate scattering, can be generalized to

$$u_j\varepsilon_j + H_{j,j+1}u_{j+1} + H_{j,j-1}u_{j-1} = u_j\varepsilon \tag{8.1}$$

for a case where couplings and state energies vary from site to site. We look for a state with an electron incident from the left, partially reflected back and partially transmitted. The most convenient way to construct a state from which we may obtain the transmission is to write $u_j = Te^{ikdj}$ to the right of the impurity [T is a transmitted-wave amplitude] and use Eq. (8.1) to work back through the impurity. We then fit the results to a $Ie^{ikdj} + Re^{-ikdj}$. [I and R are also amplitudes.] This correctly applies the defining condition, that there is no incident electron wave from the right, and R^*R/I^*I is the reflectivity. In this problem we are not concerned with boundary conditions in the sense we were when we wanted to obtain quantized energy eigenstates. If we were, we would need to apply them at the outer boundaries. It would not be permitted for example to use periodic boundary conditions to the right of the impurity and let the left region "take care of itself".

For this case we let again all couplings be the same, $H_{j,j+1} = H_{j,j-1} = V_{ss\sigma}$. We let all ε_j be the same ε_s except at one site which we number as $j = 0$. Then we see that Eq. (8.1) is satisfied (with $\varepsilon = \varepsilon_s + 2V_{ss\sigma}\cos kd$) by $u_j = Te^{ikdj}$ for $j = 1, 2, \ldots$ if indeed $u_0 = T$. Similarly, Eq. (8.1) will be satisfied for $j = -1, -2, \ldots$ by $Ie^{ikd_j} + Re^{-ikd_j}$ if $u_0 = I + R$. Thus one condition for this simple case is

$$I + R = T. \tag{8.2}$$

The other condition is Eq. (8.1) for $j = 0$, which is

$$T(\varepsilon_s + \delta\varepsilon_s) + (Te^{ikd} + Ie^{-ikd} + Re^{ikd})V_{ss\sigma} = T\varepsilon \tag{8.3}$$

or

$$(-2T\cos kd + Te^{ikd} + Ie^{-ikd} + Re^{ikd})V_{ss\sigma} = -T\delta\varepsilon_s\,. \tag{8.4}$$

We may substitute $R = T - I$ from Eq. (8.2) into Eq. (8.4) and solve for T as

$$\frac{T}{I} = \frac{2iV_{ss\sigma}\sin kd}{2iV_{ss\sigma}\sin kd + \delta\varepsilon_s}. \tag{8.5}$$

For comparison with our result from the Golden Rule, we may substitute this into Eq. (8.2) and solve for R/I to obtain a reflectivity

$$\text{Reflectivity} = \frac{R^*R}{I^*I} = \frac{\delta\varepsilon_s^2}{4V_{ss\sigma}^2\sin^2 kd + \delta\varepsilon_s^2}, \tag{8.6}$$

proportional to $\delta\varepsilon_s^2$ at small $\delta\varepsilon_s$ as for our Golden Rule calculation. In order to make a detailed comparison we may drop the small $\delta\varepsilon_s^2$ in the denominator and note that the velocity is $v = (1/\hbar)d\varepsilon/dk = -2d\,V_{ss\sigma}\sin kd/\hbar$ so that the reflectivity can be written $d^2\delta\varepsilon_s^2/(\hbar v)^2$. When the perturbation is small most of the wave is transmitted so that the beam is equally distributed over the length Nd of the system and the probability of striking the impurity per unit time is v/Nd which is to be multiplied by the reflectivity to obtain the transition rate,

$$\frac{1}{\tau} = \frac{d\,\delta\varepsilon_s^2}{v N\,\hbar^2}, \tag{8.7}$$

exactly as we obtained in Eq. (7.12). Eq. (8.6) is a more exact result, containing terms of all orders in $\delta\varepsilon_s{}^2$. It is a nice confirmation of our derivation of the Golden Rule, Eq. (7.9), for the second-order term in the transition rate.

The other limit, when $\delta\varepsilon_s{}^2$ is large, corresponds to electron tunneling. Then we evaluate one minus the reflectivity as the transmission,

$$Trans. = \frac{4V_{ss\sigma}{}^2\sin^2 kd}{4V_{ss\sigma}{}^2\sin^2 kd + \delta\varepsilon_s{}^2} \approx \frac{\hbar^2 v^2}{\delta\varepsilon_s{}^2 d^2}, \tag{8.8}$$

where in the last form we have dropped the first term in the denominator and substituted in the numerator in terms of the velocity. It is quite interesting that we can think of this limit as a transition rate of electrons from the left side of the barrier to the right. It is not easy to see how the matrix element for such a transition is to be evaluated, but we can see by writing both expressions for the rate. For transmission near zero the electron state is a standing wave on the left. If that length is written L_1 the electron will strike the barrier at a rate $v/(2L_1)$ so that multiplying by the transmission gives the rate

$$\frac{1}{\tau} = \frac{v}{2L_1} \cdot Trans. = \frac{\hbar^2 v^3}{2L_1 \delta\varepsilon_s{}^2 d^2}. \tag{8.9}$$

We may calculate the rate directly from the Golden Rule using the *unknown* matrix element T_{12} and taking the length of the region on the right as L_2 . It is

$$\begin{aligned}\frac{1}{\tau} &= \frac{2\pi}{\hbar}\Sigma_{k2} T_{21}T_{12}\delta(\varepsilon_2 - \varepsilon_1) = \frac{2\pi}{\hbar}\frac{L_2}{\pi}\int dk_2 T_{21}T_{12}\delta(\varepsilon_2 - \varepsilon_1) \\ &= \frac{2L_2}{\hbar^2 v} T_{21}T_{12}.\end{aligned} \tag{8.10}$$

In the first step we noted that the change in wavenumber between successive states on the right *for standing waves* is π/L_2, and in the final step we wrote $dk_2 = d\varepsilon_2/(\hbar v)$ and performed the integral over energy. Equating the rates obtained in these different ways we find that the tunneling matrix elements must be given by

$$T_{21}T_{12} = \frac{(\hbar v)^2}{4L_1L_2} \cdot Trans. = \frac{(\hbar v)^4}{4L_1L_2\delta\varepsilon_s{}^2 d^2} = \frac{4V_{ss\sigma}{}^4\sin^4 kd}{N_1N_2\delta\varepsilon_s{}^2} \quad . \tag{8.11}$$

The proportionality of each matrix element to $\sqrt{1/N_1N_2}$ is as expected due to the normalization of the states coupled. Even in much more complicated situations, such as we shall discuss, it is possible to calculate a barrier transmission and see what tunneling matrix elements are needed to obtain the correct formula from the Golden Rule. In Section 9.1 we shall see a case where such tunneling matrix elements can be derived more directly in terms of perturbation theory.

8.2 More General Barriers

The first generalization from the one-dimensional chain is to a three-dimensional crystal in which there is a plane of defect atoms, or several planes, of different atoms. We can see that this only slightly changes the analysis which we made for the one-dimensional chain. The simplest case would be a simple-cubic crystal with the defect plane parallel to a cube face, say an xy-plane. We still retain the translational periodicity parallel to that plane so we can apply periodic boundary conditions (or we could put vanishing boundary conditions) on the lateral faces and construct tight-binding states with $u_{j,i} = e^{i\mathbf{k}_t \cdot \mathbf{r}_{i(j)}}u_j$ for the coefficient on the i'th atom in the j'th plane. $\mathbf{k}_t$ is the *transverse wavenumber* for the state with components k_x and k_y, but no z-component. For each such transverse wavenumber we have an energy for an isolated plane $\varepsilon_j(\mathbf{k}_t) = \varepsilon_s(j) + 2V_{ss\sigma}(\cos k_xd + \cos k_yd)$. For that transverse wavenumber Eq. (8.1) applies with ε_j replaced with $\varepsilon_j(\mathbf{k}_t)$. Thus at each transverse wavenumber the problem reduces exactly to a one-dimensional problem. There can be slight complications for more complicated structures, more orbitals per atom, or different orientation of planes, but the essential point remains: for planar defects in crystals, transverse wavenumbers $\mathbf{k}_t$ can be constructed and the tunneling problem for each $\mathbf{k}_t$ is essentially a one-dimensional problem such as described by Eq. (8.1). We avoid those complications here. We continue with the tight-binding formulation as we generalize the barriers, but then consider a formulation based upon free-electron, or effective-mass states.

A second generalization is for multiple planes, or multiple defects in the one-dimensional part of the problem. One such system is illustrated in Fig. 8.1, showing a double-humped barrier. We have chosen now to have the transmitted wave on the left and flowing to the left as $u_j = Te^{-ikdj}$, rather than to the right as before. We may take $T = 1$ since in the end we evaluate T^*T/I^*I or R^*R/I^*I as in the preceding section. Then any constant factor in the transmitted wave cancels. Thus we may use Eq. (8.1) to obtain successive u_j values in order of increasing j. We need to set a wavenumber for the transmitted beam, including any transverse wavenumber $\mathbf{k}_t$ and longitudinal wavenumber k_l corresponding to an energy $\varepsilon = \varepsilon_j(\mathbf{k}_t) +$

$2V_{ss\sigma}\cos k_l d$. Then successive u_j are obtain from Eq. (8.1) which can be solved for u_{j+1} as (writing $\varepsilon_j(\mathbf{k}_t)$ as ε_j for each $\mathbf{k}_t$)

$$u_{j+1} = [u_j(\varepsilon - \varepsilon_j) - H_{j,j-1}u_{j-1}]/H_{j,j+1} . \tag{8.12}$$

In the simplest case all $H_{j,j-1}$ are taken as $V_{ss\sigma}$ and only the ε_j vary from site to site. They will be real in any case. Only u_j and u_{j-1} are needed to determine u_{j+1} so only two terms in the transmitted-beam region are needed. They will have real and imaginary parts, $u_j = x_j + iy_j$, so both x_{j+1} and y_{j+1} need to be evaluated, both using Eq. (8.12). It is quite simple to write a computer program to work through the barrier in terms of a specified set of ε_j to obtain two successive sets of coefficients in the incident region to the right, as in Problem 8.1. Then we must determine the longitudinal wavenumber k in the incident region from the starting energy. $\mathbf{k}_t$ remains the same, but in the illustration Fig. 8.1 the k will be different from the starting k_l. Only two successive sets, j and $j + 1$ are needed to determine the transmission. The derivation of the form proceeds as in Section 8.1. After some algebra it leads to a transmission T^*T/I^*I equal to

$$Trans. = \frac{4 \sin^2 kd}{(x_{j+1}-x_j\cos kd+y_j\sin kd)^2+(y_{j+1}-x_j\sin kd-y_j\cos kd)^2} . \tag{8.13}$$

Had we kept a transmitted amplitude T , the denominator would have been multiplied by T^*T and a T^*T would also have appeared in the numerator. The k_l entered in the starting $u_j = e^{-ik_l dj}$, and therefore affected the x_j and y_j which enter Eq. (8.13), but it is only the k for the right which enters this equation explicitly.

Problem 8.1 sets up a system for such a calculation, leading to a transmission as a function of energy for a double-humped barrier. In general this is a very good way to gain an understanding of various tunneling situations. It is easy to do very complex systems, keeping only the features one wishes to explore, as in the illustration here where we kept all $V_{ss\sigma}$ the same, but it is simple to allow those to change if one wishes, or to model a set of energy bands of interest. It is not even difficult to include additional orbitals and couplings which allow accurate modeling of particular band structures. We do not carry through examples except for Problems 8.1 and 8.2, but will note some results in connection with more general descriptions.

It is ordinarily less convenient to proceed numerically with the full Schroedinger Equation when modeling a system, but there are some analytical results which are of interest and it may also be simpler to

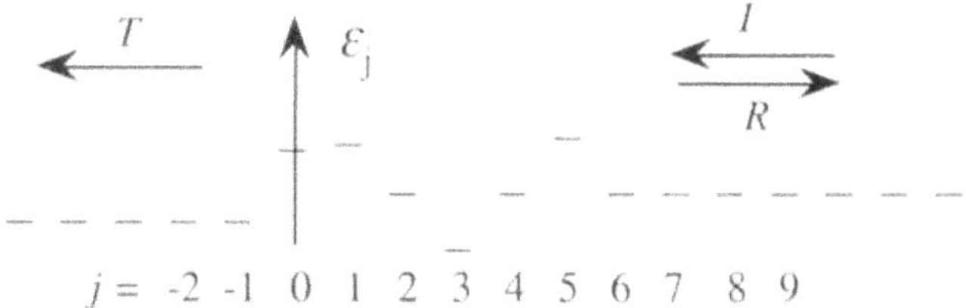

Fig. 8.1. Successive energy levels ε_j through a barrier. The coefficients u_j are written Te^{-ik_1dj} to the left, and each successive u_{j+1} determined from Eq. (8.12). Beyond the barrier to the right the transmission is obtained from the calculated coefficients and k on the right, using Eq. (8.13).

understand the procedure than for the tight-binding analysis we just gave. In doing this we are extending the analysis which we began in Section 2.2.

We imagine a free-electron Schroedinger Equation with a square barrier lying in an xy-plane, as illustrated in Fig. 8.2. As for the tight-binding representation we may take periodic boundary conditions on the lateral boundaries (xz- and yz-planes) so that outside the barrier the electron states can be written $e^{i(k_zz + k_xx + k_yy)}$. Here k_x and k_y are components of the transverse wavenumber and in the region where the potential is zero the energy is $\varepsilon = \hbar^2(k_x{}^2 + k_y{}^2 + k_z{}^2)/2m$, as in Section 2.2. Taking the transmitted beam on the left side, as in Fig. (8.1), the transmitted beam is Te^{-ik_zz}, with k_z positive, and with a factor $e^{i(k_xx + k_yy)}$ which will appear in all beams due to the matching, as for the tight-binding case, and we do not write it explicitly. On the right side the incident beam is Ie^{-ik_zz} and the reflected beam is Re^{ik_zz}. We match wavefunctions, as in Fig. 8.3, through the boundaries since the wavefunction is single-valued (this conserved k_x and k_y), and match normal derivatives at each interface so the kinetic energy operator does not give an infinite value at that point.

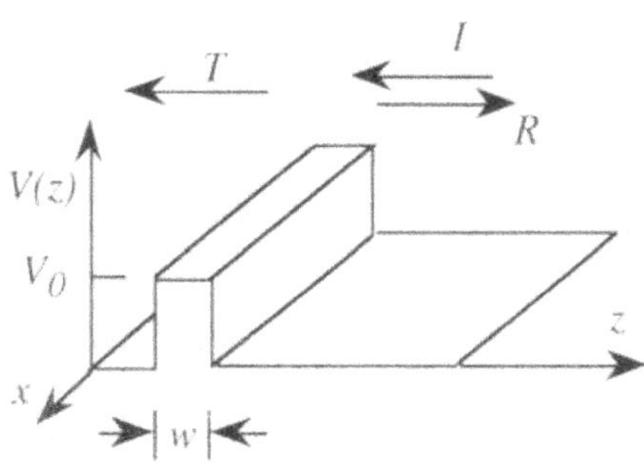

Fig. 8.2. A simple planar square tunneling barrier to free electrons.

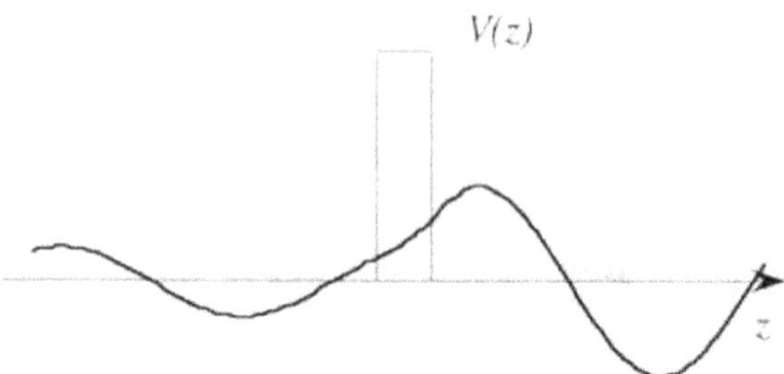

Fig. 8.3. Matching the wavefunction through the barrier, starting with a transmitted wave on the left, matching it to exponentially decaying and growing states in the barrier, and finally to an incoming and reflected wave on the right.

In the barrier region, if the height of the potential V_0 exceeds the energy of the state ε , the energy eigenvalue equation will give us an imaginary k_z with magnitude κ given by $\varepsilon = V_0 + \hbar^2(k_x{}^2 + k_y{}^2)/2m - \hbar^2\kappa^2/2m$ (which follows directly from the eigenvalue equation, Eq. (2.2)). Thus the wavefunction within the barrier has the general form

$$\psi = [Ae^{\kappa z} + Be^{-\kappa z}]e^{i(k_x x + k_y y)}. \tag{8.14}$$

We specify the transverse wavenumber, (k_x, k_y), and the energy which specifies κ as well as k_z for the incident, reflected, and transmitted waves. We match the value and slope to the incident wave at the left surface of the barrier by adjusting A and B. This allows determination of the wavefunction and slope at the right surface of the barrier, which in turn is matched to the incident and reflected wave by adjusting I and R. In the matching at the left surface we obtain one term which grows exponentially to the right and one which drops exponentially. Ordinarily we can neglect the one which drops. The resulting transmission, obtained after some algebra, is

$$Trans. = \frac{T^*T}{I^*I} = \frac{(4\kappa k_z)^2}{(k_z{}^2 + \kappa^2)^2}\, e^{-2\kappa w} \;\; + O(e^{-4\kappa w}) \tag{8.15}$$

The exponential factor comes from the wavefunction dropping exponentially from the incident to the transmitted side a distance w away. The factors in front come from the matching. It is in fact interesting to compare this form with the tight-binding form for tunneling through one atom, Eq. (8.8). The numerator in that equation contained a velocity squared, which corresponds to $(\hbar k_z/m)^2$ for free electrons and explains the $k_z{}^2$ factor in Eq. (8.15). The denominator in Eq. (8.8) contained a factor $\delta\varepsilon_s^2$, which corresponds to the barrier height squared, $(\hbar^2\kappa^2/2m)^2$ and explains the denominator (with the velocity term in k_z^2 dropped as in the final form of Eq. (8.8)). For one atom as in Eq. (8.8), there is no exponential term but otherwise the two forms are quite closely related, as we would expect.

This same formula would apply if the free-electron had an effective mass m^* as in semiconductors, discussed in Section 14.2, but there is a difficulty if the mass changes at an interface. We may see that if we then match ψ and $\partial\psi/\partial z$ at the boundary charge is not conserved; it disappears or appears at the interface. To see this we must derive the current operator, which we may do using the *continuity equation*, the requirement that charge not disappear locally. The probability density is $\psi^*\psi$ so its rate of change is obtained using the Schroedinger Equation (1.16).

$$\frac{\partial\psi^*\psi}{\partial t} = \psi^*\frac{\partial\psi}{\partial t} + \frac{\partial\psi^*}{\partial t}\psi = i\hbar\psi^*H\psi + (i\hbar H\psi)^*\psi$$
$$= \frac{i\hbar}{2m}(\psi^*\nabla^2\psi - \psi\nabla^2\psi^*) = \frac{i\hbar}{2m}\nabla\cdot(\psi^*\nabla\psi - \psi\nabla\psi^*). \tag{8.16}$$

In the second step we noted that the term in the potential canceled and kept only the kinetic-energy term. The continuity equation can be written $\partial\rho/\partial t = -\nabla\cdot\mathbf{j}_e$ with ρ the charge density and $\mathbf{j}_e$ the electric-current density, the statement that the only way the local charge density can change is from net current flowing in or out. Canceling a factor of $-e$ this becomes a relation between the time derivative of the probability density $\partial\psi^*\psi/\partial t$ and the divergence of the probability-current density. We may then identify the probability-current density $\mathbf{j}$ from Eq. (8.16) as

$$\mathbf{j} = -\frac{i\hbar}{2m}(\psi^*\nabla\psi - \psi\nabla\psi^*). \tag{8.17}$$

We may confirm that for a free electron with wavefunction $(1/\sqrt{\Omega})e^{i\mathbf{k}\cdot\mathbf{r}}$ this is $(\hbar\mathbf{k}/m)\psi^*\psi = \mathbf{v}\psi^*\psi$ as we would expect, but it is now much more general.

We now see that if we match ψ and $\partial\psi/\partial z$ at an interface, but the mass which appears on the two sides (the effective mass) is different, the current flowing into the interface will differ from that flowing away from it, so the description is incorrect. We could use the requirement of current conservation, matching $(\psi^*/m^*)\,\partial\psi/\partial z$ as one condition, but that leaves the other uncertain. A full solution of the eigenvalue equation, or Schroedinger Equation, in the solid, using the full potential in the solid will of course give the correct answer, and using that answer for an interface describable by effective-mass bands would tell the condition for that particular case. A more general, but approximate, way is to perform a tight-binding calculation to determine the states near the bottom of a band and see what matching condition would give the correct result. This was done by Harrison and Kozlov (1992) for a simple band using Eq. (8.1) with $H_{j,j+1}$ different on the two sides so the effective masses at the bottom of the bands (determined from $\partial^2\varepsilon_k/\partial k^2$) were different. The result was more complicated than anticipated, giving matching conditions

$$\psi^+ = A\psi^- + B\,\partial\psi^-/\partial z,$$
$$\partial\psi^+/\partial z = C\psi^- + D\,\partial\psi^-/\partial z, \tag{8.18}$$

with the plus and minus indicating the wavefunction to the right and to the left, and with the coefficients A B, C, and D depending upon what was chosen for the matrix element $H_{j,j+1}$ coupling the states *across* the interface. The only way the result became simple, with B and C equal to zero, was if the matrix element $H_{j,j+1}$ coupling the states across the interface was the geometric mean $\sqrt{(H_{j,j+1}{}^+ H_{j,j+1}{}^-)}$ of those to the left and to the right, a perfectly plausible choice. The matching conditions then became

$$\sqrt{\frac{1}{m^+}}\psi^+ = \sqrt{\frac{1}{m^-}}\psi^-,$$
$$\sqrt{\frac{1}{m^+}}\frac{\partial\psi^+}{\partial z} = \sqrt{\frac{1}{m^-}}\frac{\partial\psi^-}{\partial z}, \tag{8.19}$$

This would appear to be the only simple choice which can be correct. However, for real simulations of semiconductors, where the question of effective masses arises, it would seem preferable to use Eq. (8.1) directly as we did above, with parameters chosen to fit the system in question.

8.3 Tunneling Systems

We note that once we have calculated a transmission, one way or another, the argument which led to Eq. (8.11) can be used to write the matrix elements which will give the correct transmission *Trans.* when used in the Golden Rule as

$$T_{12}T_{21} = \frac{\hbar^2 v_1 v_2}{4L_1L_2}\,Trans. \tag{8.20}$$

with of course $v_{1,2}$ the velocities on the two sides and $L_{1,2}$ the length of the system on the two sides. This could for example be used with Eq. (8.15).

The discussion of tunneling in terms of transition rates is often the most appropriate way to analyze tunneling systems of current interest. An important system is a metal such as aluminum, allowed to oxidize in air which frequently leaves an insulating oxide some 20 Å thick, upon which a second metallic layer is deposited. Current between the two metals arises from tunneling through the oxide. With no voltage applied the states on both sides are filled to the same Fermi energy and no net current flows. An applied voltage ϕ raises the Fermi energy on one side by $V = -e\phi$ relative to the other, and tunneling in that energy range goes only one way. The net current may be calculated (assuming zero temperature) using the Golden Rule as in Eq. (8.10), usually assuming conservation of transverse wavenumber. Usually one will expect the matrix elements T_{12} to be insensitive to energy for applied voltages small compared to the Fermi energy. The number of coupled states occupied on one side and empty on the other will be approximately proportional to the applied voltage, so the current will be approximately proportional to applied voltage, as a simple and perhaps uninteresting resistor. However, in special cases, such as with a superconductor on one or both sides, there may be a proportionality to the density of excited states, with a gap at the Fermi energy, as observed by Giaever (1960), providing a powerful test of the theory of those excitations. The same system led to the discovery of superconducting tunneling when both metals were superconducting, the Josephson Effect (Josephson (1962)).

The transition-rate approach can be particularly useful in treating the scanning tunneling microscope where a metallic tip is held over a crystal surface, as in Fig. 8.4, sufficiently close that electrons tunnel into, or out of, the substrate. As it scans across the surface the matrix element T_{12} between

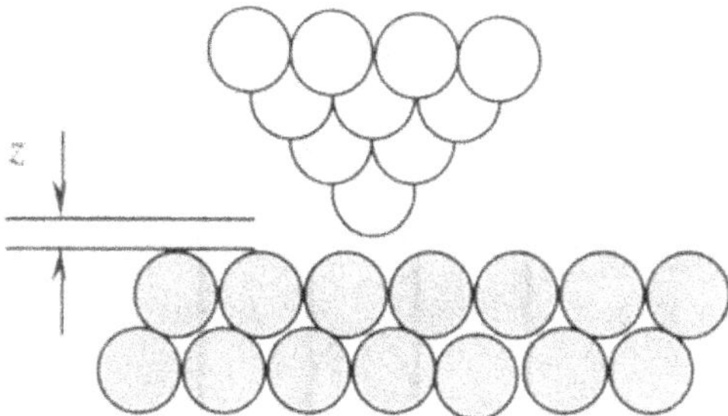

Fig. 8.4. In a Scanning Tunneling Microscope (STM) a metallic tip, shown above, is brought sufficiently close to a substrate, shown below, that electrons can tunnel between. By applying a voltage, and causing the separation z to vary such that the current is constant as the tip moves across the surface, one traces out the surface, like a phonograph needle on an atomic scale.

the tip and the substrate varies giving information about the geometry of the surface atoms. The tip can be modeled as a crystal corner, as described at the end of Section 6.1 with the matrix element between the tip wavefunction and a surface atom given by $\sqrt{2/N}\ \sin k_x d \sin k_y d \sin k_z d\ V_{ss\sigma}$ for a simple-cubic crystal corner and all s-states. One can build upon such a model to construct a more accurate representation of the tunneling spectroscopy for any system of interest. In a more complete analysis one may construct approximate eigenstate wavefunctions for the entire system and use the current operator of Eq. (8.17) to evaluate the current.

8.4 Tunneling Resonance

The system for which the transmission was calculated in Problem 8.1 is illustrated schematically in Fig. 8.5, a double-humped potential which is almost high enough to form a local, bound state - called a *resonant state* - within the barrier. For a finite barrier height the electron can tunnel out on either side as shown and as we have seen. It has finite kinetic energy, corresponding to approximately a half wavelength equal to the width of the well, also as shown. The very remarkable result in such a case, and found in the problem, is that the transmission rises to one at an energy equal to the resonant-state energy, and then drops at higher energies. This system illustrates a number of quantum effects which we shall explore.

The first concerns transitions out of such a localized state. If this were a spherical shell, rather than a one-dimensional system, the state could be constructed just as we did in Section 2.4. There could be a resonant state of, for example, spherical symmetry. We can calculate the rate electrons would leave such a state just as we calculated the transition rate between two sides of a barrier in Eq. (8.9), and we shall carry out such a calculation in Section 9.2. From that rate we can deduce the half-life of the state, the time in which it would have a 50% probability of having escaped.

Such a theory describes for example the decay of a nucleus by the

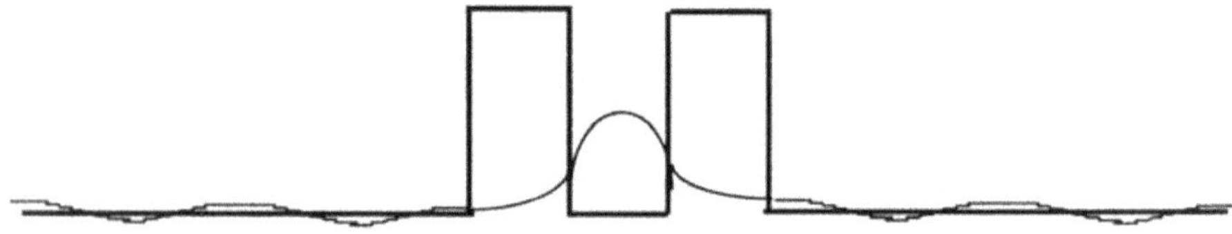

Fig. 8.5. A double barrier may very nearly form a state, called a resonant state, out of which, however, a particle can tunnel.

emission of an *alpha particle*, α-particle, which itself is a helium nucleus. We saw in Section 4.4 that the helium nucleus, with two protons and two neutrons, all in the lowest one-nucleon 1s-state, has particularly low energy and is therefore particularly stable. In the heavier nuclei the binding energy per nucleon is reduced by the Coulomb repulsion of the many protons confined to the nucleus, and at the same time the binding-energy per nucleon is rather insensitive to the exact number of nucleons just as the cohesive energy per atom of a metal becomes independent of the size of the metal for bulk crystals. Thus it becomes energetically favorable to remove one helium nucleus, an α–particle, from a heavy nucleus. There is however a barrier to this removal because as the alpha particle moves just outside of the nucleus which initially had Z_{nuc} protons, beyond the attractive square well which we described for the shell model in Section 4.4, it has a very large electrostatic potential $2(Z_{nuc}\text{-}2)e^2/r$. Thus a plot of the energy of the α-particle as a function of radial distance, relative to its energy at infinite distance should appear as in Fig. 8.6. The transmission of the barrier can be calculated as we did in Section 8.2 (using the tight-binding description as an approximate calculation on a grid of the continuous radial wavefunction, or by the WKB method, which we have not described in this text), and the tunneling rate deduced. For any case Z_{nuc} is known and one has a good idea of the nuclear radius, as indicated in Section 4.4, so one can reliably calculate the transition-rate out, as a function of the energy of the emergent α-particle, and the results are in good accord with experiment.

A particle outside the barriers of a system such as this will have very strong scattering if its energy is near that of the resonant state. This is then called a *scattering resonance* and is analogous to strong scattering of sound by a resonant body when the sound wave has frequency near that of the resonator. In this case it would be scattering of α-particles by a nucleus.

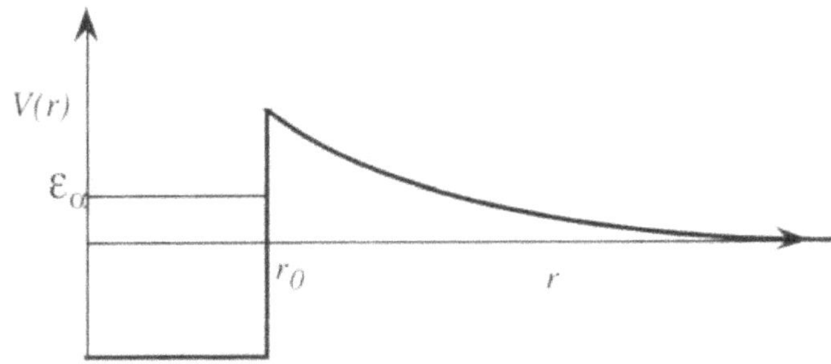

Fig. 8.6. The potential energy of an α-particle as a function of distance r. from the center of the nucleus. r_0 is the radius of the nucleus, and ε_α is thought of as the ground-state energy of the α-particle within the nucleus.

We return next to the one-dimensional case shown in Fig. 8.5 and ask how it can be that an incident electron could have transmission near unity as if it did not see the pair of barriers. The answer suggested by Fig. 8.5 is essentially correct. The probability density inside the barrier is very high for the resonant state and its tunneling out may not be symmetric so that it corresponds to a tiny transmitted beam on one side and a tiny incident and reflected wave on the other. The calculated transmission is high because in that circumstance the probability within the well is high and it feeds the transmitted wave. It is not that the incident electron does not see the barrier. If the individual barriers are thick and high, the resonance becomes extremely narrow in energy; the transmission is high only over a very narrow energy range. Thus if we constructed a wave packet for an electron approaching the double barrier, it would include states of low transmission and the packet would be largely reflected as we would expect physically.

The small-amplitude packet which does tunnel through spends a time of the order of $\hbar$ divided by the resonance width before proceeding beyond. If there is more than one electron tunneling, the probability of both being in the barrier at the same time is greatly reduced by the Coulomb repulsion between them, an effect called the *Coulomb blockade.*

We can construct states in this energy range in detail, and find that over the resonance width (the energy range with high transmission) the states are closely spaced, as without a well in the barrier, but that an extra state is crowded in within this energy range, as illustrated in Fig. 8.7. There is no single resonant state and each state has only a small probability of being within the well, a probability of order the reciprocal of the product of N, the density of states, and the resonance width.

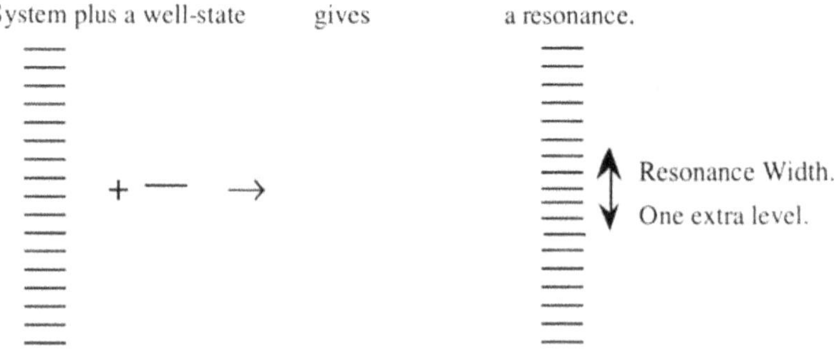

Fig. 8.7. If a potential well is inserted within a barrier, a resonant state which arises will be an additional state crowded in among those already present.

Chapter 9. Transition Rates

The case of tunneling resonances, with one state coupled to two different continua of states as in Fig. 8.5, is an interesting one to explore. We have seen the nature of the states which arise in an exact calculation and have seen the resonant transmission going to unity at the center of the resonance. It will also be useful to proceed in an approximate way to understand better the properties of such systems.

9.1 Second-Order Coupling

It is of interest to consider the effect of the resonance on tunneling for states well removed in energy from the resonance, as illustrated in Fig. 9.1 We proceed in perturbation theory using the matrix elements given in Eq. (8.20). In zero-order, the resonant state of energy ε_2 is an independent energy eigenstate, as is our starting state of energy ε_1 outside the barrier. We may correct our starting state using the first-order perturbation theory of Eq. (5.26) to obtain

$$|1\rangle \rightarrow |1\rangle + \frac{T_{21}}{\varepsilon_1 - \varepsilon_2}|2\rangle \qquad (9.1)$$

Even if there is no direct coupling between the state $|1\rangle$ on the left and a state $|3\rangle$ on the right, there will be a coupling of this first-order state to the state on the right through the coupling T_{32} between the resonant state $|2\rangle$ and the state to the right $|3\rangle$. Thus there is a *second-order coupling* through the resonant state given by

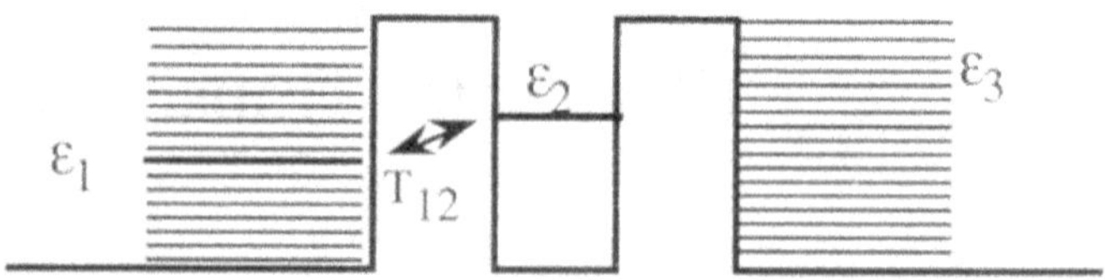

Fig. 9.1. A resonant state is treated as a zero-order energy eigenstate |2>, coupled to a continuum of states |1> and to a second continuum of states |3>.

$$<3|H^{2nd}|1> = \frac{T_{32}T_{21}}{\varepsilon_1 - \varepsilon_2} \ . \qquad (9.2)$$

There is actually an additional term arising from the first-order correction to the state |3> and a cross term from the two first-order corrections. They turn out to be of the same magnitude when $\varepsilon_1 = \varepsilon_3$ and the cross term is of opposite sign so for the calculation of transition rates between |1> and |3> we need only the term given in Eq. (9.2). If there were more that one state in the well, the terms would add in the second-order matrix element and if the terms were opposite in sign they would cancel each other. We shall make use of this feature that there is interference between terms in Section 23.4.

The matrix elements with the local state, T_{12} and T_{23}, contain a single factor of $1/\sqrt{L_1}$ or $1/\sqrt{L_3}$ so that if we write the transmission combining Eqs. (8.20) and (9.2) as

$$Trans. = \frac{4L_1L_3|T_{12}T_{23}|^2}{\hbar^2 v_1 v_3(\varepsilon_1 - \varepsilon_2)^2} , \qquad (9.3)$$

all of these length factors cancel, as they should. As the energy ε_1 of the incident (and transmitted) electron approaches the energy ε_2 of the resonant state, the transmission becomes large. In this form it would diverge but the perturbation theory is only valid when $\varepsilon_1 - \varepsilon_2$ is large. We proceed in the following section to describe tunneling in the other limit, that in which the incident energy is approximately equal to that of the resonant state.

A case where such transmission through a resonant state is important is shown in Fig. 9.2 where electrons tunnel through an oxide between two metals. If there are impurity atoms in the oxide, they can dominate the tunneling. This can be seen experimentally if vibrations in impurity molecules are excited (as described in Section 23.2), so that the transmitted electron has reduced energy.

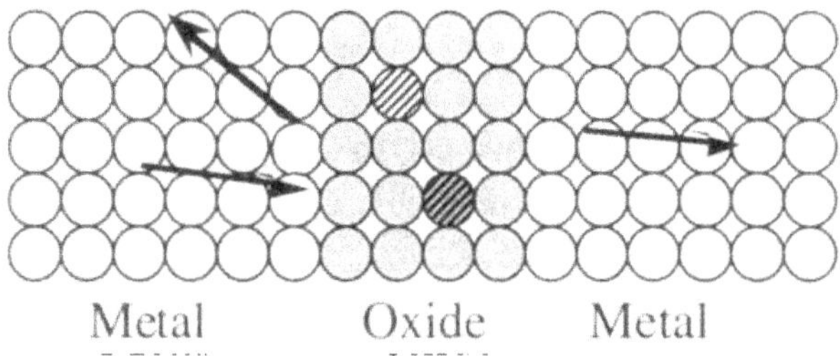

Fig. 9.2. A tunnel junction between two metals. Impurities, indicated by striped circles, can provide resonant states which contribute to, or even dominate, the tunnel current.

9.2 Carrier Emission and Capture

We must proceed differently in order to discuss tunneling when the incident electron has energy approximately equal to that in the resonant state. Then the possibility of transitions into and out of that state arises. In deriving Fermi's Golden Rule we emphasized the necessity of having a continuum of final states, which we do not have for capturing electrons, but we can use the Golden Rule to calculate emission. We could proceed as we just have to obtain matrix elements T_{23} between the local state of energy ε_2 and an external state of energy ε_3 to the right. Then the emission rate is obtained as

$$\frac{1}{\tau} = = \frac{2\pi}{\hbar}\frac{L_3}{\pi}\int dk_3 T_{23}T_{32}\,\delta(\varepsilon_2 - \varepsilon_3) = \frac{2L_3T_{23}T_{32}}{\hbar^2 v_3} . \tag{9.4}$$

v_3 is the speed in the final state. Again the size of the system to the right cancels out in $L_3T_{23}T_{32}$.

Although we cannot directly use the Golden Rule to obtain a capture rate, such capture into discrete states must occur since, for a system in equilibrium, electrons in an impurity state in the barrier will tunnel out with a rate given by Eq. (9.4) and there must be electrons tunneling in at the same rate to replace them. This is called *detailed balance.* In Chapter 10 we shall define probabilities of occupation of such an impurity state as $f(\varepsilon_2)$ and probabilities of occupation of the states into which they tunnel as $f(\varepsilon_3)$. For a transition to occur the internal state must be occupied and the outside state empty, so the transition rate out is given by $f(\varepsilon_2)(1 - f(\varepsilon_3))/\tau$. Similarly there must be a transition rate in equal to $f(\varepsilon_3)(1 - f(\varepsilon_2))/\tau'$ with a τ' which we wish to know. However, $\varepsilon_3 = \varepsilon_2$ for the transition and in equilibrium we shall see in Chapter 10 that f depends only upon energy. Thus $f(\varepsilon_3) = f(\varepsilon_2)$, so $\tau' = \tau$ and we have the rate also for capture and can use the capture rate $f(\varepsilon_3)(1 - f(\varepsilon_2))/\tau$ also when we do not have equilibrium distributions.

We return to the system shown in Fig. 9.1 in which tunneling occurs through a resonant state, but now when the incident electrons have energy ε_1 near ε_2. For the moment we let all such states $|1>$ be occupied, and none of the transmitted states $|3>$ of equal energy be occupied. Then electrons will be captured into the resonant state $|2>$ from $|1>$ and the occupation will increase until the transition rate out to $|3>$ equals the rate in, steady state. This will be with the probability of occupation of $|2>$ equal to one half if the $1/\tau$ is the same between $|2>$ and $|3>$ as it is between $|1>$ and $|2>$ and the tunneling rate through the resonant state would be half the rate given in Eq. (9.4). We look at more general steady-state cases in Section 11.1.

This would be called *sequential tunneling* rather than the resonant tunneling treated in Section 9.1. The full transmission calculation in Chapter 8, and Problem 8.1 in particular, gives transmission for all energies and includes both types without distinguishing. However, in a real system we may make the distinction and it can be very useful to consider both and to see which dominates. For example, in a semiconductor tunnel junction, with a resonant state at energy ε_2 well above the conduction-band edge E_c, the contribution of sequential tunneling will be reduced relative to resonant tunneling by the much lower occupation of electrons at the required energy, by a factor $e^{-(\varepsilon_2 - E_c)/k_BT}$ as we shall see in Chapter 10, than the occupation of levels near the conduction-band edge. However, we see from Eq. (9.4) that the rate for sequential tunneling has only two factors of T_{12} or T_{23} while the resonant tunneling rate is seen in Eq. (9.3) to have four such factors. Each of these factors is seen from Eqs. (8.15) and (8.20) to contain a factor of $e^{-\kappa w}$ which may be extremely tiny. It is not clear from the outset which type of tunneling will dominate. It will differ for different systems and it can be extremely useful to learn for any one system which mechanism is dominant and therefore what its dependence upon parameters such as temperature will be.

9.3 Time-Dependent Perturbations

We have treated transitions arising from perturbations which do not depend upon time, such as the coupling between a resonant state and continuum states, and this led to the conservation of energy. If we now allow perturbations which vary with time we shall find that they can do work on the electron and shift its energy in the transition. The derivation of the corresponding rate follows closely the derivation of Fermi's Golden Rule in Section 7.2.

We introduce a perturbation, which might be a potential $V(\mathbf{r},t)$ or other small term added to the starting Hamiltonian which had eigenstates $|j>$. Then the coupling between these states H_{ij} will also depend upon time. It is convenient to expand the time dependence of the perturbation in a Fourier series, $V(\mathbf{r},t) = \Sigma_\omega V_\omega(\mathbf{r})e^{-i\omega t}$, so that

$$H_{ji}(t) = \sum\nolimits_\omega H_{ji}(\omega)e^{-i\omega t}, \tag{9.5}$$

and each term is a separate perturbation. This however leads to a persistent ambiguity since $V(\mathbf{r},t)$ is certainly real so that $V_{-\omega}(\mathbf{r})$ must equal $V_\omega(\mathbf{r})^*$. This restriction on the Fourier components is compensated for by the fact that $V_\omega(\mathbf{r})$ has independent real and imaginary parts. We shall deal with this systematically in Chapter 15 for lattice vibrations and in Chapter 17 for light, where we find that we may think of each component as an independent term, and it is best to do that here. The ambiguity will resolve itself for any real problem if we are careful that our expansion represents the potential in that system, and that we have correctly included all the terms in the perturbation theory. For the present we treat a single term $H_{ji}e^{-i\omega t}$ and add the other terms in Eq. (9.5) at the end.

We again expand in the eigenstates $|j\rangle$ of the starting Hamiltonian, $|\psi\rangle = \Sigma_j u_j(t)e^{-i\omega_j t}|j\rangle$, as in Section 7.2 and let $u_0 = 1$ at $t = 0$. We may then proceed exactly as in Eqs. (7.6) through (7.8), but keeping the time dependence so

$$i\hbar\frac{\partial u_j(t)}{\partial t} = H_{j0}e^{-i(\omega_0 - \omega_j + \omega)t} + O(H_{j0}^2). \tag{9.6}$$

The final term indicates the second-order terms which we neglect. We may integrate directly from t=0 to get

$$u_j(t) = \frac{H_{j0}}{\hbar(\omega_0 - \omega_j + \omega)}\left(e^{-i(\omega_0 - \omega_j + \omega)t} - 1\right) \tag{9.7}$$

or

$$u_j^*(t)u_j(t) = \frac{H_{0j}H_{j0}}{\hbar^2}\,\frac{4\sin^2[(\omega_0 - \omega_j + \omega)t/2]}{(\omega_0 - \omega_j + \omega)^2}\ . \tag{9.8}$$

The second factor on the right is an energy delta function exactly as shown in Fig. 7.2, but with an additional term in the argument. Thus we obtain exactly the result, Eq. (7.9), with the argument of the delta function replaced by $\varepsilon_0 - \varepsilon_j + \hbar\omega$. We may now add all of the other terms in Eq. (9.5) to obtain the time-dependent perturbation-theory result

$$\frac{1}{\tau} = \frac{2\pi}{\hbar} \sum_j \sum_\omega H_{0j}(\omega) H_{j0}(\omega)\, \delta(\varepsilon_0 - \varepsilon_j + \hbar\omega)\,. \tag{9.9}$$

If there were only one term in the sum over ω, with $\omega = 0$, the equation reduces to Eq. (7.9) as it should. For each term in the sum over ω for which ω is positive the final-state energy is higher than the initial-state energy by $\hbar\omega$, as if a quantum of energy $\hbar\omega$ was absorbed. For the perturbation to be real, for each such term there must be another for which ω is negative, and the final-state energy for that term is lower, as if a quantum of energy was given up by the electron. When we finally quantize the vibrational states or optical energy in Chapters 16 and 17 we shall see that this is exactly the case, but for the present we have introduced a classical perturbation depending upon time and the classical system can add or subtract energy from the electrons to which it is coupled. We continue this classical treatment of the electromagnetic field in the following section and see how it can ionize an atom by exciting an electron from a bound state to the continuum.

9.4 Optical Transitions

The first step in the calculation of transitions caused by an electromagnetic field, and the ionization of an atom in particular, is to obtain the perturbation. It would be possible, and simpler, to carry out a derivation in terms of a perturbation $-(-e)\mathbf{E}\cdot\mathbf{r}$ with $\mathbf{E}$ the time-varying electric field of the light at the atom. However, that formulation is not convenient for other problems and we proceed more generally to obtain results which we shall use later. As we indicated in Section 3.3, the electric field can be included by adding a vector potential $\mathbf{A}$, in terms of which the electric field is given by $\mathbf{E} = -(1/c)\, \partial\mathbf{A}/\partial t$. Then the interaction with the electron is obtained by adding $-(-e)\mathbf{A}/c$ to the momentum operator in the Schroedinger Equation. The $\mathbf{p}^2/(2m)$ is written out and the cross term $e\mathbf{A}\cdot\mathbf{p}/(mc) = -i\hbar e\mathbf{A}\cdot\nabla/(mc)$ is the interaction, thought of qualitatively as the velocity times the field, divided by frequency since $\mathbf{E}$ is proportional to $\partial\mathbf{A}/\partial t$. We can expand the vector potential as we shall do in Eq. (18.1) but making the time-dependence explicit in analogy with Eq. (9.5), keeping two terms which will lead to real fields

$$\mathbf{A}(\mathbf{r},t) = \sqrt{\frac{4\pi}{\Omega}}\, (\mathbf{u_q} e^{i(\mathbf{q}.\mathbf{r} - \omega t)} + \mathbf{u_{-q}} e^{-i(\mathbf{q}.\mathbf{r} - \omega t)}). \tag{9.10}$$

This corresponds to a wave propagating in the direction of $\mathbf{q}$ and only the first term can raise the energy of the electron according to Eq. (9.9). We shall find that the wavelengths of the light of interest are so long (thousands of Angstroms) compared to the size of an atom that we can neglect the variation of Eq. (9.10) over one atom. Then we may take our origin of coordinates at the atom, we take our z-axis along the light wavenumber **q** and the x-axis along the polarization direction, the direction of the electric field, $\mathbf{u_q}$. Then the perturbation representing the electron-light interaction becomes

$$H_{el} = -\sqrt{\frac{4\pi}{\Omega}}\frac{i\hbar e}{mc}(u_q e^{-i\omega t} + u_{-q} e^{i\omega t})\frac{\partial}{\partial x} \quad . \tag{9.11}$$

which will become Eq. (18.10) when we quantize the light field. As we indicated, for the absorption of light we keep only the first term.

We may use this form for coupling between atomic states to learn which states are coupled. We look first at the matrix element between an s-state $|0\rangle$ and some atomic state of angular-momentum quantum number l, the matrix element $\langle l|H_{el}|0\rangle$. All of the factors in Eq. (9.11) except $\partial/\partial x$ can be taken out of the integral in the matrix element, and for a spherically symmetric state $|0\rangle$, $\partial/\partial x|0\rangle = (x/r)\partial/\partial r|0\rangle$ is of p-symmetry. Thus an s-state is coupled by the electron-light interaction only to a p-state since all other states are orthogonal to $(x/r)\partial/\partial r|0\rangle$. This generalizes to the statement that if the initial state has angular-momentum quantum number l, there will only be nonzero matrix elements with states of quantum number $l \pm 1$. We shall make this important generalization in Section 16.3. Such rules are called *Selection Rules*. In a similar way matrix elements of x between harmonic-oscillator eigenstates $\phi_n(x)$ are only nonzero if the two states differ in n by ± 1.

In the case of the electron-light interaction of Eq. (9.11) it has the important physical interpretation that a photon has a spin angular momentum of one unit, $\hbar$, since the absorption or emission of a quantum of light energy always changes the angular momentum of the system it interacts with by one unit. Indeed this is more than an interpretation, it is a proof that photons have unit angular momentum.

For excitation to the continuum, we seek matrix elements $\langle l|H_{el}|\mathbf{k}\rangle$ between an atomic state $\langle l|$ and a freely propagating state $|\mathbf{k}\rangle = (1/\sqrt{\Omega})e^{i\mathbf{k}\cdot\mathbf{r}}$. It might seem preferable to represent the freely propagating state by a plane wave which has been made orthogonal to the atomic states, $|\mathbf{k}\rangle\rangle = |\mathbf{k}\rangle - \Sigma_{l'}|l'\rangle\langle l'|\mathbf{k}\rangle$, called an *orthogonalized plane wave*, since the true propagating states *are* orthogonal to the atomic states. This would lead to cross terms for $l' = l \pm 1$, but such terms are usually small and we neglect them here. We have chosen the matrix element with the plane wave to the right so the derivative $\partial/\partial x|\mathbf{k}\rangle = ik_x|\mathbf{k}\rangle$. The matrix element we seek is then

$$<l|H_{el}|\mathbf{k}> = -\sqrt{\frac{4\pi}{\Omega}}\frac{i\hbar e}{mc}u_q e^{-i\omega t}<l|\frac{\partial}{\partial x}|\mathbf{k}>$$
$$= \sqrt{\frac{4\pi}{\Omega}}\frac{\hbar k_x e}{mc}u_q e^{-i\omega t}<l|\mathbf{k}>. \tag{9.12}$$

The matrix element $<\mathbf{k}|H_{el}|l>$ is the complex conjugate of this, which can be shown by making a partial integration on the $\partial/\partial x$.

The plane-wave states themselves can be expanded in spherical terms as (e. g., Schiff (1968), p. 119)

$$e^{i\mathbf{k}\cdot\mathbf{r}} = \Sigma_{l'}(2l'+1)i^{l'}j_{l'}(kr)P_{l'}(\cos\theta) \tag{9.13}$$

with θ the angle between $\mathbf{k}$ and $\mathbf{r}$, where the P_l are Legendre polynomials P_l^0 introduced in Eq. (2.27), and where the j_l are the spherical Bessel functions introduced in and after Eq. (2.34). For an atomic state of angular-momentum l there will be terms for $l' = l$ in $<l|\mathbf{k}>$ or terms for $l' = l \pm 1$ in $<l|\partial/\partial x|\mathbf{k}>$.

We may complete the evaluation of the matrix element using the final form in Eq. (9.12), which we do for an atomic s-state,

$$<0|\mathbf{k}> = \sqrt{\frac{1}{\Omega}}\int 2\pi r^2 dr \int \sin\theta d\theta \psi_s(r)e^{ikr\cos\theta} = \sqrt{\frac{1}{\Omega}}\int 2\pi r^2\psi_s(r)dr\,\frac{e^{ikr\cos\theta}}{-ikr}\Big|_0^\pi$$
$$= \frac{4\pi}{\sqrt{\Omega}}\int r^2\psi_s(r)\frac{\sin kr}{kr}dr. \tag{9.14}$$

This can be evaluated from tabulated wavefunction or for a simple hydrogen 1s-state, or a state approximated by that form, $\psi_s(r) = \sqrt{\mu^3/\pi}e^{-\mu r}$ with energy $\varepsilon_s = -\hbar^2\mu^2/(2m)$. For that case it becomes

$$<0|\mathbf{k}> = \sqrt{\frac{\mu^3}{\pi\Omega}}\frac{8\pi\mu}{(\mu^2+k^2)^2}. \tag{9.15}$$

Then the time-dependent matrix element becomes

$$<0|H_{el}|\mathbf{k}> = \frac{\hbar k_x\mu^{3/2}e}{mc\Omega}\frac{16\pi\mu}{(\mu^2+k^2)^2}(u_q e^{-i\omega t} + u_{-q}e^{i\omega t}). \tag{9.16}$$

We may now evaluate the rate that such transitions to an ionized state are made using the time-dependent perturbation theory of Eq. (9.9). It is

$$\frac{1}{\tau} = \frac{2\pi}{\hbar} \sum_{\mathbf{k}} \langle 0|H_{el}|\mathbf{k}\rangle \langle \mathbf{k}|H_{el}|0\rangle \, \delta(\, \varepsilon_0 - \varepsilon_k + \hbar\omega), \tag{9.17}$$

and $\langle \mathbf{k}|H_{el}|0\rangle$ is the complex conjugate of $\langle 0|H_{el}|\mathbf{k}\rangle$. We may see already from the k_x factor in the matrix element of Eq. (9.16) that the probability of exciting the electron into a state of wavenumber **k** is proportional to $\cos^2\theta$, with θ the angle which **k** makes with the direction of polarization of the light wave. This makes physical sense since that polarization direction is the direction of forces on the electron. Our neglect of the variation of the field over the atom has made the result completely independent of the direction of propagation of the light, the direction of **q**.

As usual, we replace the sum over final states (of the same spin as the initial state since the H_{el} we use does not couple electron states of different spin) by an integral, in this case with angles measured from the direction of polarization,

$$\sum_{\mathbf{k}} = \frac{\Omega}{(2\pi)^3} \int 2\pi d\theta \, \sin\theta \int dk \, k^2 = \frac{m\Omega}{(2\pi)^2 \hbar^2} \int d\theta \, \sin\theta \int d\varepsilon_k \, k \ . \tag{9.18}$$

Integrating over angle and energy gives

$$\frac{1}{\tau} = \frac{256\pi e^2}{3mc^2 \hbar\Omega} \frac{\mu^5 k^3}{(\mu^2 + k^2)^4} u_{\mathbf{q}} u_{-\mathbf{q}} \, . \tag{9.19}$$

The evaluation of the delta function relates the final-state energy to the photon energy as

$$\frac{\hbar^2 k^2}{2m} = \hbar\omega + \varepsilon_s, \tag{9.20}$$

with $\varepsilon_s = -\hbar^2\mu^2/(2m)$ measured relative to an electron at rest far from the atom.

It is difficult to interpret a result in terms of the amplitudes $u_{\mathbf{q}}$, but we can also write the light-energy flux in terms of the $u_{\mathbf{q}}$ using Eq. (9.10). The ratio of the two is meaningful and the $u_{\mathbf{q}}$ cancel out. We may return to Eq. (9.10) which gives the vector potential, and then the electric field. We can square it and divide by 8π to obtain the energy density, which averages over space or time to $q^2 u_q u_{-q}/\Omega$. There is an equal energy density from the magnetic field so we multiply by two, and by the speed of light to obtain the energy flux. We may then divide by $\hbar\omega$ to obtain a photon flux of $2qu_qu_{-q}/(\hbar\Omega)$. Finally we set the rate of ionization from Eq. (9.19) equal to this photon flux times the cross-section for ionization $\sigma_\times$ and solve for $\sigma_\times$ as

$$\sigma_\times = \frac{128\pi e^2}{3mc\omega}\frac{\mu^5 k^3}{(\mu^2+k^2)^4} = \frac{128\pi\hbar e^2}{3mc|\varepsilon_s|}\frac{\mu^7 k^3}{(\mu^2+k^2)^5}\ . \tag{9.21}$$

The leading factor in the final form is a constant, equal to 0.55 Å^2 if ε_s = -13.6 eV, and the second factor is dimensionless and can be written in terms of $w = \hbar\omega/|\varepsilon_s| = (\mu^2 + k^2)/\mu^2$as $(w - 1)^{3/2}/w^4$. This is plotted Fig. 9.3, showing how the cross-section rises above the threshold $\hbar\omega = -\varepsilon_s$. We note that the leading factor is similar in magnitude to the area, 0.91 Å^2, one would get from the Bohr radius ($1/\mu$ = 0.54 Å) squared times π, and does vary with the depth of the s-state as appropriate to that. However, the final factor is quite small, as seen in Fig. 9.3, so the correspondence between the cross-section and the "area" of the atom is not close numerically.

This evaluation of the flux in the analysis above is tricky because of factors of two and here we have been careful to use the same vector potential, Eq. (9.10), for both the transition rate and the flux. In Chapter 17 we proceed more generally and then the algebra should take care of the consistency there. It would be dangerous to use different approaches for two

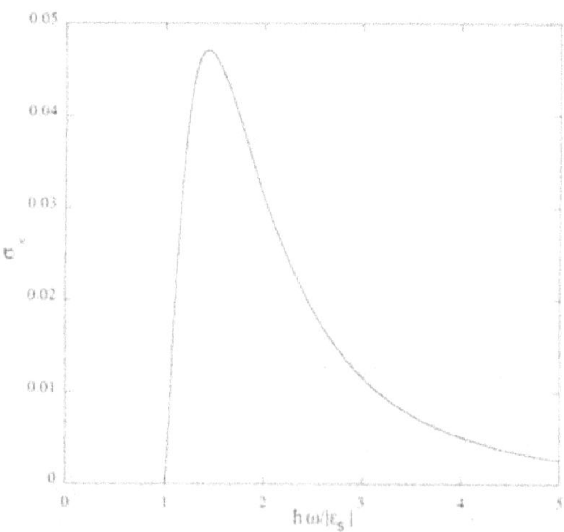

Fig. 9.3. The final factor in Eq. (9.21) giving the cross-section for photo-ionization as a function of photon energy for a hydrogenic s-state.

parts of the same problem.

In Problem 9.1 this calculation is redone for excitation of an electron out of a state in a semiconductor quantum well, localized in one dimension, but propagating in the other two. The procedure is very much the same, but with different geometry. This could be quite accurately done by constructing the excited states for the system as states were constructed in Section 8.2, but for the purposes of the problem simpler states were chosen analogous to those we have used in this section. It is interesting that for a single electron and light propagating parallel to the plane of the well, with polarization perpendicular to that plane, the ionization rate at a given electric-field strength does not depend upon the area of the well, only the thickness. Thus we can again define a cross-section for ionization σ_X, which in this case is considerably larger. When it is plotted it leads to a curve similar to that shown in Fig. 9.3.

9.5 Beta-Ray Emission from Nuclei

We turn now to a very different type of transition, but one which can be described using the Golden Rule. We indicated in Section 4.4 that a neutron could decay into a proton by emission of an electron. Such an event could only occur in conjunction with some other effect since the neutron has spin $1/2$, as do the proton and the emitted electron. With necessarily a half-integral spin change, and orbital angular momentum in integral units of $\hbar$, angular momentum could not be conserved. The other effect is the emission of a *neutrino*, which has spin $1/2$, so it could cancel the spin of the electron and leave the proton with the same spin state as the neutron. The neutrino clearly has no charge, and it has very tiny, if any, mass - we shall treat it as vanishing - so that we can almost think of it as pure spin without substance leaving the scene with the emission. However, it does have wavenumber $\mathbf{q}$ and thus momentum $\hbar\mathbf{q}$, and if it has negligible mass, its energy from the relativistic formula given in Problem 1.1 will be $\hbar qc$.

The emission of the electron and neutrino can be considered a transition, calculated with the Golden Rule. It is an interesting problem to consider since we have no way of knowing what the matrix element is. In this case, and in most others, we can nevertheless proceed by making the simplest sensible guess for the matrix element, and learn the consequences. This is what Fermi (1934) did for beta-decay and it turned out to describe the process well. If it had not, the failure would have taught him how to correct the matrix element. The theory applies to the isolated neutron, or to a neutron contained in a nucleus.

Fermi chose the matrix element to be independent of the magnitude of the momentum of the electron, the neutrino, the neutron, or the proton. However, we recall that momentum conservation is always enforced by the matrix element, so that the matrix element must only be nonzero if the sum of the momenta, or wavenumbers, for the electron, the neutrino, and the proton equals that of the starting neutron. Restating this, we may take the starting wavenumber of the neutron (or nucleus) as zero, and then the sum of the electron wavenumber $\mathbf{k}$ and the neutrino wavenumber $\mathbf{q}$ must be the negative of the proton wavenumber $\mathbf{K}$ (or that of the nucleus after the decay). There would be a relativistic correction if the starting neutron or nucleus had velocities in the laboratory frame comparable to the speed of light. We do not consider that case, but write the matrix element $\langle\mathbf{k},\mathbf{q},\mathbf{K}|H|0\rangle$ as independent of the magnitudes of $\mathbf{k}$, $\mathbf{q}$, and $\mathbf{K}$, and the condition $\mathbf{K} = -\mathbf{q} - \mathbf{k}$ remains true for a relativistic system.

We also need to write the energy delta function. We use the relativistic formula (given in Problem 1.1) for the electron and neutrino energies, which we indicated was $\hbar qc$ for the neutrino, and we keep only the first two terms for the electron. Thus we require that the sum of the electron and neutrino energies $mc^2 + \hbar^2k^2/(2m) + \hbar qc$ plus the recoil kinetic energy of the nucleon or nucleus be the negative of the change in the internal energy ΔE of the nucleus or nucleon. In fact the nucleon is so heavy in comparison to the electron and neutrino that this recoil energy $\hbar^2K^2/(2M_n)$ is negligible. However, we must retain the $\mathbf{K}$ since it allows $\mathbf{k} + \mathbf{q}$ to be nonzero. Thus we neglect the recoil energy in the energy delta function and the Golden Rule gives us a transition rate

$$\frac{1}{\tau} = \frac{2\pi}{\hbar}\sum |\langle\mathbf{k},\mathbf{q},\mathbf{K}|H|0\rangle|^2\, \delta(mc^2 + \hbar^2k^2/(2m) + \hbar qc - \Delta E), \tag{9.22}$$

with the sum over all final states.

The final states are now specified by two wavenumbers, $\mathbf{k}$ for the electron and $\mathbf{q}$ for the neutrino, since $\mathbf{K} = -\mathbf{k} - \mathbf{q}$, so we will sum over both wavenumbers, writing each sum as $\Omega/(2\pi)^3$ times a three-dimensional integral over wavenumber, with an additional factor of two for the two spin states of the electron (with the neutrino of opposite spin, though a more complete analysis of the interactions allows the electron and the neutrino to have parallel spin, and the spin of the nucleon is then flipped). The matrix element must contain a factor of the reciprocal volume of the system $1/\Omega$, just as did the matrix element of the electron-electron interaction, so that the final rate does not depend upon the volume of the system. We think of this factor of $1/\Omega$ as coming from the $1/\sqrt{\Omega}$ normalization of the neutrino and electron wavefunctions and an integration over a local interaction with the nucleon states. Thus we write $\langle \mathbf{k},\mathbf{q},\mathbf{K}|H|0\rangle = M/\Omega$, with M having units of volume times energy. We proceed by finding the rate into a particular electron wavenumber state $\mathbf{k}$ by summing over $\mathbf{q}$ and afterward convert to a distribution as a function of $\mathbf{k}$. With this simple matrix element the rate is independent of the direction of $\mathbf{k}$ or $\mathbf{q}$ or the angle between them, so the probability of a final state with a particular $\mathbf{k}$ and spin is

$$\frac{1}{\tau(\mathbf{k})} = \frac{2\pi}{\hbar}\frac{M^2}{(2\pi)^3\Omega}\int 4\pi q^2\, dq\, \delta(mc^2 + \frac{\hbar^2k^2}{2m} + \hbar qc - \Delta E)$$

$$= \frac{M^2}{\pi\Omega\hbar^4c^3}\int (\hbar cq)^2\, d(\hbar cq)\, \delta(mc^2 + \frac{\hbar^2k^2}{2m} + \hbar qc - \Delta E) \tag{9.23}$$

$$= \frac{M^2}{\pi\Omega\hbar^4c^3}(\Delta E - mc^2 - \frac{\hbar^2k^2}{2m})^2 .$$

Then the probability per unit time of emission of an electron into one of the $2\times4\pi k^2dk\;\Omega/(2\pi)^3$ times $(m/\hbar^2k)(d\varepsilon_k/dk)$ states in the energy range $d\varepsilon_k$ is

$$P_{\varepsilon k}\, d\varepsilon_k = \frac{mM^2}{\pi^3\hbar^7c^3}\sqrt{2m}\,\sqrt{\varepsilon_k}(\Delta E - mc^2 - \varepsilon_k)^2\, d\varepsilon_k, \tag{9.24}$$

shown in Fig. 9.4.

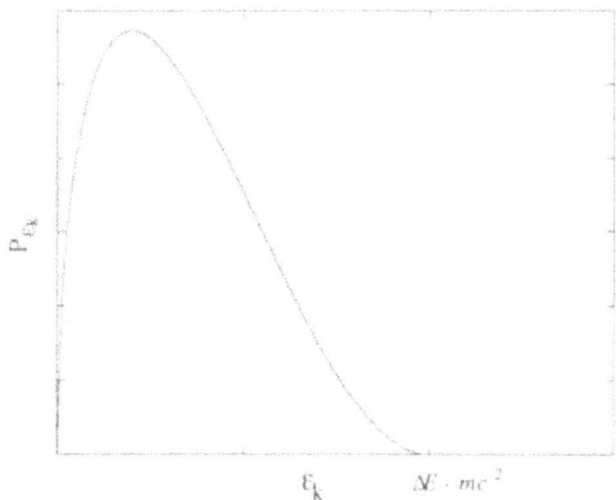

Fig. 9.4. Probability of emission of beta-rays (electrons) as a function of their energy. ΔE is the magnitude of the change in energy of the nucleus due to the transition. The rest of the energy is given to the neutrino.

The scale depends upon the matrix element M which must be obtained from experiment, such as the total half-life of the free neutron. This theory of β-decay is an impressive accomplishment of quantum theory. We do not know at the outset what nature has given us in the way of particles and interactions, but learn that a neutron has somewhat more mass-energy than a proton. Thus energy can be conserved if a neutron decays into a proton and an electron with the excess energy going to the electron. However, such a process cannot conserve angular momentum. When we learn that the process nevertheless occurs, we deduce that another particle - the neutrino - must be created (or perhaps could have been absorbed).

We of course do not know the matrix element for the process but can make the simplest guess and see if it fits experiment. The same approach was taken for a wide range of particle decays, and collision processes. One assumes that all processes are possible but each with an unknown but nonzero matrix element. If a particular process does not occur, one concludes that it is ruled out by some other conservation law. This is the path which led to the Standard Model of particle physics, mentioned in Section 4.4.

The matrix element for β-decay was found to be small, and called a *weak interaction*, small compared to the electromagnetic interaction and the strong interactions between nucleons arising from π-mesons as described in Section 17.4. At the same time the weak interactions are strong compared to gravitational interactions.

A principal goal of contemporary particle theory is to understand the ratio between these four interactions in terms of one theory. That has not been accomplished, and in any case for us it would only explain what nature has given us. It is still true that given what we have, quantum theory tells us how to understand its behavior.

IV. Statistical Physics

To a large extent in our analysis we have been able to treat individual particles or waves, based upon a one-particle or one-electron approximation introduced already in Section 2.2. Occasionally we have introduced collections of particles, as for quantized conductance in Section 2.3. It is appropriate to take time to organize the subject a little more completely, discussing first statistical distributions in equilibrium, then transport theory when systems are out of equilibrium, and finally some aspects of the theory of noise. These are separate subjects from quantum mechanics, but are absolutely necessary if one is to make quantum-mechanical studies of systems involving many particles. They are not always incorporated in undergraduate programs, so it may be necessary to include them in a course on quantum theory. They also lead to quite different results in the context of quantum mechanics. We proceed as we have in other discussions by treating the simplest cases carefully, and seeing how they generalize to more intricate contexts.

Chapter 10. Statistical Mechanics

Statistical mechanics is concerned with systems in thermal equilibrium, the state of any many-particle system if it is left by itself long enough. It does not depend on the mechanisms which bring the system to equilibrium. There may be a vast number of possible states of the entire system, and a major aspect of statistical physics is dealing with this complexity by formulating the right questions. We discuss this first in describing the distribution functions which are sought. The analysis is then based upon the statistical assumption that *in thermal equilibrium the probability of any particular quantum state depends only upon the energy of that state.* It will then follow from our first example that the dependence of that probability on its energy ε_j is given by the Boltzmann factor, $e^{-\varepsilon_j/k_BT}$, in term of the temperature T. It may seem odd to use a detailed model to obtain this very general result, but the model is easy to understand and in terms of it the derivation of the general result becomes obvious. This first example is a very large number of identical, independent harmonic oscillators, such as we discussed in Section 2.5. For it we carry out the intricate calculation of the most likely distribution. The remainder of the analysis in the subsequent sections generalizes the solution to normal modes of vibration, light waves or photons, Bose-Einstein particles, and finally fermions, particles such as electrons which obey the Pauli Exclusion Principle.

10.1 Distribution Functions

The quantum-mechanical state of a set of N_T harmonic oscillators, with displacement coordinates $\{u_j\}$, is a many-particle wavefunction $\Psi(\{u_j\},t)$.

[As in Chapter 3 we are using the brackets { } to designate a collection of values.] If the oscillators are independent, as we assume ($H = \Sigma_j H(u_j,p_j)$), we saw in Section 2.2 that the wavefunction could be factored as $\Psi(\{u_j\},t)= \Pi_j\psi(u_j,t)$. We found further in Section 2.5 that each harmonic oscillator has eigenstates numbered by an integer n_j with energy $\hbar\omega(n_j + {}^1/_2)$. Thus energy eigenstates of a collection of N_T oscillators can be specified by giving the n_j of each one. If we measure the energy of this system, we shall find it to be in one of these eigenstates.

We in fact may not care which ones of the harmonic oscillators are excited to which level and might ask only for the number of these oscillators $N(n)$ which are in the n'th level of excitation. This would be called a *distribution function.* It is a simplification of the description, obtained by asking only for the information which will prove useful. There are still a very large number of possible sets $\{N(n)\}$ and what we might really prefer is an average over many observations, on the system in equilibrium, of each, $\langle N(n)\rangle$, called an *ensemble average*. For a given total energy, the resulting $\langle N(n)\rangle/N_T$ would be some well-defined set of numbers which we would call an *equilibrium distribution function* for the excitations of each oscillator. This distribution function is enough to answer the principal questions we want to ask, and it is what we seek from a statistical analysis.

Our starting point is saying that the probability a system in equilibrium is in any particular quantum-mechanical state depends only upon the energy of that state, that all states of the same energy are equally likely. This is not an obvious statement. If for example we had $N_T = 2$ oscillators, and a total excitation energy of $2\hbar\omega$, we might incorrectly apply first an excitation randomly to one of the two oscillators and then, apply the second randomly to one of the two. Fifty percent of the times we did this the second excitation would be on the other oscillator from the first, giving a fifty-percent probability of the state with one excitation on each oscillator. The other fifty-percent of the times they would be both on the same oscillator, 25% for the state with oscillator-one doubly excited and oscillator-two in the ground state, and 25% for the reversed state. This would be inconsistent with our starting point of equal, $33^1/_3\%$, probability for each of the three states and is not correct. The error arose because the probability of excitation of an oscillator *does* depend upon whether or not it is already excited. We should avoid saying that the assumed equal probability for each state is justified by there being no reason to favor one over another since we have just, incorrectly, found one. The assumption is much deeper, but we can be as confident of it as we can of thermodynamics since variations from that assumption lead to violations of the Second Law.

Given that every quantum state of the same energy has equal probability we can calculate the probability of any particular distribution function

$\{N(n)\}$ occurring by counting the number of quantum states to which such a set corresponds, and dividing by the total number of quantum states. Thus if we had three oscillators, and a total of three units of excitation, there are three quantum states, or arrangements of the excitation energy, with $N(3) = 1$, $N(2) = N(1) = 0$, (again $N(3)$ is the number of oscillators triply excited) depending upon which is excited. There is only one arrangement with $N(1) = 3$, and six arrangement with $N(2) = N(1) = 1$. These are all of the ten quantum states of the three oscillators of this energy. We would say that with three units of excitation the probability of finding the first arrangement is 3/10, the second arrangement is 1/10, and the third is 6/10. This is a detailed calculated distribution for this energy, based upon the statistical assumption that all states of the same energy are equally likely.

We may do the same calculation for a large number N_T of oscillators. For a particular set $\{N(n)\}$, we might first assign the $N(1)$ in $N_T!/[(N_T - N(1))!N(1)!]$ ways [that is, $N_T(N_T-1) \ldots (N_T-N(1)+1)/N(1)!$, where the numerator is the number of ways $N(1)$ objects can be placed in N_T bins, and the denominator is the number of different ways those same $N(1)$ objects can be arranged in the same $N(1)$ bins]. Note that this is looking for distinct quantum states as we did above for the $N_T = 3$ case, not applying excitations to the system as we incorrectly did for the $N_T = 2$ case before. Similarly, for each such assignment of single excitations, there are$(N_T-N(1))!/[(N_T-N(1)-N(2))!N(2)!]$ ways of assigning the $N(2)$ oscillators in the second state of excitation among the $N_T - N(1)$ remaining oscillators. Continuing on for the entire series of n values, we obtain the total number of equally likely ways this set can be accomplished as

$$W = \frac{N_T!}{(N_T - N(1))!N(1)!}\,\frac{(N_T-N(1))!}{(N_T-N(1)-N(2))!N(2)!}\cdots = \frac{N_T!}{N(1)!N(2)!N(3)!\ldots}. \quad (10.1)$$

We may confirm for the example given above with $N_T=3$ that we obtain the correct answer from this for the three cases, $\{N(0),N(1),N(2),N(3)\}= \{2,0,0,1\}$, $\{0,3,0,0\}$, and $\{0,1,1,0\}$, respectively, which we discussed above, noting of course that $0! = 1! = 1$. Similarly for the case discussed above with $N_T = 2$ we correctly obtain $W = 1$ for $\{0, 2, 0\}$ and $W = 2$ for $\{0, 0, 1\}$.

These W-values would enable us to find the probability of any particular distribution function $\{N(n)\}$ as we did for the $N_T=2$ and $N_T=3$ cases, but it will be much more useful to find the *most-likely* distribution function. It turns out that when N_T is very large, the vast majority of the huge number of possible distribution functions are very close to the most probable one. We may find the most-likely distribution by maximizing W subject to the two conditions,

$$\Sigma_n \; N(n) = N_T \tag{10.2}$$

and

$$\hbar\omega \; \Sigma_n \; nN(n) = E_{exc.} \tag{10.3}$$

where $E_{exc.}$ is the total excitation energy, over the zero-point energy of $^1/_2\hbar\omega \, N_T$.

There are two ways to greatly simplify the calculation. The first is to maximize not W but equivalently to maximize the logarithm of W, first since the logarithm of the product is a *sum* of logarithms of the factorials, and second since we then can use *Stirling's formula* for the logarithm of the factorial of large numbers,

$$\ln(N!) = N \ln(N) - N + \ldots, \tag{10.4}$$

neglecting the remaining terms, which are much smaller when N is large. Then

$$\begin{aligned} \ln W &= \ln N_T! - \Sigma_n \ln N(n)! \\ &= N_T \ln N_T - N_T - \Sigma_n \, [N(n) \ln N(n) - N(n)] + \ldots \end{aligned} \tag{10.5}$$

Second, we use the method of Lagrange multipliers which we derived in Section 5.2 to apply the constraints. We subtract from $\ln W$ a Lagrange multiplier α times $\Sigma_n \; N(n) - N_T$ (which is zero by Eq. (10.2)) and a corresponding term with Lagrange multiplier β for the second constraint, Eq. (10.3). Then we can obtain the maximum of $\ln W$ subject to the two constraints by maximizing $\ln W - \alpha[\Sigma_n \; N(n) - N_T] - \beta[\hbar\omega \, \Sigma_n \, nN(n) - E_{exc.}]$ without any constraint and adjusting the α and β to fit the constraints. Setting the derivative of this expression (using the second form in Eq. (10.5)) with respect to $N(n)$ equal to zero gives immediately

$$-\ln N(n) - 1 + 1 - \alpha - \beta n\hbar\,\omega = 0. \tag{10.6}$$

We take the $\ln N(n)$ to the right, and exponentiate both sides noting that $e^{\ln N(n)} = N(n)$, to obtain

$$N(n) = e^{-\alpha} \, e^{-\beta n\hbar\omega} \,. \tag{10.7}$$

$e^{-\alpha}$ and β are to be determined by fitting the conditions, Eqs. (10.2) and (10.3).

With increasing total energy, corresponding to higher temperature, β must decrease. In fact temperature is *defined* for any equilibrium system by $T = 1/(\beta k_B)$, with k_B called the Boltzmann constant, and β the Lagrange multiplier used to fix the total energy. Then T can be shown to have all the properties we associate with temperature. [For example, the pressure of a contained gas can be used as a thermometer, and an analysis similar to that here used to show that temperature is proportional to $1/\beta$.] We use the definition in the reverse sense to write β in terms of temperature as

$$\boxed{\beta = \frac{1}{k_B T} \ . \qquad (10.8)}$$

The derivation of Eqs. (10.7) and (10.8) can be generalized to the proof of the statement that the probability of any system in equilibrium being in any accessible quantum state of energy ε_j is proportional to the Boltzmann factor, $e^{-\varepsilon_j/k_B T}$, with the same proportionality constant for every state. We do this by duplicating exactly this system many times and letting all duplicates be in equilibrium with each other. This is called an ensemble. Then each duplicate does not need to have the same energy, but the average energy - or the temperature - must be the same for the ensemble. With respect to one system, all of the others have become a *thermal reservoir.* We can even include a variety of other systems within this thermal reservoir. Then all of the energies $\varepsilon_j = \Sigma_n n\hbar\omega$ of the duplicates replace the energies $n\hbar\omega$ of the individual oscillators from our previous derivation, and the probability of any particular ε_j is proportional to the number of duplicates having that energy, $N(\varepsilon_j)$, obtained from Eq. (10.7) by replacing $n\hbar\omega$ by ε_j to obtain $e^{-\varepsilon_j/k_B T}$, with the same proportionality constant $e^{-\alpha}$. This is used in Problem 10.1, noting that the same derivation applies to the excitation of electrons from a defect, to obtain the relative probability of different charge states. It is accomplished by evaluating $e^{-\alpha}$ as we do here for the set of harmonic oscillators.

The condition Eq. (10.2) states that

$$\Sigma_n N(n) = e^{-\alpha} \Sigma_n e^{-\beta n\hbar\omega} = N_T. \qquad (10.9)$$

The sum is of the form $1 + x + x^2 + x^3 + ... = 1/(1 - x)$ so $e^{-\alpha} = N_T(1 - e^{-\beta\hbar\omega})$ and the probability of an excitation number n for any particular oscillator is

$$P(n) = \frac{N(n)}{N_T} = (1 - e^{-\beta\hbar\omega})\, e^{-\beta n\hbar\omega}. \qquad (10.10)$$

Of more interest is the average excitation per mode,

$$<n> = \Sigma_n P(n)n = (1\text{-}e^{-\beta\hbar\omega})\Sigma_n\, n e^{-\beta n\hbar\omega} = -\,(1\text{-}e^{-\beta\hbar\omega})\frac{\partial}{\partial\beta\hbar\omega}\Sigma_n\, e^{-\beta n\hbar\omega}$$

$$= -\,(1\text{-}e^{-\beta\hbar\omega})\frac{\partial}{\partial\beta\hbar\omega}\frac{1}{1 - e^{-\beta\hbar\omega}} = \frac{e^{-\beta\hbar\omega}}{1 - e^{-\beta\hbar\omega}} = \frac{1}{e^{\beta\hbar\omega} - 1}. \tag{10.11}$$

At the third equal sign we used a term-by-term derivative to write the sum, and at the fourth we performed the sum as for Eq. (10.9). The final form is the important result.

We may evaluate this for high temperatures, where $<n>$ is large and classical physics applies. Then the average excitation $<n> = 1/(e^{\beta\hbar\omega} - 1) \approx 1/(1 + \beta\hbar\omega + ... - 1) = k_B T/(\hbar\omega)$. This gives the classical result for the average excitation energy of a harmonic oscillator $\varepsilon_{exc.} = <n>\hbar\omega = k_B T$ (as the classical average energy of an ideal gas atom is $^3/_2 k_B T$). We have derived it for the quantized oscillator, but these relations reassure us that we have made the proper definition of temperature with Eq. (10.8).

10.2 Phonon and Photon Statistics

Eq. (10.11) may be immediately applied to the energy in vibrations of a crystalline lattice, which we shall consider more completely in Part V, or to sound waves in a gas, which we introduced in Section 1.8. Normal modes in a pipe of length L with closed ends could be obtained from vanishing boundary conditions so that the wavenumber q satisfies $qL = \pi n$. The amplitude $u(z)$ for each of these normal modes satisfies a classical equation of motion which may be rewritten from Eq. (1.30) as

$$\rho\frac{\partial^2 u}{\partial t^2} = -\,q^2 B\, u \tag{10.12}$$

at any position z, or in particular $u(z)$ with z at the antinode. This is exactly the harmonic oscillator equation (spring constant $\kappa = q^2 B/\rho$) and our treatment in Section 10.1 applies directly. It may somehow be even more remarkable that wave-particle duality applies to the system made up of this disordered fluid than when it applied to the center of mass of the weight in the harmonic oscillator but the statement of wave-particle duality specified that it was true of *everything*. The average energy in a mode of frequency ω is then obtained from Eq. (10.11).

This is an important quantum-mechanical result when applied to a solid. For N atoms bonded together in a solid there are $3N$ normal modes. Einstein (1907), (1911) imagined $3N$ modes of the same frequency ω_E, and therefore a thermal energy given by

$$E_{\text{therm.}} = \frac{3N\hbar\omega_E}{e^{\beta\hbar\omega_E} - 1} . \tag{10.13}$$

At high temperature, where β is small, the exponential in the denominator can be expanded and the leading term in $E_{\text{therm.}}$ is $3Nk_BT$, the classical result as discussed at the end of Section 10.1. This corresponds to a specific heat, $C_V = \partial E_{\text{therm.}}/\partial T = 3Nk_B$. However, at low temperatures β becomes large and the thermal energy, and the specific heat, go to zero, as they do experimentally. Prior to this treatment it was impossible to understand this experimental result. The observed specific heat does not drop off exponentially, as predicted by Eq. (10.13), because all of the frequencies are not the same. Debye (1912) redid the problem representing the normal modes as sound waves, with frequency equal to the wavenumber times the speed of sound v, restricting wavenumber to be less than a q_D which limits the number of each of the three (longitudinal and two transverse) modes to N. Then the thermal energy is obtained, by replacing the sum over modes by an integral as in Eq. (2.9), as

$$E_{\text{therm.}} = \frac{3\Omega}{(2\pi)^3} \int_{0,q_D} \frac{4\pi q^2 dq \;\hbar v q}{e^{\beta\hbar v q} - 1} . \tag{10.14}$$

This again leads to the classical result at high temperatures, small β. At low temperatures we may change variables in the integral to $x = \beta\hbar vq$, bringing a factor of $1/\beta^4$ outside an integral over x. At low temperatures the integral can be extended to infinity, becoming a constant, so the thermal energy is proportional to T^4 and the specific heat proportional to T^3, in good accord with experiment. This appears to be the first time quantum theory was applied to such a macroscopic system as a sound wave. In Problem 10.1 we use this same Debye Approximation to calculate the vibrational zero-point energy.

Eq. (10.11) also leads immediately to the familiar *Planck Distribution* of the energy of light. We first note the equation of motion for the amplitude of light, Eq. (1.20) based upon the vector potential, is of the same form as the equation of motion for sound given in Eq. (1.30), so again each light mode in a cavity is an independent harmonic oscillator of frequency $\omega = cq$, now with c the speed of light. We will carry this analysis out in detail in Chapter 18. The average energy in each mode, $\hbar\omega(\langle n \rangle + 1/2)$ is again obtained from

Eq. (10.11). We write the energy in a frequency range $\delta\omega = \delta q/c$, adding the two directions of polarization and dropping the zero-point energy, as

$$\delta E_{\text{light}} = \frac{2\Omega}{(2\pi)^3} 4\pi q^2 \delta q <n> \hbar\omega = \frac{\hbar\omega^3\,\Omega}{\pi^2 c^3} \frac{1}{e^{\beta\hbar\omega} - 1} \delta\omega, \qquad (10.15)$$

which is the Planck distribution of light in thermal equilibrium obtainable only with the use of quantum theory. It is shown in Fig. 10.1.

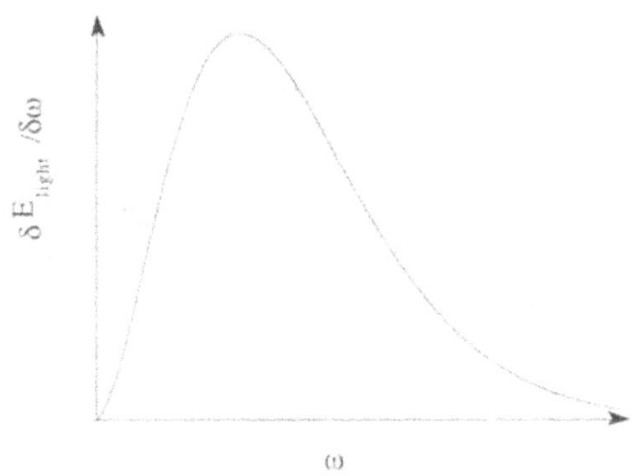

Fig. 10.1. The Planck distribution of the energy density of light in thermal equilibrium as a function of frequency, from Eq. (10.15).

10.3 Bosons

The photons which we treated in Section 10.2 are particles with energy $\hbar\omega$ and momentum $\hbar q$, and one unit of spin as we saw in Section 9.4 when we saw that they added or subtracted one unit of angular momentum to an atom when they were absorbed or emitted. Phonons, the quantized vibrational energies of the sound waves in the solid, are also particles. It would not mean anything to distinguish two photons since they simply represent the second state of excitation of the oscillator if they are the same mode, or excitations of two different oscillators if they are different modes. We would therefore say that "interchanging" two photons does not change the state of the system and we shall take this to mean that the wavefunction representing the state is the same. The field theory of other particles with integral spin similarly leads to particles for which the wavefunction of the system "goes into itself" under interchange, and all are called *bosons*. Their statistics are called Bose-Einstein statistics.

When bosons have mass, the statistical calculation is the same except that there is an additional constraint: the total number of bosons does not change when they collide. When we change the temperature of a photon gas, the number of photons, by Eq. (10.10) automatically changes, but energy is not available to create or destroy massive bosons when the temperature changes. Thus for bosons with mass, we cannot treat each **q** independently because the number conservation applies to the collection of states, and we must therefore redo the calculation.

We do this by dividing the spectrum into groups of N_T states allowed by the boundary conditions (rather than modes) in a range of energy $\delta\varepsilon$ near ε (that ε replaces $\hbar\omega$ for modes of frequency ω). The number of ways a distribution of occupations $\{N_\varepsilon(n)\}$ can be made is again given by Eq. (10.1), with a different set of N_ε for each range. We should multiply these W's together, which adds their logarithms as in Eq. (10.5). We add the same constraint for the number of states $\sum_n N_\varepsilon(n) = N_T$ for each range with Lagrange multipliers α_ε, but now a global (summed over ε as well as n) $\sum_{n,\varepsilon} N_\varepsilon(n)n\varepsilon = E_{tot.}$ with a Lagrange multiplier β and in addition a global $\sum_{n,\varepsilon} N_\varepsilon(n)n = N_{tot.}$ with a new Lagrange multiplier $-\beta\mu$. fixing the total number of particles. [The use of the product $-\beta\mu$ turns out convenient in the end.] Then setting the derivative with respect to $N_\varepsilon(n)$ equal to zero gives

$$-\ln N_\varepsilon(n) - \alpha - \beta\varepsilon n + \beta\mu n = 0. \tag{10.16}$$

in place of Eq. (10.6) and solving this gives

$$N_\varepsilon(n) = e^{-\alpha_\varepsilon} e^{-\beta(\varepsilon-\mu)n}. \tag{10.17}$$

We can again fix α_ε so that $\sum_n N_\varepsilon(n)n = N_T$ and again evaluate the average n for a single state, at each ε, to obtain

$$\langle n(\varepsilon)\rangle = \frac{1}{e^{\beta(\varepsilon-\mu)} - 1}, \tag{10.18}$$

in place of Eq. (10.11). At high energies, where $\langle n(\varepsilon)\rangle$ is small, this approaches the classical Boltzmann distribution proportional to $e^{-\beta\varepsilon}$. Again $\beta = 1/k_BT$ and μ is called the *chemical potential* for these bosons. It is adjusted to obtain the correct number $N_{tot.} = \Sigma_\varepsilon N_T\langle n(\varepsilon)\rangle$ of particles.

At high temperatures, μ is large and negative. As the temperature drops, μ increases to keep the number N of particles fixed. If we find $\mu > 0$ (the energy of the lowest states is zero), $\langle n \rangle$ in Eq. (10.18) becomes negative. This is not meaningful and in fact a phase transition must occur. To obtain the condition for this we evaluate the number of particles using Eq. (10.18) with $\mu = 0$ and an energy $\varepsilon = \hbar^2 k^2 (2M)$. The condition for finding $\mu > 0$ becomes, using a sum of Eq. (10.18) over states rather than ranges ε,

$$N > \frac{\Omega}{(2\pi)^3} \int dk\, 4\pi k^2 \frac{1}{\exp(\beta\hbar^2 k^2/(2M)) - 1} = \left(\frac{2M}{\hbar^2 \beta}\right)^{3/2} \frac{\Omega}{(2\pi)^2} \int du \frac{\sqrt{u}}{e^u - 1}. \quad (10.19)$$

The integral can be evaluated numerically (Kittel and Kroemer (1980) p. 204) giving $1.306\pi^{1/2}$. Then we may solve for $k_B T = 1/\beta$ to obtain the condition in Eq. (10.19) as $k_B T < (2\pi\hbar^2/M)(N/(2.612\Omega))^{2/3}$ given by Kittel and Kroemer (1980), and called the *Einstein Condensation Temperature*. We may see that at this point $\langle n \rangle$ for the lowest state has increased to the point that a sizable fraction of the bosons are in the lowest-energy one-particle state, which is called Bose-Einstein condensation. Using the M and N/Ω for liquid helium (^{4}He), this gives 3°K for ^{4}He, approximately equal to the temperature at which helium becomes superfluid. This is generally agreed to be the nature of superfluid helium, though treating the helium atoms as noninteracting is a very crude approximation. There are a number of peculiar properties of helium in this state, discussed by Kittel and Kroemer, op. cit., such as a vanishing viscosity. In recent years alkali-metal vapors were found to undergo Bose-Einstein condensation (Ensher, Jin, Matthews, Wieman, and Cornell (1996)). In these atoms the valence-electron spin combines with the nuclear spin to form an atom of vanishing or integral spin, which can then condense into the Bose-Einstein ground state.

10.4 Symmetry Under Interchange

We have seen that there is no meaning to interchanging two photons since they are simply degrees of excitation of an oscillator, and that this is also true of other bosons. Thus if we are to regard them as particles, and write a wavefunction for two photons (we use a capital Ψ for a state of more than one particle), $\Psi(\mathbf{r}_1, \mathbf{r}_2)$, it must be same state as $\Psi(\mathbf{r}_2, \mathbf{r}_1)$. This would seem to require that $\Psi(\mathbf{r}_2, \mathbf{r}_1) = \Psi(\mathbf{r}_1, \mathbf{r}_2)$ for all identical particles but when we discuss half-integral spins shortly we shall see that when the two particles have the same spin the wavefunctions with coordinates interchanged are of opposite sign. It is said of bosons, with $\Psi(\mathbf{r}_2, \mathbf{r}_1) = \Psi(\mathbf{r}_1, \mathbf{r}_2)$, that the wavefunction is symmetric under interchange of particles.

For noninteracting bosons, for which we saw in Section 2.2 that we can find product solutions, the products must take the form

$$\Psi(\mathbf{r}_1, \mathbf{r}_2) = \frac{[\psi_1(\mathbf{r}_1)\psi_2(\mathbf{r}_2) + \psi_1(\mathbf{r}_2)\psi_2(\mathbf{r}_1)]}{\sqrt{2}} \tag{10.20}$$

One consequence of this symmetry of the wavefunction is the Bose-Einstein statistics which we have developed above, and the Bose-Einstein condensation which can occur at low temperatures.

There is another quite remarkable consequence for the rotational states of molecules for which the nuclei are identical bosons. If we wished to discuss the rotational states of a molecule such as O_2, with nuclear positions $\mathbf{r}_1$ and $\mathbf{r}_2$, we would ordinarily change variables to a center-of-mass coordinate $\mathbf{R}$, and a relative coordinate $\mathbf{r} = \mathbf{r}_2 - \mathbf{r}_1$, and further write $\mathbf{r}$ in spherical coordinates $\{r, \theta, \phi\}$. With no external torques, the eigenstates will be of the form $\psi_c(\mathbf{R})\psi_l(r)Y_l^m(\theta,\phi)$ as for other spherically symmetric systems as discussed in Section 2.4. In particular, there will be rotational states with energy proportional to the square, $l(l+1)\hbar^2$, of the total angular momentum. There is however a difficulty with this if both atoms in the molecule are oxygen-16. We saw in Section 4.4 that this nucleus with eight protons and eight neutrons completes a shell so, as for electrons in an inert gas, there is no net angular momentum and a total spin of zero. Thus these nuclei are bosons. With zero-spin they must be in the same spin state, and interchanging them does not change the state, so the wavefunction (the state of the system) must be the same. If we imagine the molecule in a rotational state of angular-momentum quantum number l , written $\psi_c(\mathbf{R})\psi_l(r)Y_l^m(\theta,\phi)$, for odd l the spherical harmonic changes sign when $\mathbf{r}$ is replaced by $-\mathbf{r}$, violating the condition $\Psi(\mathbf{r}_2, \mathbf{r}_1) = \Psi(\mathbf{r}_1, \mathbf{r}_2)$. We conclude that only even l are allowed states and that is found to be true in experimental infra-red spectra. *No such odd-integer rotational states are observed* for $O^{16}O^{16}$, though they are observed for $O^{16}O^{17}$ since the extra neutron in O^{17} makes the nucleus distinguishable from the O^{16} (Hilborn and Yuca(1996)).

It seems truly remarkable that these two nuclei, which have negligible possibility of direct interaction with each other, can "know" the symmetry of the other nucleus. As with the Aharanov-Bohm Paradox of Section 1.4, this is not resolved by imagining some tiny interaction, but is to be recognized as a consequence of wave-particle duality, our starting assumption.

We would seem to have shown that only integral values of angular momentum were allowed under any circumstances when we formulated the rotational states of an $O^{16}O^{16}$ or $O^{16}O^{17}$ molecule. We simply apply the same analysis to any other rotating object. This may be clearer if discussed relative to the angular-momentum axis where with circular symmetry the wavefunction will have the form $e^{im\phi}$. If we say that the object is in the same state if rotated 360°, we conclude that the wavefunction must be the same and m must be an integer.

Let us consider, however, an object such as the ball rolling without sliding which we treated in Problem 3.1. In particular, let it roll around the inside of a pipe of diameter $^3/_2$ that of the ball, as illustrated in Fig. 10.2. We may see that when it completes one path around the pipe, the ball has only rotated 180°. Two circuits are required to return to the same state. We could write the total kinetic energy, including both the spin of the ball and the revolution of the center of gravity around the center of the pipe, in terms of $\dot{\phi}$ and define the canonical momentum conjugate to ϕ (defined in Section 3.1) from the derivative with respect to $\dot{\phi}$. Again the operator for that canonical momentum is $(\hbar/i)\partial/\partial\phi$ but with a change in ϕ of 4π required return to the same state, the phase factor can be $e^{i\phi/2}$ and the canonical momentum is then $\hbar/2$. This same rolling ball can also have any other half-integral multiples (but not full-integral multiples) of $\hbar$ as states of higher energy. The actual angular momentum of each state, the sum of the angular momentum of the spinning ball and of the center of gravity revolving around the center of the pipe (in the opposite direction), is equal to this half-integral canonical momentum conjugate to ϕ, just as it is integral multiples in simple rotating bodies independent of the moment of inertia of the body. We have found, contrary to our first guess, that a simple mechanical system can have half-integral angular momentum.

We note further, that if the ratio of the two radii were an irrational number, there would have been no quantization of the angular momentum at all. The point is that the simple rule of angular momentum quantized in units of $\hbar$ need not apply in all systems, but we retain the concept that the wavefunction goes into itself when the system is rotated into the same physical state.

We might never have thought of the possibility of half-integral angular momentum had it not arisen experimentally. However, we accept whatever particles nature provides, and it provides an electron with mass m, charge $-e$ and angular momentum $\hbar/2$, appropriately called *spin* since it is intrinsic to the particle. We accommodate to the spin as we did to $E = p^2/2m$ when we

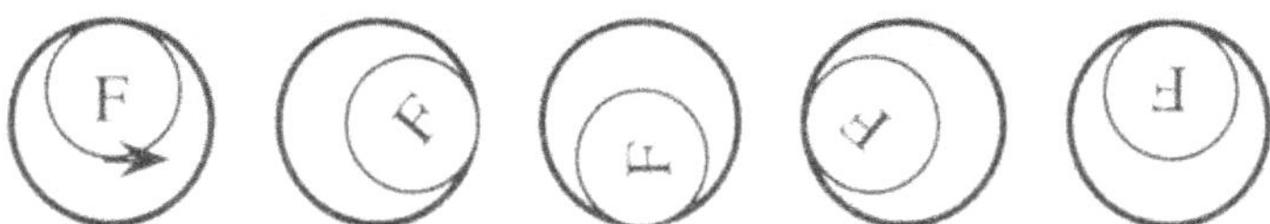

Fig. 10.2. A ball, with "F" written on its side, roles without slipping inside a pipe of diameter $^3/_2$ as large. It has only rotated 180° by the time it has rolled completely around the inside of the pipe, and like a fermion must roll around twice to return to its initial state. Its canonical angular momentum can be half-integral multiples of $\hbar$.

invented the Schroedinger Equation in Section 1.2, even though we continue to think of the electron as a point particle. If nature gives us a particle of spin $\hbar/3$ we will know how to proceed, and one could even be constructed using a scheme such as that in Fig. 10.2.

We can make a general connection between spin of particles and the symmetry of their wavefunctions under interchange, just as we did for the $O^{16}{}_2$ molecule above. We replace the oxygen molecule by a hydrogen molecule, for which the two nuclei are protons, particles of spin $^1/_2$ just as electrons are. We again write an energy eigenstate $\Psi(\mathbf{r}_1,\mathbf{r}_2)$ in terms of the coordinates $\mathbf{r}_1$ and $\mathbf{r}_2$ of the two proton nuclei. Now, however, the two protons have spin and the wavefunction contains another factor with the spin coordinate for the first proton $\psi_1(\phi_1) = e^{i\phi_1/2}$ if the first proton has its spin parallel to our z-axis. If the other proton has parallel spin (called orthohydrogen), the state contains also a factor for the second electron $\psi_2(\phi_2) = e^{i\phi_2/2}$. [We have neglected any interaction between the spin and the other coordinates for this discussion, allowing, as in Section 2.2, the wavefunction to be factored as $\Psi(\mathbf{r}_1,\mathbf{r}_2)\psi_1(\phi_1)\psi_2(\phi_2)$.] Now we imagine rotating this entire wavefunction around the z-axis such that the two protons interchange positions, replacing $\mathbf{r}$ by $-\mathbf{r}$, or equivalently interchanging $\mathbf{r}_1$ and $\mathbf{r}_2$. In addition, the rotation changes $\psi_1(\phi_1)$ and $\psi_2(\phi_2)$ each by a factor $e^{-i\pi/2}$ so that the rotated state becomes $-\Psi(\mathbf{r}_2,\mathbf{r}_1)\psi_1(\phi_1)\psi_2(\phi_2)$. The two protons are identical so that, as for the oxygen, this is the same state as before rotation, $\Psi(\mathbf{r}_1,\mathbf{r}_2)\psi_1(\phi_1)\psi_2(\phi_2)$, and we conclude that $\Psi(\mathbf{r}_2,\mathbf{r}_1) = -\Psi(\mathbf{r}_1,\mathbf{r}_2)$.

This antisymmetry applies in general for half-integral-spin particles of parallel spin, including protons, neutrons, and electrons. They are antisymmetric with respect to interchange as illustrated in Fig. 10.3. If this is two electrons of parallel spin, and we neglect any interaction between

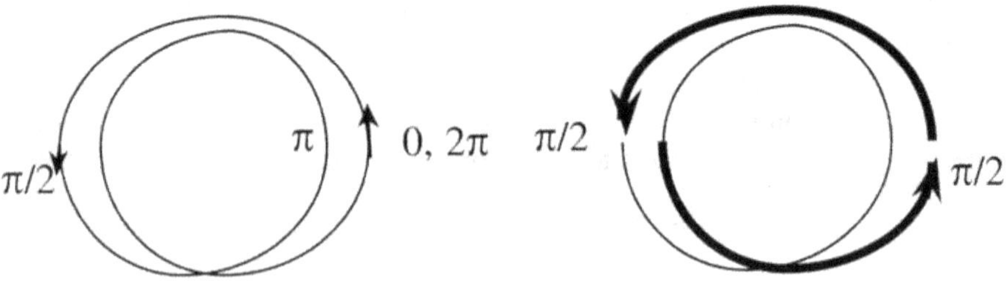

Fig. 10.3. To the left is a schematic representation of a state of half integral spin, $e^{i\phi/2}$, which changes sign with a single full rotation and must be rotated twice to return to itself. The state of two such particles, $e^{i\phi_1/2}e^{i\phi_2/2}$, will change sign if the particles are interchanged through a rotation of π radians, as illustrated to the right.

them so that they can be written as a product wavefunction, that product can be written, in analogy with Eq. (10.20), as

$$\Psi(\mathbf{r}_1, \mathbf{r}_2) = \frac{[\psi_1(\mathbf{r}_1)\psi_2(\mathbf{r}_2) - \psi_1(\mathbf{r}_2)\psi_2(\mathbf{r}_1)]}{\sqrt{2}}. \tag{10.21}$$

For noninteracting electrons of antiparallel spin the wavefunction is again given by Eq. (10.20). We say that if the state is symmetric with respect to spins (parallel), the spatial wavefunction is antisymmetric, but if the state is antisymmetric with respect to spins (antiparallel), the spatial wavefunction is symmetric.

If we again look for rotation states of the H_2 molecule, by writing $\Psi(\mathbf{r}_1,\mathbf{r}_2)= \psi_c(\mathbf{R})\psi_l(r)Y_l^m(\theta,\phi)$, we see that rotational states for orthohydrogen (parallel nuclear spins) are allowed only for l an odd integer, since only then does the wavefunction change sign when $\mathbf{r}$ is replaced by $-\mathbf{r}$. If we redo the analysis with the two proton nuclei of opposite spin (parahydrogen), $e^{i\phi_1/2}$ and $e^{-i\phi_2/2}$, we find that $\Psi(\mathbf{r}_2,\mathbf{r}_1) = \Psi(\mathbf{r}_1,\mathbf{r}_2)$, and rotational states are allowed only for even angular-momentum quantum number l, as it was for $O^{16}{}_2$. The allowed states for the tumbling of hydrogen molecules are just as remarkable as those for oxygen. In the case of hydrogen only even or odd values of rotational quantum numbers are allowed depending upon whether the spin of the two nuclei is antiparallel or parallel. Parahydrogen can have lower (zero) rotational kinetic energy than orthohydrogen, providing an effective interaction favoring antiparallel nuclear spins in the ground state. A recent publication, with references, by Bertino, et al., (1998) described additional effects on the bouncing of these molecules from crystal surfaces.

10.5 Fermions

Such particles of half-integral spin are called *fermions*. The effect of the antisymmetry of states of the same spin is even more profound than the effect of symmetry on bosons. It tells us that two electrons of the same spin cannot occupy the same orbital, because if we take $\psi_1(\mathbf{r}) = \psi_2(\mathbf{r})$ in Eq. (10.21), then $\Psi(\mathbf{r}_1, \mathbf{r}_2) = 0$, meaning as always that there is no such state. This is the *Pauli Principle*, which we have used in all of our discussions of many electrons, but now the origin is clearer. It applies to any pair of identical particles of half-integral spin. It is again a direct consequence of the wave-particle duality we assumed at the outset.

The separation out of the spin state, as parallel or antiparallel, is only possible if there is no term in the Hamiltonian coupling spin and orbital motion so that we can make the factorization, as in Section 2.2. There is such a coupling, which we shall discuss in Section 22.5, and a proper treatment of the states requires the relativistic theory, Dirac (1926) theory, which is seldom used in molecular and solid-state problems. It is usually adequate to ignore spin-orbit coupling as we have done here and specify each electron state as spin-up or spin-down. The Pauli Principle allows one of each to occupy each orbital. For half-integral spins of $s = {}^1/_2, {}^3/_2, {}^5/_2, \ldots$, the total spin angular momentum squared, as for integral spin, is $L^2 = s(s + 1)\hbar^2$, and the z-component is $s_z\hbar$ with $s_z = s, s\text{-}1, s\text{-}2 \ldots \text{-}s$. We will only be concerned here with spins of ${}^1/_2$ and thus $s_z = \pm{}^1/_2$.

The generalization of the antisymmetric state of two electrons in Eq. (10.21) to many particles is called the *Slater Determinant*, and is written

$$\Psi(\{\mathbf{r}_i\}) = \frac{1}{\sqrt{N!}}\mathrm{Det}\begin{pmatrix} \psi_1(\mathbf{r}_1)\,\psi_1(\mathbf{r}_2)\,\psi_1(\mathbf{r}_3) & \ldots \\ \psi_2(\mathbf{r}_1)\,\psi_2(\mathbf{r}_2)\,\psi_2(\mathbf{r}_3) & \ldots \\ \psi_3(\mathbf{r}_1) \quad \ldots \quad \ldots & \ldots \\ \ldots & \end{pmatrix}. \tag{10.22}$$

Recall that determinants have the property that they change sign when two columns are interchanged, here corresponding to interchanging two electrons. Each $\psi_j(\mathbf{r}_i)$ is imagined to contain a spin-state factor, providing the overall antisymmetry.

The antisymmetry of fermions profoundly affects the statistics, allowing only two electrons in each orbital state, and we must redo the analysis we did for bosons. We again divide the states into sets of N_T states (counting different spins as different states) in a small range of energies $\delta\varepsilon$ near ε, and say N_ε of these are occupied. The number of ways to do this is

$$W(\varepsilon) = \frac{N_T!}{(N_T - N_\varepsilon)!N_\varepsilon!}, \tag{10.23}$$

We do same for every energy range, take the logarithm of the product of all $W(\varepsilon)$, use Stirling's formula, to obtain $\Sigma_\varepsilon \ln(N_T!) - \Sigma_\varepsilon (N_T - N_\varepsilon)\ln(N_T - N_\varepsilon) - N_\varepsilon \ln(N_\varepsilon) + (N_T - N_\varepsilon) + N_\varepsilon$. The condition fixing the total energy is $\Sigma_\varepsilon N_\varepsilon \varepsilon = E_{tot.}$, applied with a Lagrange multiplier β. The condition fixing the total number of particles is $\Sigma_\varepsilon N_\varepsilon = N_{tot.}$, applied with a Lagrange multiplier, again $-\beta\mu$. Setting the derivative with respect to N_ε equal to zero yields

$$\ln(N_T - N_\varepsilon) - \ln(N_\varepsilon) + 1 - 1 - \beta\varepsilon + \beta\mu = 0. \tag{10.24}$$

We take the logarithms to the right and exponentiate to obtain $N_\varepsilon/(N_T - N_\varepsilon) = e^{-\beta(\varepsilon - \mu)}$ and finally, the fraction of states occupied is

$$f_0(\varepsilon) = \frac{N_\varepsilon}{N_T} = \frac{1}{e^{\beta(\varepsilon - \mu)} + 1}, \tag{10.25}$$

which is the *Fermi (or Fermi-Dirac) distribution function* with again $\beta = 1/k_BT$ and μ the chemical potential, or *Fermi energy*. It is plotted in Fig. 10.4. Well below the Fermi energy it goes to one; each state is occupied for both spins. It goes to zero well above the Fermi energy. For any system the Fermi energy is shifted to correspond to the correct number of electrons. When we treated free-electron metals in Section 2.2 we had one, or a few, free electrons per atom, which gave a Fermi energy μ measured relative to the band minimum ε_0 as $E_F = \mu - \varepsilon_0 \approx 5$ eV $>> k_BT$. The Fermi function could be taken as a sharp cut-off, $f_0 = 1$ for $\varepsilon < \mu$, $f_0 = 0$ for $\varepsilon > \mu$. At finite temperature electrons are excited to energies of the order of k_BT from this cut-off and the step is smoothed out as in Fig. 10.4. However, the total excitation energy for the electrons is clearly very much less than the classical $^3/_2k_BT$ so the predicted electronic specific heat (derived in most solid-state texts) is greatly suppressed in comparison to classical theory, as is the experimental value, another of the early achievements of quantum theory.

For semiconductors there is a gap E_g in energy between the states occupied and the states empty in the ground state, as illustrated in Fig. 10.5, and as we shall discuss in more detail in Chapter 14. The Fermi energy ordinarily lies in that gap, allowing a small occupation of the upper band from the tail of the distribution to the right in Fig. 10.4, and a small number of empty states (holes) in the valence band below due to the deviation of the distribution from one to the left in Fig. 10.4. Again the Fermi energy adjusts

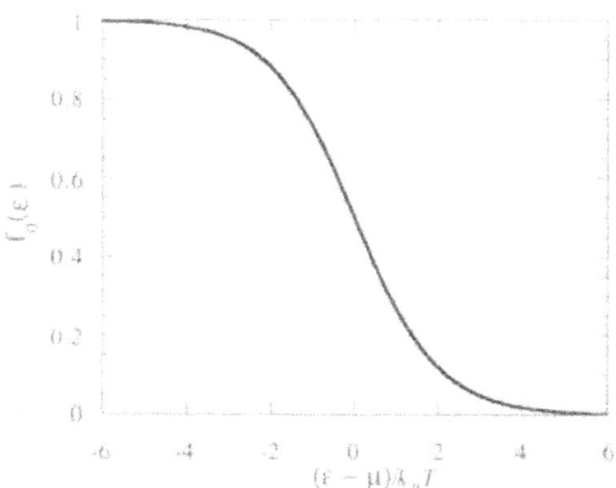

Fig. 10.4. The Fermi distribution function from Eq. (10.25), as a function of the energy measured from the Fermi energy, in units of k_BT.

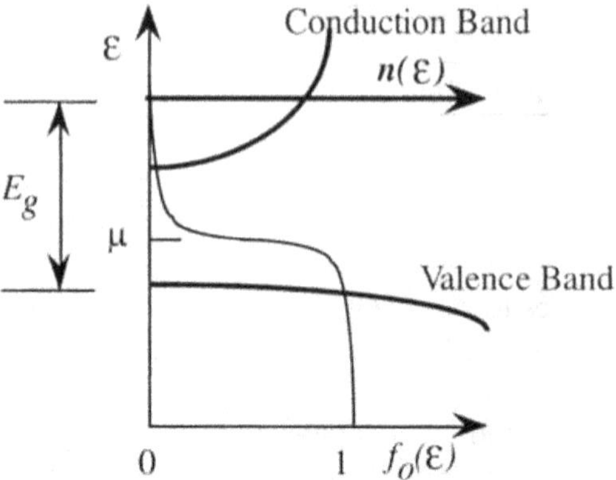

Fig. 10.5. Semiconductors have a density of states $n(\varepsilon)$ for the conduction band, shown as the heavy curve above, and a density of states for the valence band below it and separated by a gap in energy E_g. The Fermi distribution is shown as the light line, with a Fermi energy in the gap.

to correspond to the appropriate number of electrons present, equal numbers of electrons and holes for the undoped system. If extra electrons are added by doping (e. g., substituting a few phosphorous atoms, each with an extra electron, for silicon atoms in silicon) the Fermi energy moves to higher energy to account for those electrons.

In most circumstances in semiconductors the Fermi energy lies in the energy gap, well removed from either band edge. Then $\beta(\varepsilon - \mu) = (\varepsilon - \mu)/k_BT$ is large for all ε in the conduction band and the one in the denominator in Eq. (10.25) can be neglected. Then we may measure energies from the conduction-band minimum ε_c and Eq. (10.25) becomes

$$f_0(\varepsilon) \approx e^{-(\varepsilon_c - \mu)/k_BT} e^{-(\varepsilon - \varepsilon_c)/k_BT} . \tag{10.26}$$

The final factor is a simple classical *Boltzmann distribution*, and the leading factor is a constant which sets the number of electrons. f_0 is small because of this leading factor and so there is almost no effect from the Pauli Principle which required no more than single occupancy of a state. The electron gas is then considered to be *classical*. With very heavy doping, when f_0 becomes a sizable fraction, the electron gas is called *degenerate* and we need to return to Eq. (10.25) to describe its distribution.

In a similar way $\beta(\varepsilon - \mu) = (\varepsilon - \mu)/k_BT$ becomes large and negative for all states in the valence band. We may expand the denominator taking $e^{\beta(\varepsilon - \mu)}$ as small to obtain $f_0 \approx 1 - e^{\beta(\varepsilon - \mu)}$ with $\varepsilon - \mu$ large compared to k_BT and negative. This may be stated as a very small concentration of missing electrons, or of *holes*, in the valence band, again with a classical distribution as a function of depth into the valence band.

In either case, the carriers lie very near the band edge, where the bands may be treated as carriers having an effective mass m^*, as we shall see in Chapter 14. Thus the density of states for the conduction bands may be written in terms of the energy measured from ε_c using Eq. (2.11) as

$$n(\varepsilon) = \frac{\delta N}{\Omega \delta \varepsilon} = \frac{1}{2\pi^2}\left(\frac{2m^*}{\hbar^2}\right)^{3/2} \sqrt{\varepsilon - \varepsilon_c}\,. \tag{10.27}$$

Then we can calculate the total number of electrons per unit volume in the conduction band as

$$\begin{aligned} N_c &= e^{-(\varepsilon_c - \mu)/k_B T} \int_{\varepsilon_c,\infty} n(\varepsilon) e^{-(\varepsilon - \varepsilon_c)/k_B T}\, d\varepsilon \\ &= e^{-(\varepsilon_c - \mu)/k_B T}\, \frac{(k_B T)^{3/2}}{\pi^2} \left(\frac{2m^*}{\hbar^2}\right)^{3/2} \int_{0,\infty} x^2 e^{-x^2}\, dx \\ &= \frac{1}{\sqrt{2}}\left(\frac{m^* k_B T}{\pi \hbar^2}\right)^{3/2} e^{-(\varepsilon_c - \mu)/k_B T}, \end{aligned} \tag{10.28}$$

where we have changed variables in the integration to $x = \sqrt{\varepsilon / k_B T}$. The factor preceding the exponential in the last form can be thought of as an effective conduction-band density of states, which when multiplied by the Boltzmann factor for the band edge, gives the density of electrons. The corresponding expression for holes in the valence band contains the hole mass and the exponential $e^{-(\mu - \varepsilon_v)/k_B T}$ in place of that in Eq. (10.28). We see in Problem 10.2 how this equation is used to determine the Fermi energy and carrier densities in GaAs, as well as the average kinetic energy of the carriers.

Chapter 11. Transport Theory

Transport theory, like statistical mechanics, is not really a part of quantum mechanics - in fact, the Boltzmann Equation we shall derive in Section 11.2 will be seen to rest on a classical description. However, transport also is modified by quantum effects and transport equations are needed to make use of many of the quantum processes we have discussed. While statistical mechanics described systems in thermal equilibrium, transport theory specifically deals with systems out of equilibrium. We begin with the simplest generalization, obtaining an equation for the time-dependence of a distribution function. For that we use the sequential tunneling problem we discussed in Section 9.2. We then move on to full transport theory with distribution functions which depend upon momentum, position, and time. By far the most important transport properties concern electrons and we shall formulate each step in terms of electrons.

11.1 Time-Dependent Distributions

When we discussed sequential tunneling in Section 9.2 we found a rate $1/\tau$ at which an electron occupying a resonant state $|2\rangle$ (local state in the absence of coupling with the continua) would make a transition to a continuum of states $|1\rangle$. If the probability of that resonant state being occupied is $f_2(\varepsilon_2)$, and the probability of the states $|1\rangle$ being occupied is $f_1(\varepsilon_1)$, then the rate electrons make transitions from $|2\rangle$ to $|1\rangle$ is $f_2(\varepsilon_2)(1 - f_1(\varepsilon_1))/\tau$ for $\varepsilon_1 = \varepsilon_2$. We then used detailed balance, the fact that in equilibrium, where the distribution functions are a single function of energy $f_0(\varepsilon)$, all processes go in both directions at the same rate. This allowed us to show that capture of carriers into state |2> from the continuum |1> is given by $f_1(\varepsilon_1)(1 - f_2(\varepsilon_2))/\tau$ with again $\varepsilon_1 = \varepsilon_2$, whether or not the system is in equilibrium. We then applied the corresponding relation for a second continuum |3> coupled to |2> and found that if $f_1(\varepsilon_2)$ was held at one and $f_3(\varepsilon_2)$ was held at zero, $f_2(\varepsilon_2)$ would come to the steady-state value at which electrons were arriving and leaving at the same rate. We generalize this steady-state calculation to time-dependent situations.

We are considering transitions which conserve energy so we can drop the ε_2 for each of the three distributions. Then the *net* rate electrons are added to |2> from the continuum |1> is $(f_1(1 - f_2) - f_2(1 - f_1))/\tau = (f_1 - f_2)/\tau$, with any effect of occupation of the final states canceling out, as we shall note also in the next section. We may allow the rate $1/\tau'$ to be different for transfer between the state |2> and the continuum |3> so the net transfer from |3> to |2> is similarly $(f_3 - f_2)/\tau'$. Thus the net rate the occupation of |2> is changing with time is

$$\frac{\partial f_2}{\partial t} = \frac{f_1 - f_2}{\tau} + \frac{f_3 - f_2}{\tau'} , \tag{11.1}$$

which is a simple equation for the time-dependence of f_2. This equation allows us to predict the change in f_2 with time, in terms of the distributions $f_1, f_2,$ and f_3 , just as the Schroedinger Equation allowed us to find how the wavefunction changed with time.

The equilibrium state $f_1 = f_2 = f_3 = f_0(\varepsilon_2)$ is a trivial solution of Eq. (11.1). A more interesting case is the steady-state solution, which we discussed in Section 9.2. If we apply the condition that $\partial f_2/\partial t = 0$, we obtain

$$f_2 = \frac{\tau' f_1 + \tau f_3}{\tau + \tau'} . \tag{11.2}$$

In the simple case we assumed for sequential tunneling ($f_1 = 1, f_3 = 0, \tau' = \tau$) this gave $f_2 = {}^1/_2$ and a transfer rate of $1/(2\tau)$. We move directly to the development of a time-dependence equation for an electron gas with an f which depends upon momentum, position and time.

11.2 The Boltzmann Equation

For electrons each state, including spin, is occupied by one electron, or no electrons and the distribution function we introduced is $f_0(\varepsilon)$, the probability of occupation of a state of energy ε in equilibrium. For a system in equilibrium it does not depend upon where in the system we are or whether the electron is moving to the right or left. Often a system is near equilibrium, but the chemical potential (Fermi energy) or the temperature are different in different parts of the system. Then also there will be flow of carriers or energy from place to place and the distribution depends upon position and upon direction of motion and may depend upon time. In such a case we write a distribution function $f(\mathbf{p},\mathbf{r},t)$, depending upon both momentum and position. This becomes intrinsically classical because in a quantum system we cannot specify both momentum and position, though we did both approximately when we constructed wave packets in Section 1.2. It nevertheless retains some quantum features since we restrict f to be always between zero and one, according to the Pauli Principle. Such a discussion is sometimes called *semiclassical*. The physics of the analysis is quite clear, whatever we choose to call it.

In order to proceed we need an equation from which we can determine the distribution function, the analog of the Schroedinger Equation for determining the wavefunction of a particle. We will then be able to predict observables, such as the current, from the distribution function as we calculated observables, such as the energy, from the wavefunction. The equation we use gives us the time dependence of the distribution function, just as the Schroedinger Equation, Eq. (1.16), gave us the time dependence of the wavefunction.

We imagine electrons moving in the presence of various fields so that their classical trajectories $\mathbf{p}(t)$, $\mathbf{r}(t)$ can be plotted as illustrated in Fig. 11.1. Some of these trajectories could be occupied by electrons and others not.

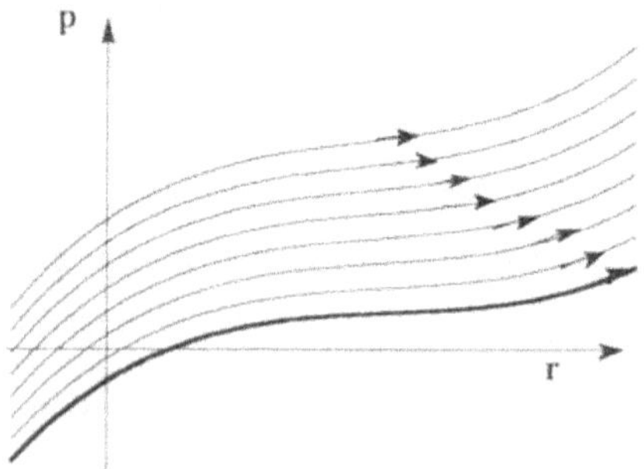

Fig. 11.1. A schematic representation of trajectories of electrons moving in the presence of fields. Different curves represent different starting points at the time $t = 0$, and t increases along the lines in the direction of the arrows.

However, in the absence of any scattering, if we follow a trajectory, that occupation $f(\mathbf{p}(t),\mathbf{r}(t),t)$ will not change with time. Alternatively, we can say that the only change in the distribution function with time, as we follow along the trajectory, will be that due to scattering, $\partial f(\mathbf{p}(t),\mathbf{r}(t),t)/\partial t|_{\text{scat.}}$. Mathematically, we write this

$$\frac{\mathrm{d}f(\mathbf{p}(t)\mathbf{r}(t)t)}{\mathrm{d}t} = \frac{\partial f(\mathbf{p}(t)\mathbf{r}(t)t)}{\partial t}\bigg|_{\text{scat.}} \tag{11.3}$$

(The different manner of writing the arguments of f in different expressions is not significant.) The total derivative, $\mathrm{d}/\mathrm{d}t$ means the total derivative including dependences upon all three of the variables, $\mathbf{p}$, $\mathbf{r}$, and t, and the partial derivative $\partial/\partial t$ includes only the dependence upon the one variable t, not $\mathbf{p}$ nor $\mathbf{r}$. We may write out the three terms on the left explicitly,

$$\frac{\partial f(\mathbf{p}\,\mathbf{r}\,t)}{\partial \mathbf{p}}\frac{\mathrm{d}\mathbf{p}}{\mathrm{d}t} + \frac{\partial f(\mathbf{p}\,\mathbf{r}\,t)}{\partial \mathbf{r}}\frac{\mathrm{d}\mathbf{r}}{\mathrm{d}t} + \frac{\partial f(\mathbf{p}\,\mathbf{r}\,t)}{\partial t} = \frac{\partial f(\mathbf{p}\,\mathbf{r}\,t)}{\partial t}\bigg|_{\text{scat.}} \tag{11.4}$$

The derivative of f with respect to $\mathbf{p}$ in the first term means that if we went to $\mathbf{r}$ at time t the derivative is the variation of f with $\mathbf{p}$ at that moment at that position. The $\mathrm{d}\mathbf{p}/\mathrm{d}t$ is the rate the momentum changes with time at that point, equal to the applied force. This is the *Boltzmann Equation* , and is exact in the semiclassical context where $f(\mathbf{p}\,\mathbf{r}\,t)$ is meaningful. However, of more use is an approximate form when the distribution is quite close to the equilibrium distribution $f_0(\varepsilon)$. Then we write the distribution as $f(\mathbf{p}\,\mathbf{r}\,t) = f_0(\varepsilon) + f_1(\mathbf{p}\,\mathbf{r}\,t)$. [Sometimes a local equilibrium distribution is used, as when the temperature varies with position, but we use the simpler form here.] f_1 is of first-order in any applied fields which exert forces $\mathbf{F}(\mathbf{r},t)$ on the electrons, and cause them to be out of equilibrium. We substitute this form for f in Eq. (11.4) and will keep only such first-order terms. There are no zero-order terms as nothing changes with time without applied forces.

The $\mathrm{d}\mathbf{p}/\mathrm{d}t$ is equal to $\mathbf{F}(\mathbf{r},t)$ so we may take only the zero-order term in $\partial f/\partial\mathbf{p}$ in that term, which is $(\mathrm{d}f_0(\varepsilon)/\mathrm{d}\varepsilon)\cdot\mathrm{d}\varepsilon/\mathrm{d}\mathbf{p} = (\mathrm{d}f_0(\varepsilon)/\mathrm{d}\varepsilon)\,\mathbf{v}$, using Eq. (1.7). In the second term, only f_1 depends upon $\mathbf{r}$ and $\mathrm{d}\mathbf{r}/\mathrm{d}t$ is $\mathbf{v}$. We also simplify the scattering term on the right. Certainly the rate electrons leave the trajectory is proportional to $f(\mathbf{p}\,\mathbf{r}\,t)$, and we write a proportionality constant $1/\tau$ though the rate really will depend upon the particle velocity also, as we have seen in other parts of this text. This is called the *relaxation-time approximation*. We must also include the rate electrons are scattered *onto* the trajectory, but if we write the rate they are scattered off as $f(\mathbf{p}\,\mathbf{r}\,t)/\tau$ we note that in equilibrium, with $f(\mathbf{p}\,\mathbf{r}\,t) = f_0(\varepsilon)$, the scattering rate onto the trajectory equals the rate they are scattered off. We take that same rate to apply in this case so that the net rate of change of $f(\mathbf{p}\ \mathbf{r}\ t)$ with time due to scattering becomes $-f(\mathbf{p}\ \mathbf{r}\ t)/\tau + f_0(\varepsilon)/\tau = -f_1(\mathbf{p}\ \mathbf{r}\ t)/\tau$. Making these approximations in Eq. (11.4) leads us to the *linearized Boltzmann Equation in the relaxation-time approximation,*

$$\frac{df_0(\varepsilon)}{d\varepsilon}\mathbf{v}\cdot\mathbf{F}(\mathbf{r},t) + \mathbf{v}\cdot\nabla f_1(\mathbf{p}\,\mathbf{r}\,t) + \frac{\partial f_1(\mathbf{p}\,\mathbf{r}\,t)}{\partial t} = -\frac{f_1(\mathbf{p}\,\mathbf{r}\,t)}{\tau}\ . \tag{11.5}$$

It is in this equation we use the scattering times from impurities and phonons which we have calculated in other parts of the text. We illustrate its application in detail only for the conductivity in the next section.

11.3 Conductivity, etc.

The simplest application of the Boltzmann Equation is to the dc-conductivity, and that will serve to illustrate the approach. For dc-conductivity the distribution function will not change with position, nor with time, so the second two terms on the left in Eq. (11.5) vanish. We have only the first term on the left, with $\mathbf{F} = -e\mathbf{E}$ and we may solve for the first-order distribution,

$$f_1(\mathbf{p}) = e\tau\mathbf{v}\cdot\mathbf{E}\,\frac{df_0(\varepsilon)}{d\varepsilon}\ . \tag{11.6}$$

We have solved for the distribution function and can evaluate the current density in terms of it. We do this by summing the current from each state $-e\mathbf{v}$ over the occupied states. The equilibrium distribution leads to no current, so only f_1 enters. We take a volume Ω for the system and the current density is the $\mathbf{j} = \Sigma_{\mathbf{p}}\, f_1(\mathbf{p})\ (-e\mathbf{v})/\Omega$. We now return to a quantum description of the states with $\mathbf{p} = \hbar\mathbf{k}$, and convert the sum over $\mathbf{p}$ to an integral over $\mathbf{k}$ using Eq. (2.9). We multiply by two for spin and have

$$\mathbf{j} = -\frac{2\Omega}{(2\pi)^3}\int d^3k\ e^2\tau\mathbf{v}\cdot\mathbf{E}\,\frac{df_0(\varepsilon)}{d\varepsilon}\,\frac{\mathbf{v}}{\Omega} = -\frac{2e^2\tau}{3(2\pi)^3}\mathbf{E}\int 4\pi k^2 v^2\,\frac{df_0(\varepsilon)}{d\varepsilon}\,dk\ . \tag{11.7}$$

In the last step we recognized that the current was parallel to $\mathbf{E}$, wrote the angle between $\mathbf{E}$ and $\mathbf{k}$ or $\mathbf{v}$ as θ and took the angular average of the integrand, to obtain the factor $1/3$. The evaluation is simplest for a metal, where f_0 drops from one to zero just at the Fermi energy so $-df_0(\varepsilon)/d\varepsilon$ is approximately a delta-function $\delta(\varepsilon - E_F)$. We take one factor of v =

$(1/\hbar)d\varepsilon/dk$ to write $vdk = (1/\hbar)d\varepsilon$ and the integral becomes $-4\pi k_F{}^2 v_F/\hbar$. Now $v_F = \hbar k_F/m$ and $(2/(2\pi)^3)4\pi k_F{}^3/3)$ is equal to the electron density n, so we have $\mathbf{j} = \sigma\mathbf{E}$ with the conductivity given by

$$\sigma = \frac{ne^2\tau}{m}. \tag{11.8}$$

Actually, one can also perform the integration by converting the final integral over k in Eq. (11.7) to an integral over energy, perform a partial integration on energy and obtain an integral which gives exactly the electron density whether or not $-df_0/d\varepsilon$ is treated as a delta function, so the same result, Eq. (11.8), applies also for a Boltzmann distribution of electrons. One virtue of doing it as we did is in showing that though we sum over all wavenumbers in a metal, only the values at the Fermi surface enter.

The treatment of other properties is quite straight-forward. (See, for example, Harrison (1970).) For the Hall effect, the magnetic force $-(e/c)\mathbf{v}\times\mathbf{H}$ is added. For thermal conductivity f_0 in the first term contains a temperature varying with position, and this yields also the other thermoelectric properties. One can also calculate the diffusion constant D, describing an electron flux $\mathbf{j}/(-e) = -D\nabla n$, with n again the electron density.

There are also simpler, less accurate, approximations for treating the transport properties. One assumes an electron density, $n(\mathbf{r},t)$, varying with position and time, and then approximates the flow locally as $\mathbf{j}(\mathbf{r},t) = \sigma\mathbf{E}(\mathbf{r},t) + eD\nabla n(\mathbf{r},t)$. Combining this with the continuity equation, $-e\,\partial n(\mathbf{r},t)/\partial t = \nabla\cdot\mathbf{j}$, and Poisson's Equation, one can frequently obtain an adequate description of transport properties. Such an approach misses *nonlocal* effects, such as the decrease in the current in the neighborhood of a surface. Such nonlocal effects *can* be calculated using the Boltzmann Equation.

Chapter 12. Noise

A familiar case of "noise" is the static on a radio. Noise exists in classical systems, but can be strongly affected by quantum effects and it is appropriate to include some discussion here. There are a wide variety of origins of noise in general, as there are for static on a radio. We enumerate the principal ones. A recent reference on noise is Kogan (1996).

12.1 Classical Noise

Thermal noise, also called Johnson-Nyquist noise, is understandable in terms of the statistical mechanics which we discussed in Sections 10.1 and 10.2. At any finite temperature the modes of light are excited in a Planck distribution, and the electromagnetic modes of any electrical circuit or transmission line are similarly excited. If there is a resistor in the circuit, it will absorb energy from these thermal fluctuations, and in equilibrium it will radiate at the same rate into the same modes. In order to understand the distribution of power radiated by a resistor, we construct electrical modes in a line, a wire with resistance R', with both its ends connected to a resistor R. There can be current fluctuations of various wavelengths, and therefore various frequencies. For a length L we apply periodic boundary conditions so that modes will have wavenumbers k such that $kL = 2\pi n$ and the frequency $\omega = ck$, with c the speed of light, which may depend upon the geometry of the line but is calculated by applying Maxwell's Equations to the line. The average energy in each mode is $[1/(e^{\hbar\omega/k_BT} - 1) + 1/2]\hbar\omega$ as we saw in Section 10.3 and it strikes the end of the line at a rate c/L times that. If there is no reflection at the resistor it will absorb at that rate, and therefore emit at that rate. Thus in a frequency range $\Delta f = \Delta\omega/2\pi = c\Delta k/2\pi$ there will be power absorbed and emitted at each end of the resistor of

$$\Delta P = \frac{L\Delta k}{2\pi}\left(\frac{1}{e^{\hbar\omega/k_BT} - 1} + \frac{1}{2}\right)\hbar\omega\ \frac{c}{L} = \left(\frac{1}{e^{\hbar\omega/k_BT} - 1} + \frac{1}{2}\right)\hbar\omega\ \Delta f. \quad (12.1)$$

In the classical, or high-temperature, limit, $\Delta P \rightarrow k_B T\,\Delta f$, and this is called *thermal noise.* It is *white noise*, meaning that the power emitted by the resistor in a frequency range Δf is independent of the frequency f, which in the case of light defines white light. It is sometimes restated by relating the power emitted in a frequency range to the square of the voltage $\langle V^2 \rangle$ and the resistance R. The resulting relation between voltage fluctuations and resistance is called the *Nyquist Theorem.* Einstein had noted such a relation between the fluctuations of a dust particle, Braunian motion, in a gas and the viscosity of the gas, and there are similar rigorous relations between every other kind of fluctuation, and the corresponding dissipation of energy. In the quantum limit, or as the temperature goes to zero, the power in Eq. (12.1) approaches $(\hbar\omega/2)\Delta f$. This is then called *quantum noise,* or zero-point noise, rather than thermal noise. The *Fluctuation-Dissipation Theorem* is the quantum generalization of the Nyquist theorem. From Eq. (12.1) we see that thermal noise changes continuously to quantum noise as we lower the temperature.

12.2 Quantum Noise and van-der-Waals Interaction

One might ask if these zero-point fluctuations of a harmonic-oscillator dipole in the ground state really generate observable fields, and to make a meaningful check we need to imagine an experiment to detect them. One way of doing that would be to bring another harmonic-oscillator dipole near, as illustrated in Fig. 12.1. If the first is generating fluctuating electric fields, the second will be polarized by those fields and the interaction between the induced dipole and the dipole which caused it would produce an attraction between the two. Measuring that force would be detecting the fluctuating fields. This force *is* observable, and is called a *van-der-Waals interaction,*

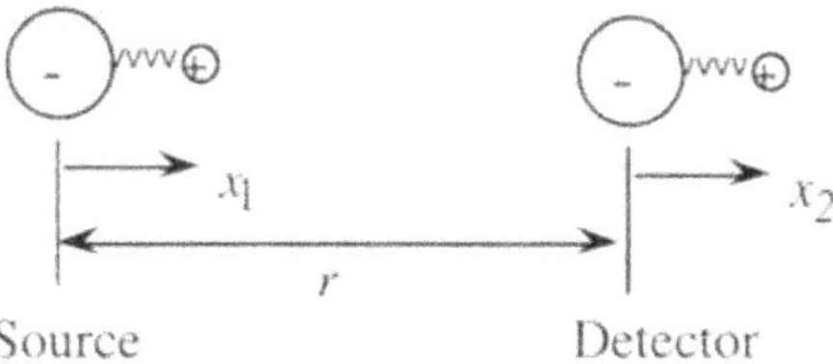

Fig. 12.1. Two coupled dipole oscillators, in the ground state, have correlated zero-point fluctuations, giving a van-der-Waals attraction between them.

named after an attractive interaction added to ideal-gas theory by van der Waals many years ago. It is quite easy to derive the force by calculating the ground state of the two interacting dipole oscillators, as done earlier by Kittel (1976), p. 78.

The calculation is simplest with two identical collinear oscillators as in Fig. 12.1, but the result is easily generalized. We let the oscillators (spring constant κ, mass M) have displacement coordinates x_1 and x_2, and therefore dipoles $p_1 = ex_1$ and $p_2 = ex_2$. The presence of a dipole p_1 produces a field $2p_1/r^3$ at the second, a distance r away, so that if the second has a dipole p_2 there is a lowering in energy of $-2p_1p_2/r^3 = -2e^2x_1x_2/r^3 = -\kappa' x_1x_2$ with $\kappa' = 2e^2/r^3$. Thus the potential-energy term in the Hamiltonian is $^1/_2\kappa x_1{}^2 + {}^1/_2\kappa x_2{}^2 - \kappa' x_1x_2$. The kinetic-energy term is $^1/_2M\dot{x}_1{}^2 + {}^1/_2M\dot{x}_2{}^2$. We may rewrite these terms in the energy in terms of normal coordinates $u_1 = (x_1 + x_2)/\sqrt{2}$ and $u_2 = (x_1 - x_2)/\sqrt{2}$ to obtain the energy $^1/_2M\dot{u}_1{}^2 + {}^1/_2M\dot{u}_2{}^2 + {}^1/_2(\kappa + \kappa')u_1{}^2 + {}^1/_2(\kappa - \kappa')u_2{}^2$ for the two coupled oscillators. This represents two oscillators with frequencies given by $\omega_1{}^2 = (\kappa + \kappa')/M$ and $\omega_2{}^2 = (\kappa - \kappa')/M$. The quantum-mechanical ground state of the system will have each in the ground state with a total energy $^1/_2\hbar(\omega_1 + \omega_2)$. With κ' equal to zero the frequencies are the same $\omega = \sqrt{\kappa/M}$ and the energy is that for the uncoupled oscillators. If we expand in κ' the two linear terms cancel out but in second order both energies are lowered by $-(1/16)(\kappa'/\kappa)^2\,\hbar\omega$. Thus, due to the interaction the energy is reduced by $-(1/8)(\kappa'/\kappa)^2\hbar\omega$.

This result may be written in terms of the polarizability of the two oscillators, defined in terms of the equilibrium dipole due to an applied field E by $p = \alpha E$. It is easily obtained as $\alpha = e^2/\kappa$ by equating the spring force to the negative of the electric-field force. Substituting also for κ' we have the interaction energy

$$E_{\rm vdW} = -\frac{\alpha^2\,\hbar\omega}{2r^6}. \qquad (12.2)$$

It is quite remarkable that we can so simply calculate the effect of the correlated motion of two coupled quantum systems in this way. In Problem 12.1 we generalize this to two three-dimensional dipole oscillators by adding the contribution of the other two vibrational directions to the interaction.

In Problem 12.2 the van-der-Waals interaction is obtained for two polarizable atoms or molecules. For two identical atoms with only a single electronic state occupied on each atom the result could immediately be guessed from Eq. (12.2) by replacing $\hbar\omega$, which is the energy of excitation to the nearest coupled harmonic oscillator state, by the energy of excitation to the first coupled atomic state. Often a harmonic oscillator provides a valid model for an atom or molecule, with $\hbar\omega$ taken equal to the electronic excitation energy, allowing estimates of properties in terms of clear simple models. For this particular case, the analysis in Problem 12.2 is also made for two different atomic types and tells the form which Eq. (12.2) would take for two dipole oscillators with different parameters,

$$E_{vdW} = -\frac{\alpha_1\alpha_2\,\hbar\omega_1\hbar\omega_2}{(\hbar\omega_1 + \hbar\omega_2)r^6}. \tag{12.3}$$

In either case, we have treated the ground state of a multibody interacting system. This is the simplest case of an extremely important set of problems, called many-body problems. Up to this point we have avoided such problems by introducing the one-electron approximation in Sections 2.2 and 4.2. The electron-electron interaction $-2e^2x_1x_2/r^3$ which we introduced here is not to be confused with the coupling between states of a single electron such as $\langle\psi_1(\mathbf{r}_1)|H(\mathbf{r}_1)|\psi_0(\mathbf{r}_1)\rangle$. We shall introduce the appropriate formulation for such electron-electron interactions in Section 16.1 and discuss a number of other many-body effects in Part VII.

Aside from being a very important type of quantum effect which we have not discussed before, it is an important physical phenomenon. The van-der-Waals interaction is an interaction between objects which do not overlap each other at all, but interact with each other through electromagnetic radiation. It is the principal attractive interaction between inert-gas atoms, which cannot form covalent bonds, and between most molecules. The extension to nuclear physics describes the interaction between nucleons through the effects of meson fields, as we discuss in Section 17.4

It is also central to the discussion of noise, which is the context in which we brought it up. The field which arises from the quantum fluctuations of the dipole, $\langle\mathbf{E}^2\rangle = 4e^2\langle x^2\rangle/r^6$, is a noise field which we can think of as power flowing from the source and there must be an equal flow inward. Similarly an atom in the ground state is emitting and absorbing noise at the frequency of its excitations. In a metal, with a partly filled band, incident light can be absorbed by transferring an electron from an occupied to an empty state. At zero temperature the corresponding absorption of zero-point light fluctuations must be balanced by the emission of noise power from the band electrons in the ground state. Thus individual electrons in metals emit quantum noise even in the ground state and even without the Coulomb interaction between different band electrons.

12.3 Shot Noise

A classical (or a quantum) charged gas will show fluctuations in current across a plane due to the individual arrival times of the electrons, like raindrops on a roof. This depends on the real discrete size of the charges, and would go to zero if the charges were subdivided into smaller and smaller particles of the same constant e/m, which thermal noise, at k_BT per frequency interval, does not. This is a fundamentally different origin for additional noise.

In a classical charged gas, such as a dilute electron gas in a semiconductor, we may ordinarily assume that the electrons are crossing a given plane at random times and calculate the current distribution illustrated in Fig. 12.2. If there are N carriers per unit volume in the system, at a root-mean-square velocity in the x-direction of $v_x = \sqrt{k_BT/m}$ (as in Problem 10.3, the average kinetic energy for one direction is $<\frac{1}{2}mv_x^2> = \frac{1}{2}k_BT$). The half moving in the positive-x direction strike an area A of a yz-plane at an average rate approximately (since the average speed $<v_x>$ is appropriate rather than the root-mean-square speed) $AN<v_x>/2= (AN/2)\sqrt{k_BT/m}$. The number crossing per unit area in the negative x-direction is equal to that in the positive x-direction at $(N/2)\sqrt{k_BT/m}$ as illustrated in Fig. 12.2.

We may select a long time period t_0 and Fourier transform the current with respect to time, using frequency components such that $\omega = 2\pi n/t_0$ with n any integer.

$$j(t) = \Sigma_j -e\delta(t-t_j) = \Sigma_{\omega'} j_{\omega'} e^{i\omega' t} = (t_0/2\pi)\int d\omega' j_{\omega'} e^{i\omega' t}. \tag{12.4}$$

Then the Fourier components j_ω are obtained by multiplying by $e^{-i\omega t}$ and integrating over time, $j_\omega = (1/t_0) \int_{0,t_0} e^{-i\omega t} j(t)\, dt = (-e/t_0) \Sigma_j \pm_j e^{-i\omega t_j}$, with t_j the arrival times and $\pm_j$ being plus for arrivals from the right and minus for arrivals from the left. The j_ω approach zero with random sign as the time period t_0 is made long and are not so interesting, but $j_\omega{}^*j_\omega = (e/t_0)^2 \Sigma_{i,j} e^{-i\omega(t_i - t_j)}$ may be evaluated noting that for random arrival times only the terms $i = j$ contribute and give $(e/t_0)^2N\sqrt{k_BT/m}\, t_0$ for the $(N/2)\sqrt{k_BT/m}\, t_0$ hits from each direction, or

$$j_\omega{}^*j_\omega = \frac{e^2}{t_0} N\sqrt{\frac{k_BT}{m}}\ . \tag{12.5}$$

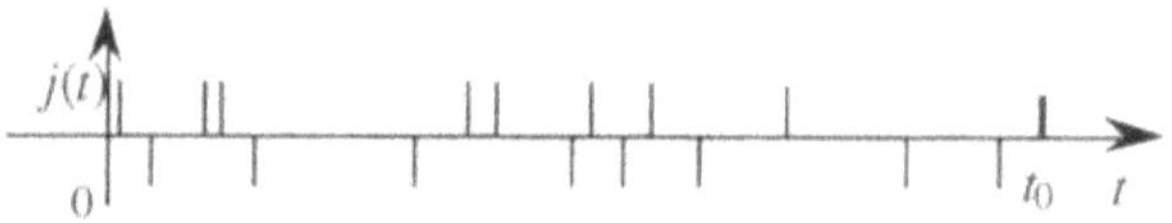

Fig. 12.2. Electrons in a classical gas cross a plane at random times giving a fluctuating current $j(t)$ as shown.

In the frequency range $\Delta f = \Delta\omega/2\pi$ there are $\delta\omega\, t_0/2\pi$ values so that the sum of values in that range, $\Sigma j_\omega{}^* j_\omega$ in Δf is $e^2 N\sqrt{k_B T/m}\,\Delta f$. Note that t_0 has canceled out. This noise depends upon the "graininess" of the system, and if we were to clump the particles together the noise would increase. Combining particles in groups of n would let $e \to ne$, $m \to nm$, $N \to N/n$, and would increase the noise by a factor $\sqrt{n}$. It may be best to think of shot noise in terms of current rather than power. If we have a resistor we can write the power absorbed and emitted, but both are reduced by the reflectivity of the surface. We note that this shot noise goes to zero for a classical system as the temperature goes to zero and the particles move more slowly.

In a quantum gas, such as electrons obeying the Pauli Principle, the situation is different. If we had a *full* band of electrons, all states occupied, we would have no currents and no noise from that band alone. We can understand what this means by returning to the polarizable molecules and their van-der-Waals interaction discussed just before Eq. (12.3). Imagine the lower, filled, level as a bond level and the excited, empty, level as an antibonding level. Had we filled *both*, the polarizability α would have been zero, and polarization in the bond state would have been canceled by that in the antibonding state. Correspondingly, in the context of this two-level system, there would be no van-der-Waals force, Eq. (12.2), the force we used to "detect" noise fields arising from one molecule. In that sense, a full band (with no coupling to empty bands) would have no current fluctuations. If we add coupling between the full-band states and those in an empty band, there will indeed be current fluctuations, which can be calculated exactly as in Problem 12.2. They will be much smaller than fluctuations in a partly-filled band, and we neglect them here.

If we have a partly-filled band as in a metal, the states well below the Fermi energy which are entirely filled do not contribute to the shot noise, except from the coupling to empty states as discussed at the end of Section 12.2 and in the preceding paragraph. At finite temperature only the electrons near the Fermi energy contribute to the shot noise, a fraction of the electrons of order $k_B T/E_F$. Thus the shot noise is suppressed by such a factor by the Pauli Principle, just as is their contribution to the specific heat, as we indicated after Eq. (10.25). In this case the velocities which enter are for electrons at the Fermi energy rather than thermal energies so the square-root factor in Eq. (12.5) is enhanced by a factor $\sqrt{E_F/k_B T}$ and the net suppression of the shot noise of Eq. (12.5) is only by a factor of $\sqrt{k_B T/E_F}$. In Bose-Einstein systems, on the other hand, quantum noise can be enhanced by the statistics. The condensation of electrons into the lowest state is analogous to the clumping of particles in classical shot noise as we discussed in connection with the graininess of a system.

In a system in which tunneling is occurring, as described in Section 8.3, we ordinarily expect the tunneling events to occur at random times, as for the carriers in a classical gas, and to have the same shot noise. There are situations where that is not the case, as in the Coulomb blockade from a tunneling resonance discussed in Section 8.4. A similar effect occurs in small junctions, with very small electrical capacity C, where a tunneling transition shifts the voltage across the junction by $-e/C$. If this drop in voltage is comparable to the applied voltage, the tunneling probability drops considerably. As current continues to flow into the junction, the voltage across the junction builds up toward the applied voltage, as illustrated in Fig. 12.3, increasing again the tunneling probability. Then the tunneling events become more evenly spaced in time. A Fourier transform such as we have constructed then concentrates the noise at frequencies near the tunneling frequency $\langle j(t)\rangle/e$, with $\langle j(t)\rangle$ the time-average current. For optical communication this can be an important effect, leaving most of the frequency domain quite free of the shot noise which might otherwise have seemed completely unavoidable. It also shows an important effect that though the tunneling events are describable as discrete events, the flow of charge into the capacitor can be regarded as continuous, as individual electrons move closer and into the capacitor.

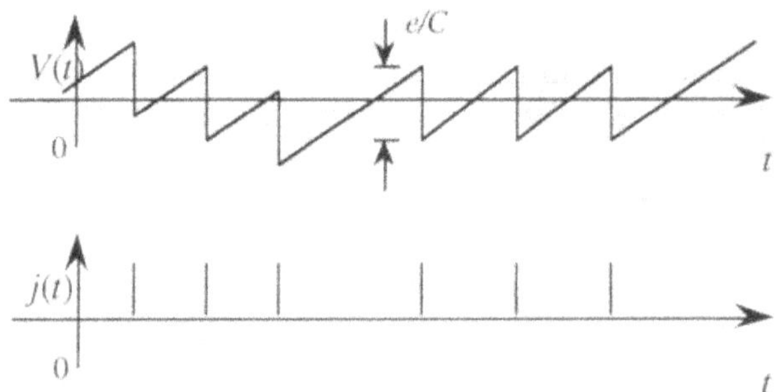

Fig. 12.3. If the capacity C of a tunnel junction is small each tunneling event, shown by a spike below, drops the voltage across the capacitor sufficiently to inhibit tunneling until the voltage rebuilds, spacing the events more uniformly and concentrating the shot noise to a narrow frequency rage.

12.4 Other Sources

With current flowing in a quantum wire, as we described in Section 2.3, we might imagine the states flowing to the right filled to a Fermi energy higher than those to the left, but a sharp cut-off at the Fermi energy in both cases. Then shot noise is suppressed, as we discussed in the last section. If then we add a scattering mechanism, represented by a transmission *Trans.*, as described in Section 8.1, some electrons moving to the right can be reflected back, producing a shot noise exactly as for tunneling in the last section. This is called *partition noise* arising from the partitioning of part of the forward current into reflected current. A similar partition noise arises in a y-shaped channel where some of the electrons move to the right and some to the left.

For the case of reflection by a defect it can be calculated just as we calculated shot noise in the preceding section. The classical shot noise was proportional to the density N of electrons involved and for small reflectivity 1 - *Trans.*, it is proportional to that reflectivity. On the other hand when the reflectivity approaches one, so that almost all reflected states are filled, the noise is again suppressed, being proportional to the transmission. Indeed over the entire range the noise is proportional to *Trans.*(1 - *Trans.*). In a similar way the partition noise in a y-shaped channel is proportional to the fraction flowing to the right times the fraction flowing to the left.

Each of these mechanisms can be regarded as shot noise, and calculated as we calculated shot noise in Eq. (12.5). If there is also inelastic scattering, so that the carriers also relax toward the low-temperature Fermi distribution, this will suppress this partition noise, just as lowering the temperature decreases the shot noise in a metal.

There is another familiar type of noise, called *1/f noise*, because the noise power varies approximately as the inverse of frequency which occurs in a wide variety of systems. In contrast to the types we have discussed it arises from a type of cooperative effect, such as illustrated in Fig. 12.4. We imagine current carried by carriers which hop from one site to the next, but can only hop if the neighboring site is empty. If we sit at one site, and note the times at which a carrier moves to the right, we obtain a current as a function of time as shown below in Fig. 12.3. If we then evaluate j_ω as in Eqs. (12.4) and (12.5) we find that noise power varies approximately as $1/\omega$. This is illustrated in Fig. 12.5 for which we have made such an evaluation for the model shown in Fig. 12.4, but for 100 sites, with periodic boundary

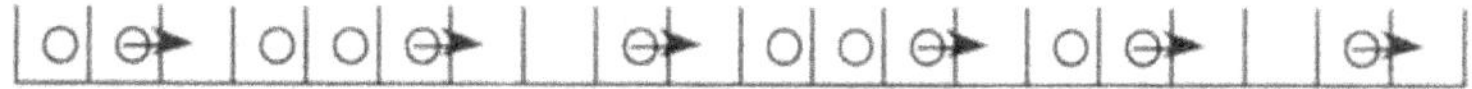

Fig. 12.4. Carriers which hop from site to site on a grid, but can only hop if the neighboring site is empty, produce a current measured at any one site which shows noise power per frequency interval, inversely proportional to the frequency, *1/f noise*. Displacements associated with a single time step are shown by arrows.

conditions, and with half the sites initially occupied at random. We did this with 800 time steps, took the Fourier transform, and averaged over eight neighboring frequencies. We repeated that calculation 100 times and averaged again to obtain the result shown in Fig. 12.5, which appears as a term proportional to $1/f$ plus some constant contribution. The noise for this calculation begins to rise again at higher frequencies(smaller $1/f$), apparently because configurations with alternate occupied and empty states move along unchanged and contribute strongly near the corresponding frequency. If one were interpreting some statistical data, the constant term seen in Fig. 12.5 might be interpreted as from some other mechanism, and the straight line in Fig. 12.5 would then represent the $1/f$ contribution. The model we used may not be worth exploring that much further.

$1/f$ noise is ubiquitous, arising from many different kinds of systems. For example, it apparently is observable in the traffic flow on busy freeways. It seems not so easy to derive the form, but it can be simulated as for Fig. 12.5.

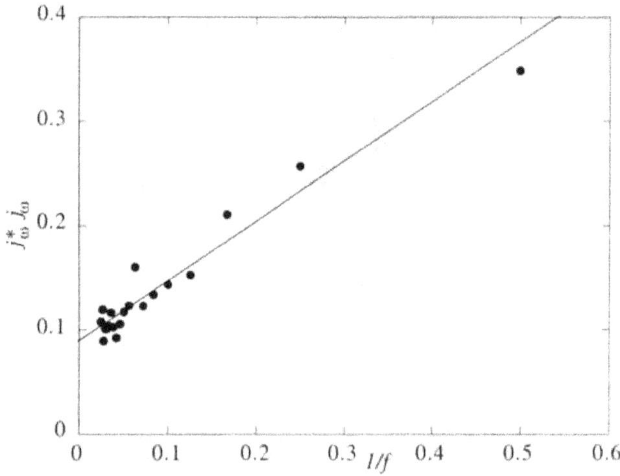

Fig. 12.5. Noise power, per frequency interval, plotted against the reciprocal of the frequency of the noise, for a periodic system such as Fig. 12.4 with half filling of 100 sites. It was calculated as indicated in the text.

V. Electrons and Phonons

We have made applications of quantum mechanics to solids throughout the text, and seen energy bands in simple systems. Crystalline systems are so important, and their understanding so heavily based upon quantum mechanics, that we should present the organization of the subject which is generally used. For many purposes the tight-binding basis is most flexible and easiest to use, as for the tunneling calculation in Chapter 8. However, the nearly-free-electron limit is also useful and provides a good introduction to Brillouin Zones, as well as formulating diffraction of waves in general, and we use it here. Our study of lattice vibrations in Chapter 15 will be closer to an energy-band formulation using tight-binding theory, but the Brillouin Zones are the same.

Chapter 13. Energy Bands

We begin by describing a procedure by which accurate energy bands could be obtained using the pseudopotential method. We do it for our approximate, empty-core form for the pseudopotential, but if we used instead one of the more rigorous forms (e. g., Harrison (1966)) the formulation would be the same and it could be a state-of-the-art band calculation. We then proceed to approximations based upon that formulation which will be more informative and allow discussion of a wider range of properties.

13.1 The Empty-Core Pseudopotential

Our beginning discussion of electron states was for free electrons, and we then saw that the effects of the potentials from the atoms constituting a crystal could be described in terms of a weak pseudopotential, which we took in the empty-core form, Eq. (4.18),

$$w(r) = \begin{cases} 0 & \text{for } r < r_c \\ v(r) & \text{for } r > r_c, \end{cases} \tag{13.1}$$

with $v(r)$ the free-atom potential, which we think of as $-Ze^2/r$ though ordinarily it included also terms in the potential arising from the valence electrons. Now in the solid we write the total pseudopotential as a *superposition* of such atomic pseudopotentials,

$$W(\mathbf{r}) = \sum_j w(\mathbf{r} - \mathbf{r}_j), \tag{13.2}$$

with $\mathbf{r}_j$ the positions of the atomic nuclei in the crystal.

For a pseudopotential band calculation we would solve the energy eigenvalue equation, Eq. (1.21),

$$-\frac{\hbar^2}{2m}\nabla^2\phi(\mathbf{r}) + W(\mathbf{r})\phi(\mathbf{r}) = \varepsilon\phi(\mathbf{r}), \tag{13.3}$$

now called the pseudopotential equation, with $\phi(\mathbf{r})$ the pseudowavefunction. It would be solved by expanding $\phi(\mathbf{r})$ in plane waves, $|\phi\rangle = \Sigma_{\mathbf{k}'}u_{\mathbf{k}'}|\mathbf{k}'\rangle$, and

$$|\mathbf{k}'\rangle = \frac{1}{\sqrt{\Omega}}e^{i\mathbf{k}'\cdot\mathbf{r}}\ , \tag{13.4}$$

with $\mathbf{k}'$ satisfying periodic boundary conditions. The expansion of $|\phi\rangle$ is substituted in Eq. (13.3), we multiply on the left by $\langle\mathbf{k}|$ and obtain

$$\frac{\hbar^2k^2}{2m}u_{\mathbf{k}} + \Sigma_{\mathbf{k}'}\langle\mathbf{k}|W|\mathbf{k}'\rangle u_{\mathbf{k}'} = Eu_{\mathbf{k}}\ . \tag{13.5}$$

For the first term on the left and the only term on the right, only terms for $\mathbf{k}' = \mathbf{k}$ were nonzero. The first step is the evaluation of the matrix elements, which is quite simple using Eq. (13.1) and would also be quite straightforward for more rigorous pseudopotentials. The important simplification comes from the use of Eq. (13.2).

$$\begin{aligned}\langle\mathbf{k}'|W|\mathbf{k}\rangle &= \frac{1}{\Omega}\int d^3r\, e^{i(\mathbf{k}-\mathbf{k}')\cdot\mathbf{r}}\Sigma_j w(\mathbf{r}-\mathbf{r}_j)\\ &= \frac{1}{\Omega}\Sigma_j e^{i(\mathbf{k}-\mathbf{k}')\cdot\mathbf{r}_j}\int d^3r\, e^{i(\mathbf{k}-\mathbf{k}')\cdot(\mathbf{r}-\mathbf{r}_j)}\, w(\mathbf{r}-\mathbf{r}_j) \\ &= \frac{1}{N}\Sigma_j e^{i(\mathbf{k}-\mathbf{k}')\cdot\mathbf{r}_j}\ \ \frac{1}{\Omega_0}\int d^3r\, e^{i(\mathbf{k}-\mathbf{k}')\cdot\mathbf{r}}\, w(\mathbf{r}) \equiv S(\mathbf{k}'-\mathbf{k})\, w_{\mathbf{k}'-\mathbf{k}}.\end{aligned} \tag{13.6}$$

In the first step we wrote out the two plane waves. In the second we interchanged the sum and integral, and multiplied under the sum by $e^{i(\mathbf{k}-\mathbf{k}')\cdot\mathbf{r}_j}e^{-i(\mathbf{k}-\mathbf{k}')\cdot\mathbf{r}_j} = 1$. In the third step we changed the variable of integration from $\mathbf{r} - \mathbf{r}_j$ to $\mathbf{r}$ and factored the volume into the number of atoms N and the atomic volume Ω_0.

This third form is factored into two terms, a structure factor and a form factor, written as $S(\mathbf{q})$ and $w_{\mathbf{q}}$ in the final form, as is usual in diffraction theory. We write $\mathbf{k}' - \mathbf{k} = \mathbf{q}$, and the *structure factor* is written

$$S(\mathbf{q}) \equiv \frac{1}{N}\sum_j e^{-i\mathbf{q}\cdot\mathbf{r}_j} \ . \tag{13.7}$$

It depends only upon the positions of the atoms, and is independent of the particular pseudopotential which has been used. The *form factor* is given by

$$w_{\mathbf{q}} = \frac{1}{\Omega_0}\int d^3r \ e^{-i\mathbf{q}\cdot\mathbf{r}} w(\mathbf{r}) = \frac{4\pi}{\Omega_0}\int dr \, r^2 \frac{\sin qr}{qr} w(r) \ . \tag{13.8}$$

For empty core pseudopotentials (with a convergence, or screening, factor $e^{-\kappa r}$) the integral may be carried out to obtain

$$\begin{aligned} w_{\mathbf{q}} &= -\frac{4\pi Ze^2}{\Omega_0 q}\int_{r_c,\infty} \sin qr \, e^{-\kappa r} \, dr \\ &= -\frac{4\pi Ze^2}{\Omega_0(q^2+\kappa^2)}\cos qr_c. \end{aligned} \tag{13.9}$$

We would anticipate taking $\kappa = 0$ in the convergence factor, but when we calculate in Section 20.2 the redistribution of the electron charge due to these pseudopotentials, self-consistently to first order in the resulting electrostatic potential, we will find that in the Fermi-Thomas approximation that the net effect is to introduce a κ^2 exactly as in Eq. (13.9), with

$$\kappa^2 = \frac{4me^2k_F}{\pi\hbar^2} \ . \tag{13.10}$$

We see incidentally from the first form in Eq. (13.9) that this "screening of the potential" has the effect of reducing the long-range Coulomb potential to $-Zee^{-\kappa r}/r$.

The pseudopotential form factor is all we need to know about an element in order to perform a band calculation, or to calculate other properties using the structure factors for whatever arrangement of atoms we wish to consider. If we are satisfied with the approximate empty-core form, we need know only the empty-core radius and the valence Z. In Problem 4.3 we calculated the pseudopotential core radius for lithium and sodium from the atomic term values given in Table 4.1. In Fig. 13.1 we show the pseudopotential form factor obtained from Eq. 13.9 with that radius for sodium, along with an earlier full calculation. The largest difference is at $q/k_F = 0$, where our value, $-2/3E_F$, would now be considered correct. For some purposes one might try

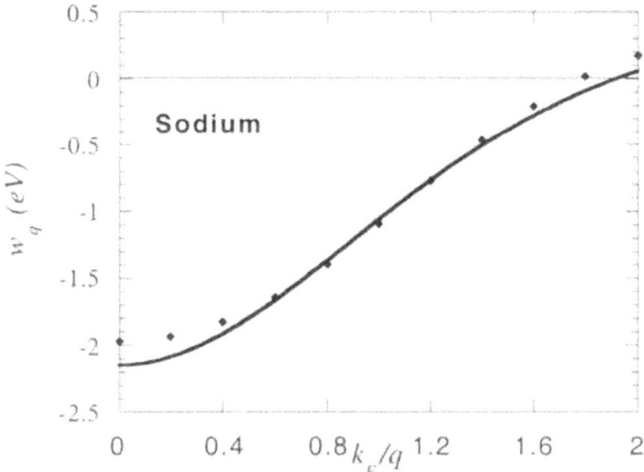

Fig. 13.1. The pseudopotential form factor for sodium from Eq. (13.9) using the empty-core radius from Problem 4.3 and k_F = 0.92 Å^{-1}. The points are the values calculated by Animalu and Heine (1965).

to improve the predictions by adjusting r_c to fit a prediction to some known property, and one can then use the same r_c for related properties.

The evaluation of structure factors may be illustrated for the one-dimensional case. For a regular chain of atoms, spaced by d as in Fig. 6.1, the atomic positions are $x_j = jd$ for $j = 0, 1, ...N-1$. Then periodic boundary conditions require that k, k' and q all are of the form $2\pi n/(Nd)$, so

$$S(q) = \frac{1}{N}\Sigma_j\, e^{-iqd_j} = \frac{1}{N}(1 + a + a^2 + ...a^{N-1})$$
$$= \frac{1}{N}\frac{1 - a^N}{1 - a}, \quad \text{with } a = \exp\left(2\pi i\frac{n}{N}\right) \tag{13.11}$$

The numerator $1 - a^N$ is always zero. We can only find nonzero $S(q)$ when n/N is an integer so the denominator is zero. Then every term is one and $S(q) = 1$. These wavenumbers, $q = 2\pi/d$ times an integer, are called *lattice wavenumbers* (or loosely called reciprocal lattice vectors, with or without the factor 2π). The same result applies for simple cubic crystals, with $q_x = 2\pi n_x/d$, $q_y = 2\pi n_y/d$, and $q_z = 2\pi n_z/d$.

For this one-dimensional case in Section 6.1 we defined a Brillouin Zone as the range of wavenumbers which gave distinct states in tight-binding theory, $-\pi/d < k \leq \pi/d$. That generalizes to three-dimensions as being the region of wavenumber space closer to $\mathbf{q} = 0$ than to any other lattice wavenumber. We shall see the significance of the Brillouin Zone in the free-electron context. We shall also evaluate the structure factor for atomic arrangements other than the perfect crystal.

13.2 A Band Calculation

For a chain of equally-spaced atoms, with nonvanishing structure factors only at the lattice wavenumbers, a plane wave $|k\rangle$ is coupled, according to Eq. (13.6), only to states $|k \pm 2\pi n/d\rangle$, states which differ from k by a lattice wavenumber. The very large number of wavenumbers which are allowed by periodic boundary conditions in a large system, N times as many as there are lattice wavenumbers for a chain of N atoms, is irrelevant since only those differing by a lattice wavenumber are coupled. If we focus on the smallest k among a set of coupled plane waves, the eigenstate can be written as a linear combination of that plane wave and all plane waves coupled to it. In an approximate treatment, we include only a limited number of such coupled states and the pseudopotential makes it possible for that number to be quite small. We can understand this in terms of the pseudowavefunction for sodium metal which we plotted in Fig. 4.3, a sum of atomic pseudowavefunctions for the $k = 0$ state. The single plane wave for $k = 0$ would be a constant. We may add to it the contributions of the two smallest lattice wavenumbers $\pm q = 2\pi/d$, so the pseudowavefunction becomes $\phi(\mathbf{r}) \approx A_0 + 2A_1 \cos qd$. . Fitting the maximum and minimum in the curve in Fig. 4.3 gives A_1/A_0 only -0.15. These small corrections reproduce the pseudowavefunction quite well, with the remaining discrepancy is largely eliminated by even smaller contributions from the next set of $q = 4\pi/d$.

With only a small number of plane waves needed for the pseudowavefunction, the problem is the same as the calculation of molecular states in terms of a small number of atomic states as we discussed in Section 5.1 and 5.2. We need to solve as many simultaneous linear equations as we have terms in the expansion, as in Eq. (5.13). The coefficients H_{ji} which enter such equations form a matrix, called the Hamiltonian matrix, and the solution of the equations is called *diagonalizing the matrix* . In these terms the pseudopotential has reduced the problem to the diagonalization of a small Hamiltonian matrix, based only on plane waves which differ by the smallest lattice wavenumbers.

This would not have been true had we sought an expansion of the full wavefunction, given approximately by a sum of the full atomic states, for sodium each with a large peak at the nuclear position and two nearby nodes on either side. An extraordinarily large set of plane waves would have been

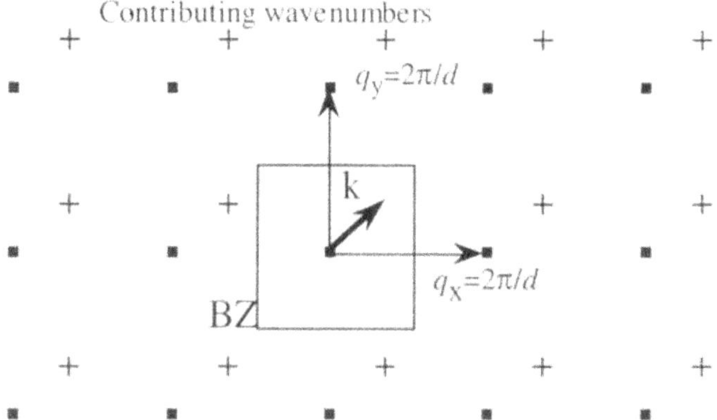

Fig. 13.2. Wavenumber space, showing the lattice wavenumbers (square dots) and a wavenumber **k** in the Brillouin Zone (BZ) and all states (+) coupled to the state |**k**>.

required, and the diagonalization of the corresponding huge matrix. The same reduction occurs in two and three dimensions. In fact, had we regarded Fig. 4.3 as the pseudowavefunction along a line in the three dimensional crystal, the corrections to the $\mathbf{k} = 0$ plane wave would have been twelve plane waves with a coefficient A_1/A_0 = -0.025. We discuss the calculation in detail first for the two-dimensional case.

For a two-dimensional square lattice, with interatomic distances d, the lattice wavenumbers are all integral linear combinations of lattice wavenumbers of length $2\pi/d$ in the x- and y-directions, as indicated in Fig. 13.2. The wavenumbers of the states to which a plane wave of wavenumber **k** in the Brillouin zone is coupled, shown by +'s, should be included in the calculation, but the ones differing by larger wavenumbers are only weakly coupled (the coupling drops as $1/q^2$ at large q, according to Eq. (13.9)) and they differ greatly in energy, also reducing their effect. The state which is calculated contains terms with all of these different wavenumbers, but we ordinarily specify it by giving the wavenumber with the smallest magnitude, the one which lies in the Brillouin Zone shown for the square lattice in Fig. 13.2. This is the two-dimensional counterpart of the one-dimensional Brillouin Zone $-\pi/d \leq q < \pi/d$ discussed in Section 6.1.

For a simple cubic lattice in three dimensions, the wavenumber lattice is simple cubic and the Brillouin Zone is a cube. Exactly the same situation obtains with respect to a band calculation, which can be performed with of the order of ten or twenty plane waves, by diagonalizing the corresponding ten-by-ten or twenty-by-twenty Hamiltonian matrix. Before discussing the results of such a straight-forward calculation, we see how this simple-cubic lattice is generalized to a more important structure, the *face-centered-cubic*, or *fcc*, lattice. This lattice is not only the crystal structure of many metals, e. g., copper and aluminum, but also has the same translational symmetry as most semiconductors and also rock salt.

The fcc lattice is illustrated in Fig. 13.3. It is based upon a simple-cubic lattice, but with an addition (identical) atom at the center of each of the six faces of every cube. The length of the cube edges is traditionally called a and is then related to the nearest-neighbor distance d by $a = \sqrt{2}d$. Traditional crystallographic notation takes a Cartesian coordinate system oriented along these cube edges, and specifies directions as [100] along the x-axis, or more generally parallel to any cube edge, [110] parallel to any face diagonal, and [111] parallel to any cube diagonal. If this lattice is extended to many cubes we see that the face-center atoms have identical arrangements of neighbors to the those at the cube corners, and the cubes could as well have had a corner at any atom in the crystal. Thus the smallest translations τ_i of the lattice which take every atom in the interior to the position previously occupied by another atom are of length d and are in [110] directions. Three such translations are indicated in Fig. 13.3, and are called *primitive lattice translations.* The density of atoms corresponds to four atoms per cube, counted by noting that the eight corner atoms are shared by eight cubes and the six face atoms are each shared by two cubes. If the spheres representing each atom are expanded till they touch their nearest neighbors, each is seen to touch twelve nearest neighbors, a close-packed lattice with the densest possible packing of spheres in an extended system.

The extension of the concept of the simple-cubic lattice wavenumbers of Fig. 13.2 to the fcc lattice is not so obvious since the primitive translations τ_1, τ_2, and τ_3 are not perpendicular to each other. We may see that the essential feature is that the primitive lattice wavenumber $\mathbf{q}_1$ be perpendicular to τ_2 and τ_3 and therefore proportional to $\tau_2 \times \tau_3$. Then the primitive lattice wavenumbers become,

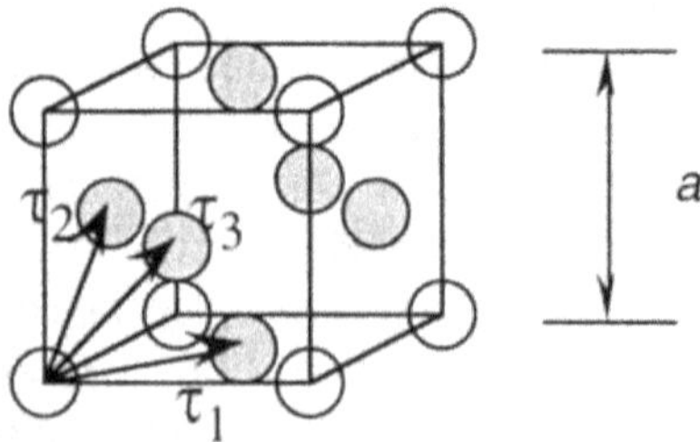

Table 13.3. One cube of a face-centered-cubic lattice, showing the atoms centered at each face, and a set of primitive translations τ_i which take the lattice into itself. The cube edge a is also indicated.

$$\mathbf{q}_1 = \frac{2\pi\tau_2 \times \tau_3}{\tau_1 \cdot \tau_2 \times \tau_3}, \text{ etc.,} \tag{13.12}$$

the other two obtained by rotating indices. For simple-cubic lattices these reduce to the primitive lattice wavenumbers we gave. In the fcc lattice these primitive lattice wavenumbers lie in [111] directions. The states coupled to any plane wave of wavenumber **k** in the Brillouin Zone have wavenumbers differing from **k** by an integral linear combination of these primitive lattice wavenumbers, the linear combinations again being called lattice wavenumbers. The wavenumber lattice made up of these lattice wavenumbers, based on primitive lattice wavenumbers in [111] directions, is called a *body-centered cubic* lattice. It is again based upon a simple-cubic lattice but has additional sites at the cube center rather than in the cube faces. It is also a common crystal lattice for elemental metals.

The Brillouin Zone for the face-centered-cubic crystal lattice again is the surface containing all wavenumbers closer to $\mathbf{q} = 0$ than to any other lattice wavenumber, and is shown in Fig. 13.4. The primitive lattice wavenumbers are shown as arrows, and the planes bisecting them, which would form a regular octagon if all eight such planes were included, form part of the Brillouin Zone. However, points inside that octahedron, but outside the cube drawn, are closer to a lattice wavenumber $4\pi/a$ along the cube direction than to $\mathbf{q} = 0$, so the cube faces truncate the points of the octahedron leading to the shape shown. It may also be regarded as the cube shown, with its eight

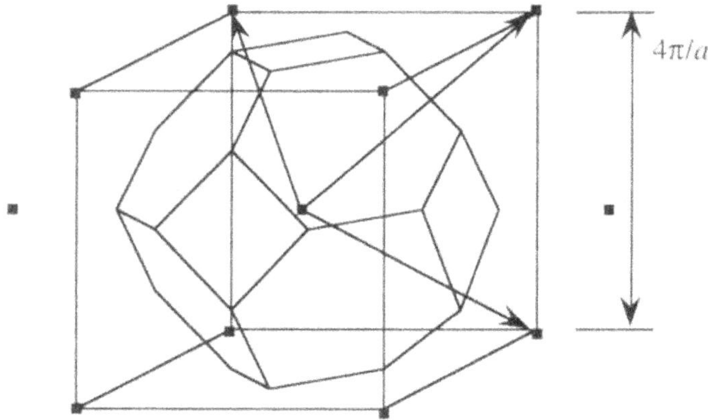

Fig. 13.4. The Brillouin Zone for the face-centered cubic crystal lattice. Lattice wavenumbers are again shown as square dots, and three primitive lattice wavenumbers are shown as arrows, leading to the corners of a circumscribed cube.

corners truncated. The Zone has half the volume of that cube. This is also the Brillouin Zone for the tetrahedral semiconductors, Si, GaAs, etc., because they have the same translational symmetry. For example, the gallium atoms in GaAs have a face-centered-cubic arrangement, with another face-centered-cubic lattice of arsenic interspersed.

For discussing free-electron bands and band calculations we return to the simple-cubic lattice. There are free-electron states at all wavenumbers, but we may represent all of them by the wavenumber in the Brillouin Zone for the plane wave to which they are coupled. In this way every wavenumber indicated by a "+" in Fig. 13.2 is replotted at the same point in the Brillouin Zone. This is also what we did in Section 6.2 for the simple-cubic lattice for matching with tight-binding sp-bands. Then for a wavenumber **k** in the Brillouin Zone, there is a state with energy $\hbar^2k^2/(2m)$ but also one of energy $\hbar^2(\mathbf{k} + \mathbf{q}_j)^2/(2m)$ for every lattice wavenumber (every integral combination of the primitive lattice wavenumbers such as the $\mathbf{q}_1$ in Eq. (13.12)). These are drawn (as in Fig. 6.6) in Fig.13.5.

States of the same wavenumber in the Brillouin Zone are coupled to each other by matrix elements of the pseudopotential. For any state well-removed in energy from the others, the coupling can be treated in perturbation theory and the shifts are small. When two coupled states

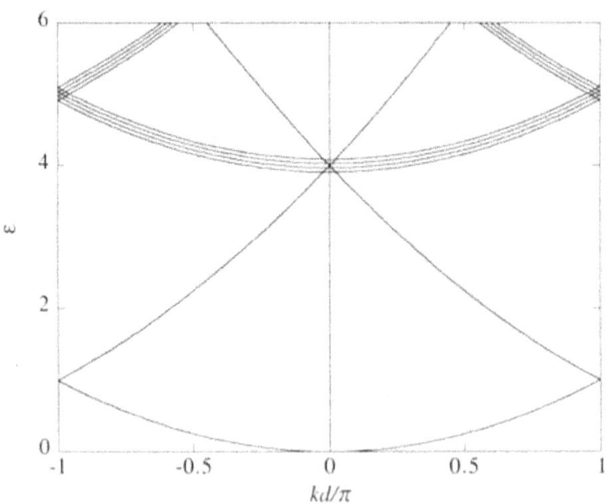

Fig. 13.5. Free-electron bands redrawn in a cube direction in the Brillouin Zone for a simple cubic lattice with spacing d. The four degenerate bands are drawn separately.

become close in energy, as for the two lowest bands at the edge of the Brillouin Zone, the two coupled states must be treated exactly, solving the two-by-two Hamiltonian matrix to obtain, in the case of those two states near the Brillouin-Zone edge ($k \approx q/2$ with $q = 2\pi/d$),

$$\varepsilon_k = \frac{\frac{\hbar^2 k^2}{2m} + \frac{\hbar^2 (k-q)^2}{2m}}{2} \pm \sqrt{\left(\frac{\frac{\hbar^2 k^2}{2m} - \frac{\hbar^2 (k-q)^2}{2m}}{2}\right)^2 + w_q^2}\,. \tag{13.13}$$

We have written the matrix element coupling the two states $\langle \mathbf{k+q}|W|\mathbf{k}\rangle = S(\mathbf{q})w_q = w_q$. Such a solution, when two coupled states are very close in energy, is sometimes called *degenerate perturbation theory.*

The resulting two bands are plotted in Fig. 13.6. A gap equal to $2w_q$ has been opened up at the Brillouin-Zone edge, where the two free-electron states are degenerate. Away from this region the states are quite free-electron-like and any effect of the pseudopotential could be treated as a perturbation. At the left edge of the Brillouin Zone it would be appropriate to use the two states k and $k + q$, rather than k and $k - q$. Then a gap would appear there. The resulting bands within the Brillouin Zone represent the electronic structure of the solid. The bands shown outside the Brillouin Zone to the right are redundant replications of the bands within the zone, as were the tight-binding bands outside the Brillouin Zone in Fig. 6.2.

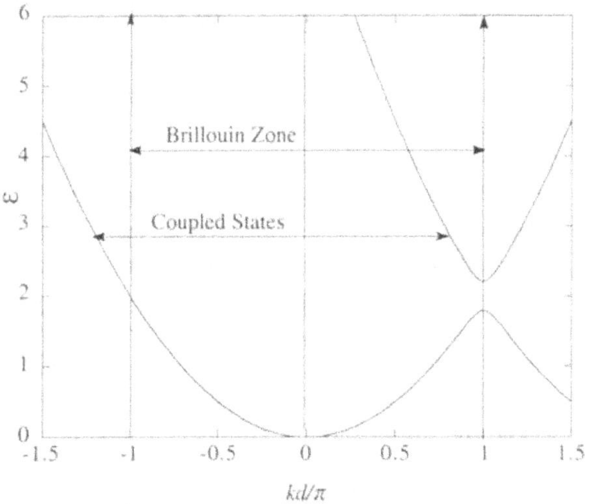

Fig. 13.6. Bands from Eq. (13.13) giving the energy of two free-electron states coupled by a matrix element w_q with $q = 2\pi/d$.

The energy bands given in Eq. (13.13) correspond to the results of a simple band calculation. In a full calculation one would include not only the three plane waves, k, $k - q$, and $k + q$, which are needed to obtain the lower bands, but all of the plane waves which have an appreciable effect on the states. One could also use a more accurate method, in comparison to the empty-core pseudopotential, for obtaining the matrix elements $\langle \mathbf{k} + \mathbf{q}|W|\mathbf{k}\rangle$ which couple the various plane waves. These are all, however, straightforward generalizations of the simple calculation we performed here. The most important approximation in any case is the one-electron approximation which we introduced in Sections 2.2 and 4.2.

In a metal the coupling is very small compared to the Fermi energy which would lie just below the gap in Fig. 13.6 for a monovalent metal (e. g., sodium in a simple-cubic structure), and just above for a divalent metal. For such a divalent metal the shifts in the band also distort the Fermi surface, which would be spherical in the absence of a perturbing pseudopotential. Experimental studies of these Fermi surfaces were important in learning how to understand metals in terms of pseudopotentials (Harrison and Webb (1960)) but by now have become a rather specialized topic. We see in Problem 13.1 how diffraction changes the electron orbits in a magnetic field, which can be interpreted in terms of Fermi surfaces made up of rearranged segments of a Fermi sphere.

13.3 Diffraction

The opening of a band gap as in Fig. 13.6 sheds further light on the diffraction of electrons by a periodic lattice. We shall see in Chapter 14 that applied electric and magnetic fields cause electrons to move continuously along the energy bands, and this was illustrated in Problem 13.1. Thus when gaps are introduced at the left, as well as at the right, edges of the Brillouin Zone in Fig. 13.6, an electron moving up to the Zone face at the right must continue on to the right, or equivalently emerge from the left face of the Zone. The electron has changed the direction of its wavenumber and its velocity, which physically corresponds to a diffraction of the electron by the periodic pseudopotential of the lattice. Indeed the Bragg condition for diffraction is that two states of the same energy are coupled, exactly the condition which causes us to solve the two-by-two degenerate-perturbation-theory equation, Eq. (13.13). An approximate description of Fermi surfaces in polyvalent metals (in these metals the Fermi sphere always crosses diffraction planes) is possible simply by taking these diffractions into account, again illustrated in Problem 13.1. This corresponds to the real Fermi surfaces obtained from a band calculation, but in the limit as the pseudopotential becomes small. We shall see how this affects the electron dynamics in the following chapter.

13.4 Scattering by Impurities

We discussed scattering of electrons by impurities in the context of tight-binding theory in Section 7.3. It is useful to understand it also in the context of weak pseudopotentials.

We have seen that the matrix elements $\langle\mathbf{k}'|\Sigma_j w(\mathbf{r} - \mathbf{r}_j)|\mathbf{k}\rangle$ are zero for a perfect crystal if the wavenumber $\mathbf{k}$ does not lie on a Bragg plane. Thus if we were to change one atomic pseudopotential at $\mathbf{r}_i$ in the sum by $\delta w(\mathbf{r}-\mathbf{r}_i) = w'(\mathbf{r}-\mathbf{r}_i)-w(\mathbf{r}-\mathbf{r}_i)$, for all of these states away from the Bragg planes the matrix element will be zero plus the matrix element of the change. Let us again write $\mathbf{k}'= \mathbf{k}+\mathbf{q}$ (as after Eq. (13.6)), and then

$$\langle\mathbf{k}+\mathbf{q}|\Sigma_j w(\mathbf{r} - \mathbf{r}_j)|\mathbf{k}\rangle = \langle \mathbf{k}+\mathbf{q}|\delta w(\mathbf{r} - \mathbf{r}_i)|\mathbf{k}\rangle$$
$$= \int d^3r(w'(\mathbf{r}-\mathbf{r}_i)-w(\mathbf{r}-\mathbf{r}_i))\frac{e^{-i\mathbf{q}\cdot\mathbf{r}}}{\Omega} = (w_q' - w_q)\frac{e^{-i\mathbf{q}\cdot\mathbf{r}_i}}{N}. \tag{13.14}$$

This is closely analogous to the matrix element $\delta\varepsilon_s e^{i(k-k')di}/N$ obtained for tight-binding theory, in one dimension, following Eq. (7.10). As in that case we can proceed with the Golden Rule but now the matrix element depends upon the difference in wavenumber between the two coupled states.

$$\frac{1}{\tau} = \frac{2\pi}{\hbar}\Sigma_{\mathbf{k}'}\left(\frac{w_q' - w_q}{N}\right)^2 \delta(\varepsilon_{\mathbf{k}'} - \varepsilon_{\mathbf{k}}) \tag{13.15}$$

For the evaluation we replace the sum over $\mathbf{k}'$ by an integration as we indicated at the end of Section 7.3, and take a coordinate system as shown in Fig. 13.7. For a given magnitude of $\mathbf{q} = \mathbf{k}' - \mathbf{k}$, the states in a circular ring of radius $k'\sin\theta$ and small cross-section dk' by $k'd\theta$ have the same matrix element. The density of states in wavenumber space is $\Omega/(2\pi)^3$ with Ω the volume of the system so

$$\frac{1}{\tau} = \frac{2\pi}{\hbar}\frac{\Omega}{(2\pi)^3}\int 2\pi d\theta\sin\theta\int dk'k'^2\left(\frac{w_q' - w_q}{N}\right)^2\delta(\varepsilon_{\mathbf{k}'}-\varepsilon_{\mathbf{k}}). \tag{13.16}$$

We multiply by $(d\varepsilon_{k'}/dk')/(\hbar^2 k'/m)$ which is equal to one, and integrate over energy $\varepsilon_{k'}$ to obtain

$$\begin{aligned}\frac{1}{\tau} &= \frac{mk}{\hbar^3}\frac{\Omega}{2\pi N^2}\int d\theta \sin\theta\left(\frac{w_{q'} - w_q}{N}\right)^2 \\ &= \frac{m\Omega_0}{2\pi\hbar^3 kN}\int_{0,2k} dq\, q\,(w_{q'} - w_q)^2.\end{aligned} \tag{13.17}$$

with Ω_0 the volume per atom. In the final step we noted that for an isosceles triangle ($k' = k$) we have $q^2 = 2k^2(1\text{-}\cos\theta)$ so $\sin\theta d\theta = q dq/k^2$. This form is convenient to use with known form factors w_q as in Fig. 13.1.

We may compare this with the tight-binding result in Eq. (7.13), if we replace the velocity in that expression by $v = \hbar k/m$, which leads to

$$\frac{1}{\tau} = \frac{\delta\varepsilon_s{}^2 m\, k\, \Omega_0}{\pi\hbar^3 N}. \tag{13.18}$$

We see that $\delta\varepsilon_s{}^2$ has been replaced by $^1/_2\int_{0,2k} d(q/k)\,(q/k)(w_{q'} - w_q)^2$. The integral could readily be performed using the empty-core form for the pseudopotential, Eq. (13.9).

For calculating the conductivity as in Section 11.3 we should use the *momentum relaxation time* τ in which each scattering event is weighted by the fractional loss of initial momentum, $1 - \cos\theta = q^2/(2k^2)$. This factor can be directly inserted in the integrand in Eq. (13.17).

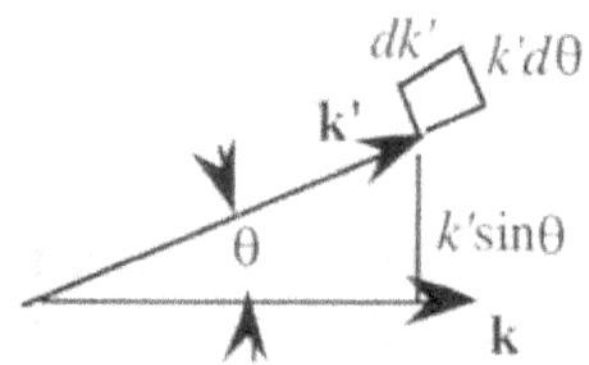

Fig. 13.7. The coordinate system for summing over states in cylindrically-symmetric systems.

13.5 Semiconductor Bands

Perhaps the most important energy bands are those for semiconductors, and we will be discussing specifically electron dynamics in those bands in the following chapter. We introduce them briefly here in the context of the

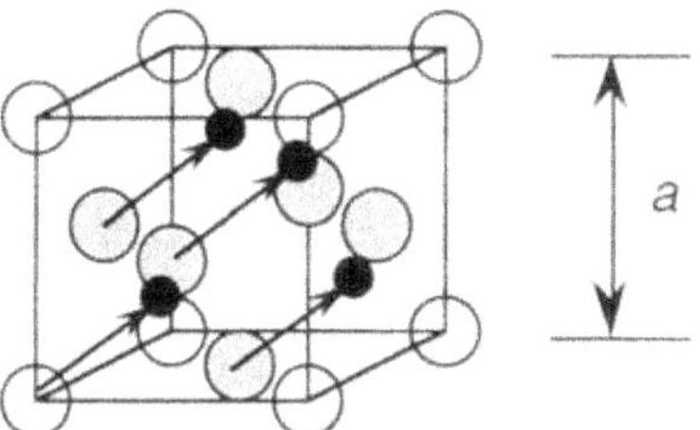

Fig. 13.8. The diamond structure is obtained from the face-centered-cubic structure (empty and lightly-shaded circles, as in Fig. 13.3) by adding a second atom, displaced by one quarter of a cube body diagonal from each of the original sites.

free-electron bands.

The semiconductors silicon and germanium are in the diamond structure, which is based upon the face-centered-cubic structure which we discussed in Section 13.2. One cube of that structure is redrawn in Fig. 13.8 and a second atom added for each original atom, as indicated. Note that each added atom is surrounded be a regular tetrahedron of the original atoms. Similarly, each original atom is surrounded by a regular tetrahedron of added atoms of inverted geometry (compared to the tetrahedra of original atoms). Most compound semiconductors, such as gallium arsenide, are in this structure with for example the original face-centered-cubic atoms gallium and the added atoms arsenic. In either case the translational symmetry is that of the face-centered-cubic structure, with two atoms in each primitive cell.

The free-electron bands for this structure are shown to the left in Fig.

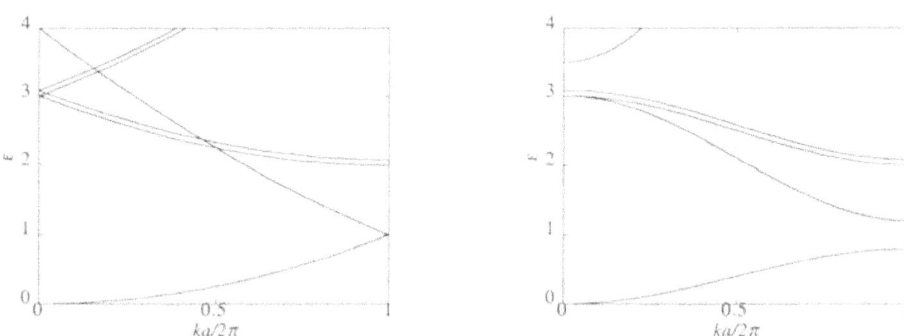

Fig. 13.9. The free-electron bands, on the left, for the face-centered-cubic (or Si or GaAs) structure for **k** along a cube-axis direction, analogous to the free-electron bands for the simple-cubic structure shown in Fig. 13.5. ε is given in units of $(\hbar^2/2m)(2\pi/a)^3$. To the right are the corresponding bands for a semiconductor such as GaAs.

13.9, for wavenumbers along a [100] direction in the Brillouin Zone of Fig. 13.4. A total of eight free-electron bands meet at $\mathbf{k} = 0$ with energy $3(\hbar^2/2m)(2\pi/a)^3$ (the bands shown double represent four bands each). As the pseudopotential is introduced, these split into two sets of three-fold-degenerate bands and two single bands. One of these three-fold sets is lowest and the three bands emerging from this point are the upper *valence bands* as shown to the right in Fig. 13.9 (here the double line represents two bands). With eight electrons per atom pair the lowest four bands are filled and are collectively called the valence bands. The lowest band above these is the empty *conduction band* and in most compound semiconductors has its minimum energy at $\mathbf{k} = 0$. These are the bands for which we discussed the statistics of occupation in Section 10.5. We shall discuss the dynamics of electrons in such bands in the following chapter.

These bands are also understandable in terms of tight-binding theory, as are the simple-cubic bands discussed earlier in this chapter. It was in fact the comparison of the free-electron and tight-binding simple-cubic bands in Section 6.2 which gave us our universal matrix elements. In the case of covalent semiconductors we proceed from the atomic states to form sp^3-hybrids, as indicated in Section 6.3, and then form bonding and antibonding states, four each for each atom pair. The four valence bands shown to the right in Fig. 13.9 arise from coupling between neighboring bond states just as the coupling between atomic states broadens them into bands. Each of the states in the band is a linear combination of bond orbitals (and in a more accurate calculation some admixed antibonding orbitals). The lowest state at $\mathbf{k} = 0$ turns out to be the sum of every bond orbital with an equal coefficient. Thus on each atom it contains all four sp^3-hybrids with equal coefficients, the p-states all cancel out leaving a pure s-like state, as in the simple-cubic bands. Similarly, the three states at the top of the valence band consist entirely of p-states on the individual atoms. As wavenumber increases to the right in Fig. 13.9, the σ-oriented (parallel to $\mathbf{k}$) p-state combines with the s-states to form the three nondegenerate bands shown. The other two bands are degenerate π-bands.

These tight-binding and free-electron bands can be used to derive universal coupling parameters, as for simple-cubic bands in Section 6.2, and the resulting parameters are more appropriate for the study of covalent solids. Interestingly enough the values are quite similar. It is found that $V_{ss\sigma} = -9\pi^2/64\,\hbar^2/(md^2)$, only slightly larger than the $-\pi^2/8\,\hbar^2/(md^2)$ which we obtained here. Extensive discussions of the electronic structure and properties of covalent solids based upon this tight-binding picture are given in Harrison (1999).

Chapter 14. Electron Dynamics

It is clear that the electrons in the simple metals, with weak pseudopotentials, have the dynamics of free electrons, with the additional effect of diffraction by the pseudopotentials. They have velocity $\mathbf{v} = \hbar\mathbf{k}/m$ and acceleration given by any applied force divided by the mass m. With the more complicated energy bands, such as we discussed for semiconductors in the preceding section, the dynamics may be deduced by constructing packets much as we did in deriving the Schroedinger Equation in Section 1.2.

14.1. Dynamics of Packets

We found already in Eq. (1.6) that a packet moves with velocity $v = \partial\omega/\partial k$, which for particles became $(1/\hbar)\partial\varepsilon/\partial k$. That same result applies to energy bands since we can make wave packets of band states just as we made them of plane waves and follow the same argument. Thus for a system with energy bands $\varepsilon_{\mathbf{k}}$ we have the velocity

$$\mathbf{v} = \frac{1}{\hbar}\nabla_{\mathbf{k}}\varepsilon_{\mathbf{k}} \tag{14.1}$$

with of course $\nabla_{\mathbf{k}}$ having an x-component of $\partial\varepsilon_{\mathbf{k}}/\partial k_x$, etc.. This is the usual form for free electrons, and for bands approximated by a parabola as $\varepsilon_{\mathbf{k}} = \hbar^2k^2/(2m^*)$ it is $\hbar k/m^*$ For cosine-like bands, as shown to the right in Fig. 13.9, we note that the velocity is zero at $\mathbf{k} = 0$, but also at the Brillouin-Zone edge, $k = 2\pi/a$ (or π/d in a linear chain with spacing d).

We now imagine this packet moving in a slowly varying potential $V(\mathbf{r})$ so that it will pick up potential energy at a rate $\mathbf{v}\cdot\nabla V(\mathbf{r})$. It must therefore give up kinetic energy, or band energy $\varepsilon_{\mathbf{k}}$, at the same rate. This occurs by a change in the central wavenumber $\mathbf{k}$ of the packet, corresponding to $d\mathbf{k}/dt \cdot \nabla_{\mathbf{k}}\varepsilon_{\mathbf{k}} = -\mathbf{v}\cdot\nabla V(\mathbf{r})$. But using Eq. (14.1) for $\nabla_{\mathbf{k}}\varepsilon_{\mathbf{k}}$ we see that at least for the components of force $\mathbf{F} = -\nabla V(\mathbf{r})$ parallel to the direction of motion,

$$\hbar\frac{d\mathbf{k}}{dt} = \mathbf{F} = -\nabla V(\mathbf{r})\,. \tag{14.2}$$

This is a very simple and plausible result, which turns out to be applicable for all components of the force. The principal limitation on the validity is that it can be used only where a description of the state in terms of packets makes sense. It would not be applicable to potentials arising from an impurity which varied rapidly over one atomic distance since making a packet small enough to define the position at such a distance requires the entire band, and an energy uncertainty corresponding to the band width.

These two equations, Eqs. (14.1) and (14.2), are exactly Hamilton's Equations, Eqs. (3.5) with $\hbar\mathbf{k}$ playing the role of momentum and the Hamiltonian $H(\mathbf{p}, \mathbf{r})$ obtained from the energy bands $\varepsilon_{\mathbf{k}}$ plus a potential $V(\mathbf{r})$. Thus they describe completely the dynamics of the wave packet just as Hamiltonian mechanics described the dynamics of classical particles.

For free electrons, with momentum $\mathbf{p} = \hbar\mathbf{k}$, this is the classical $d\mathbf{p}/dt = \mathbf{F}$. For any band structure, it tells that the wavenumber changes according to the same formula. For a constant electric field the wavenumber moves through the Brillouin Zone at a constant velocity. In a complicated band structure the electron velocity itself, $\mathbf{v} = (1/\hbar)\,\nabla_{\mathbf{k}}\,\varepsilon_{\mathbf{k}}$, may have a complicated variation, but the wavenumber behaves simply. For example, for a uniform electric field parallel to $\mathbf{k}$ in the lowest cosine-like bands to the right in Fig. 13.9, the wavenumber moves to the right at a constant rate, and the electron increases its speed, reaches a maximum and then slows to a stop when the wavenumber reaches the Brillouin-Zone face. At this point we would represent the state by the equivalent wavenumber at the opposite Zone face (though we could continue on outside the Zone if we so chose) and the electron begins picking up speed in the opposite direction, reaches a maximum and comes again to rest at $\mathbf{k} = 0$. Physically we could say that the electron accelerated but made a gradual diffraction, reversing its direction and moving against the field, which brings it back to rest. If the pseudopotential were weaker, the bands would be more like those to the left in Fig. 13.9, bending over in a much shorter wavenumber range as in Fig. 13.6, and the diffraction would be much more abrupt, but still continuous. If the pseudopotential were sufficiently weak, compared to the applied forces, no diffraction would occur. In terms of bands such as those in Fig. 13.6 this would mean that the electron jumped to another band as the wavenumber crossed the Brillouin-Zone face. When this occurs as the wavenumber changes due to a magnetic field, the jump between bands is called *magnetic breakdown* of the band gap. In order to discuss that we must consider magnetic forces.

As we indicated, Eq. (14.2) remains true for all forces, whether or not they are parallel to the direction of motion, so they apply to the forces due to a magnetic field $\mathbf{F} = (-e/c)\dot{\mathbf{r}}\times\mathbf{H}$ given in Eq. (3.13). Inserting this into Eq. (14.2) we have

$$\hbar\frac{d\mathbf{k}}{dt} = \frac{(-e)\dot{\mathbf{r}}}{c}\times\mathbf{H} = -\frac{e}{c}\frac{\nabla_{\mathbf{k}}\,\varepsilon_{\mathbf{k}}}{\hbar}\times\mathbf{H}. \qquad (14.3)$$

From the final form we see that the wavenumber changes are perpendicular to the gradient of the energy with respect to wavenumber, so the energy of an electron in a magnetic field does not change with time. The change in wavenumber is also seen to be perpendicular to the magnetic field so the trajectory of an electron in wavenumber space is the intersection of a constant energy surface in a band with a plane perpendicular to the magnetic field. This is consistent with the motion we expect for free electrons, but also when the Fermi surface in a metal has a complicated shape, electrons at the Fermi energy move along the intersection of that Fermi surface and such a plane perpendicular to the magnetic field. From the first form in Eq. (14.3) we see that $\dot{\mathbf{r}}$ is proportional to $\dot{\mathbf{k}}$ (with a constant ratio $e\hbar/eH$ and a 90° rotation) so that the shape of the electron orbit in real space (projected on a plane perpendicular to the magnetic field) is exactly the same as the shape in wavenumber space. Thus experimental studies of the electron orbits in real space reveal the exact shape of the Fermi surfaces in metals (e. g., Harrison and Webb (1960)).

For simple metals, where the effects of the pseudopotential can be described as a simple diffraction, the electron orbits in a uniform magnetic field correspond to motion along circular paths between discontinuous changes in momentum at the diffraction, providing for example lens-shaped orbits when there are two diffractions, which correspond to cross-sections of the nearly-free-electron Fermi surface.

When the fields are very large and the pseudopotentials very weak, the diffraction does not occur and we have magnetic breakdown. This is a special case of the general problem in which two states $|i\rangle$ and $|j\rangle$, coupled by a matrix element H_{ij}, change their relative energies such that they cross ($\varepsilon_j - \varepsilon_i$ goes through zero and changes sign as a function of time). The probability that a particle in state $|i\rangle$ makes a transition to the state $|j\rangle$ during such a *level crossing* is given by

$$P_{ij} = \frac{2\pi H_{ij} H_{ji}}{\hbar\, d(\varepsilon_j - \varepsilon_i)/dt} \tag{14.4}$$

when that probability is small so that we may use perturbation theory. This may be derived from Eq. (7.9) by multiplying and dividing by the time-derivative of the energy difference and integrating over time. For magnetic breakdown in a simple metal, H_{ij} would be the pseudopotential matrix element between two free-electron states and 1 - P_{ij} would be the probability of breakdown occurring.

14.2 Effective Masses and Donor States

In semiconductors, we have seen that there may be small numbers of electrons, concentrated within an energy k_BT of the bottom of the conduction band. In this small energy range we may expand the energy as a function of the wavenumber measured from the conduction-band minimum at $\mathbf{k}_0$. The first derivative of $\varepsilon_\mathbf{k}$ is zero at the minimum and we obtain a quadratic form in the components of $\mathbf{k} - \mathbf{k}_0$. If the conduction-band minimum is at $\mathbf{k} = 0$ as to the right in Fig. 13.9, in the tetrahedral structure (for which x-, y-, and z-axes are equivalent) the result to second-order in $\mathbf{k}$ is

$$\varepsilon_\mathbf{k} = \varepsilon_0 + \frac{1}{2}\frac{\partial^2 \varepsilon_k}{\partial k^2} k^2 + \ldots \equiv \varepsilon_0 + \frac{\hbar^2 k^2}{2m^*} + \ldots \tag{14.5}$$

This defines the *effective mass* m^*, which is adjusted to fit the band curvature at the minimum. Clearly from the discussion in the preceding section we can see that electrons with energies near the minimum of this conduction band have the dynamics of a particle with mass m^*, rather than the true electron mass. The velocity is given by $\mathbf{v} = (1/\hbar)\partial\varepsilon_\mathbf{k}/\partial\mathbf{k} = \hbar\mathbf{k}/m^*$ and the change of its momentum $\mathbf{p} = \hbar\mathbf{k}$ with time will equal any applied force, or $m^* d\mathbf{v}/dt = \mathbf{F}$. Further, it will carry current as $-e\mathbf{v}$ and give rise to a potential in the semiconductor given by $-e/r\varepsilon$, with ε the relative dielectric constant for the semiconductor and r measured from the position of the carrier, or its packet. This allows us to carry over all of our intuition about free particles to electrons moving in such a band.

If the conduction band minimum occurs at some $\mathbf{k}_0$ away from the $\mathbf{k} = 0$, as in silicon where it occurs some $^6/_7$ths of the way to the Zone face in a [100] direction, the variation $\partial^2\varepsilon_\mathbf{k}/\partial k_x^2$ along that axis will be different from the variation $\partial^2\varepsilon_\mathbf{k}/\partial k_y^2 = \partial^2\varepsilon_\mathbf{k}/\partial k_z^2$ transverse to that axis. Then the corresponding *mass tensor* is not isotropic, but the dynamics are correctly given in terms of the corresponding longitudinal and transverse effective masses. In silicon these are of the order of m, and $0.19m$ respectively.

It is an important quantum-mechanical point that these electron packets which behave as a particle of mass m^*, also behave as a wave, just as the center of gravity of any object behaves as a wave. We have found the corresponding Hamiltonian in the preceding section and for the simplest case of an isotropic mass, the corresponding Hamiltonian of $p^2/2m^* + V(\mathbf{r})$ relative to the conduction-band minimum leads to the *effective mass equation*,

$$-\frac{\nabla^2\psi(\mathbf{r},t)}{2m^*} + V(\mathbf{r})\psi(\mathbf{r},t) = \frac{-i\hbar\partial\psi(\mathbf{r},t)}{\partial t}, \tag{14.6}$$

the Schroedinger Equation for the packet. It applies whenever the potentials vary slowly enough with position to be applicable to packets. It may seem strange to have worked through the Schroedinger Equation to obtain the behavior of electrons in an energy band as classical particles and then to reform Schroedinger's Equation for that particle. However, it may not be so different from solving the eigenvalue equation to obtain the electronic structure of the molecule and then applying quantum mechanics to the dynamics of the center of gravity of the resulting molecule.

We may apply this equation to an electron moving in the presence of a charged impurity, such as a germanium atom substituted for a gallium atom in GaAs. The extra proton in the germanium nucleus, relative to that of gallium, produces a potential energy $-e^2/(r\varepsilon)$ and its extra electron is placed in the conduction band since the valence band was full; the atom is thus called a *donor*. We predict that its ground state will be a hydrogenic 1s-state, $\sqrt{\mu^3/\pi}e^{-\mu r}$ as in Eq. (4.5), but with its "Bohr radius"

$$\frac{1}{\mu} = \frac{\hbar^2\varepsilon}{e^2m^*}, \tag{14.7}$$

larger by a factor $\varepsilon m/m^*$ than the 0.529 Å of hydrogen. In GaAs, with $\varepsilon \approx 11$ and $m^*/m \approx 0.07$ this radius is over 80 Å. This would seem large enough for the packet-like description and we expect the state to be approximately correct. Using the same parameters, the ground-state energy is $e^4m^*/(2\varepsilon^2\hbar^2) = 0.008$ eV relative to the conduction-band minimum. The electrons are so weakly bound that at room temperature almost all donors will be ionized, with the conduction electrons free to conduct.

If this were a donor in silicon, such as phosphorus, we would obtain a state which was spread out a similar distance in the two transverse directions, for which the effective mass is small, but the larger longitudinal mass would contract it in that direction, giving a pancake-shaped orbital. (One might construct it with a variational form, analogous to our treatment of the hydrogen orbital in Problem 4.1.) We would obtain similar pancake-shaped states from the conduction-band valleys in the other five [100] directions. There is coupling between the states from different valleys and the real ground state is a combination of orbitals from all valleys, with equal coefficients.

14.3 The Dynamics of Holes

It is well known that an electron in an empty state at the top of a valence band behaves dynamically as a positively charged particle. However, to demonstrate it we must associate a wavenumber **k'** with the particle which is the negative of the wavenumber of the hole, as indicated in Fig. 14.1. We consider the simplest case of a single isotropic band with maximum energy at $\mathbf{k} = 0$ so that

$$\varepsilon_{\mathbf{k}} \approx \varepsilon_V - \frac{\hbar^2 k^2}{2m^*}, \tag{14.8}$$

with ε_v the energy at the valence-band maximum, and m^* a positive number. [The true bands at the valence-band maximum are more complicated because of the three-fold degeneracy seen to the right in Fig. 13.9. Then one band drops rapidly with **k** in a [100] direction (a *light hole* with small m^*), while the other two are called *heavy holes.* The sharp curvature of the light-hole band comes principally from coupling $V_{sp\sigma}$ with the conduction band and may produce a rather isotropic band as in Eq. (14.8), but there may be major anisotropies for the heavy holes, as discussed for example in Harrison (1999), Chapter 6.]

The energy to create a hole, according to Eq. (14.8), *increases* with increasing wavenumber (as the energy of a bubble in water increases with depth) since carrying the corresponding electron to the conduction band, or elsewhere, takes more energy. Thus we may associate positive kinetic energy with the hole. It is a missing electron so the charge to be associated with it is *positive.* We may make a packet of valence-band states, as we

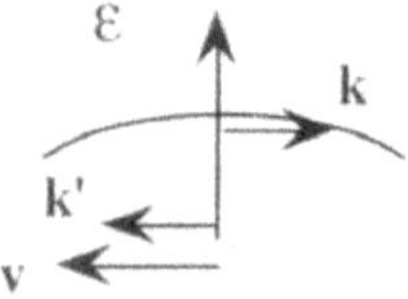

Fig. 14.1. An empty state at wavenumber **k** near the top of the valence band behaves as a positively charged particle of wavenumber $\mathbf{k}' = -\mathbf{k}$

made packets of free-particle states in Section 1.2, and that localized hole will produce an electrostatic potential e/r, with r measured from the packet location.

The packet will have a velocity again given by

$$\mathbf{v} = \frac{1}{\hbar}\frac{\partial \varepsilon_{\mathbf{k}}}{\partial \mathbf{k}} = -\frac{\hbar \mathbf{k}}{m^*} = \frac{\hbar \mathbf{k}'}{m^*}, \tag{14.9}$$

and the associated positive charge will move with that velocity. Here we have introduced the wavenumber **k'** in the direction in which the packet moves, the negative of the wavenumber of the valence-band packet we constructed. The hole then contributes to the current as a positive charge e moving with this velocity. If an electric field **E** is applied, we found that the wavenumber of the packet (made in this case of valence-band states) changes as $\hbar\, d\mathbf{k}/dt = -e\mathbf{E}$ and this remains true. However, when we associate the wavenumber $\mathbf{k}' = -\mathbf{k}$ with the hole that wavenumber changes as

$$\hbar \frac{d\mathbf{k}'}{dt} = +e\mathbf{E} \tag{14.10}$$

and similarly we may see that in changing to **k'** the deflection in a magnetic field is given by

$$\hbar \frac{d\mathbf{k}'}{dt} = +\frac{e}{c}\, \mathbf{v}\times\mathbf{H} \tag{14.11}$$

as for a positively charged particle.

In all regards the hole is behaving as a positively charged particle of wavenumber **k'**. The electron which was removed from the valence band will be attracted to the hole left behind and can form a bound state just as the donated electron could be bound to the donor atom, but the two particles are now similar to the bound positron-electron pair. Such a bound electron and hole are called an *exciton*. When it moves through a crystal it carries no current, but it carries an energy comparable to the band gap. Like a positron-electron pair, the electron and hole can annihilate each other, in this case it is just an electron dropping into the hole state, perhaps emitting a photon. We indicated that the Dirac theory of the electron led also to positrons and the physical interpretation of that theory is very close to this semiconductor band picture.

When a semiconductor is dilated (expanded), the valence-band maximum ordinarily will go up in energy since it corresponds to p-states in a bonding relationship to their neighbors. (The antibonding state, among the conduction bands, will move down in energy.) One may follow the argument we made in deriving Eq. (14.2) to see that if the dilation varies with position $\hbar\mathbf{k}'$ changes with time corresponding to accelerating the hole toward the dilated region, where the valence band is higher . For the same reason an electron tends also to be accelerated toward this dilated region and clearly an exciton will be attracted to that region, where the band gap is smaller and the exciton has lower energy. For the same reason an exciton will tend to expand the lattice in the region where it is, and this expansion can attract any other exciton which is present. Clearly an intricate theory of the dynamics of such a system can be constructed.

Finally, we should note that the simple intuitive behavior of a hole as a positive particle, with positive inertial mass, only applies at the top of the valence band where the bands are curve downward. Empty states near the bottom of the valence bands, where the bands curve upward, do not behave in this intuitive manner.

Chapter 15. Lattice Vibrations

We introduced sound waves in Section 1.8, and obtained their velocity in terms of the bulk modulus and density of the medium. We used the corresponding frequency, equal to the speed of sound times the wavenumber, to discuss the vibrational specific heat of a solid in Section 10.2. We needed there to restrict the total number of modes to the number of degrees of freedom of the vibrating solid, limiting the range of wavenumbers just as the wavenumbers for electron states were limited to a Brillouin Zone in solids. A more complete calculation of the vibrations, analogous to the tight-binding theory of electron states, makes that more natural. We do that here by calculating the vibrations in a chain of atoms, analogous to the electron states in such a chain in Section 6.1. We then generalize the result to three dimensions and introduce the electron potentials which such vibrations give rise to, the electron-phonon interaction, in preparation for a quantum treatment of the vibrations in the following chapter, and of interacting electrons and photons in Chapter 17.

15.1 The Spectrum

We may imagine a chain of atoms as masses M, each coupled to its neighbors by springs, of spring constant κ , as illustrated in Fig. 15.1, analogous to the chain of atoms considered in Section 6.1. We allow a displacement $\mathbf{x}_n$ for the n'th atom along the chain axis, and write force equals

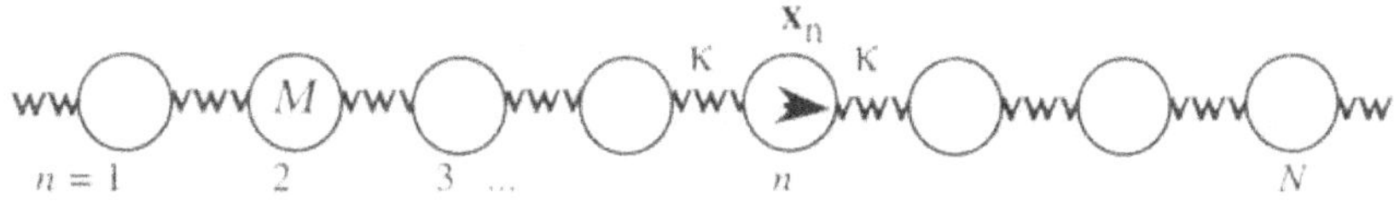

Fig. 15.1. A row of masses, representing atoms, connected by springs of spring constant κ. Vibrations are represented by displacements $\mathbf{x}_n$ of each atom, shown for the n'th atom.

mass times acceleration for each, obtaining the forces from the relative displacements of the neighboring atoms as $\mathbf{F}_n = \kappa(\mathbf{x}_{n+1} - \mathbf{x}_n)$ or $\kappa(\mathbf{x}_{n-1} - \mathbf{x}_n)$. The resulting equations are

$$M\frac{d^2}{dt^2}\mathbf{x}_n = \kappa(\mathbf{x}_{n+1} - \mathbf{x}_n + \mathbf{x}_{n-1} - \mathbf{x}_n)\,. \tag{15.1}$$

We seek *normal modes*, vibrations where every atom moves in phase as $\cos\omega t$ or $\sin\omega t$ or even more conveniently as a complex form, $e^{-i\omega t}$. For each n the left side of Eq. (15.1) becomes $-M\omega^2\mathbf{x}_n$. This is a set of equations closely analogous to Eq. (6-1) for electron states in a chain, and indeed we may again apply periodic boundary conditions on the chain, as if it were bent into a circle, and seek solutions of these equations of the form $\mathbf{x}_n = \mathbf{u}_q e^{iqdn}/\sqrt{N}$ in analogy with the solutions of Eq. (6.1), using notation which will be convenient as we proceed further. In particular, we use q for the wavenumber, rather than the k which we use for electron wavenumbers. We could take the real part of this expression to obtain real displacements, but here and in what follows it is convenient to use the complex expressions as is commonly done for alternating electric currents. With this form we have

$$-M\omega^2\frac{\mathbf{u}_q}{\sqrt{N}}e^{i(qdn - \omega t)} = \kappa(e^{iqd} - 1 + e^{-iqd} - 1)\frac{\mathbf{u}_q}{\sqrt{N}}e^{i(qdn - \omega t)}\,. \tag{15.2}$$

The factors $\mathbf{u}_q e^{i(qdn - \omega t)}/\sqrt{N}$ cancel so all N equations are identical and satisfied if

$$\omega^2 = \frac{\kappa}{M}(2\cos qd - 2) = \frac{4\kappa}{M}\sin^2\frac{qd}{2}\,. \tag{15.3}$$

The frequency depends upon the wavenumber q we have chosen, and $\omega(q)$ is called a *dispersion curve* and is illustrated in Fig. 15.2. With periodic boundary conditions q is limited such that qNd equals an integral multiple of 2π. The points shown are for N equal to twenty. Wavenumbers satisfying this condition, but lying outside the Brillouin Zone shown, produce identical displacements to those for some mode inside, so the points represent all of the twenty normal modes of the system. Each mode corresponds to an independent harmonic oscillator with displacement u_q and frequency ω_q. The mode at $q = \pi/d$ has alternate atoms displaced in opposite directions and is the highest-frequency mode. In the Debye approximation discussed in Section 10.2 this dispersion curve was approximated by straight

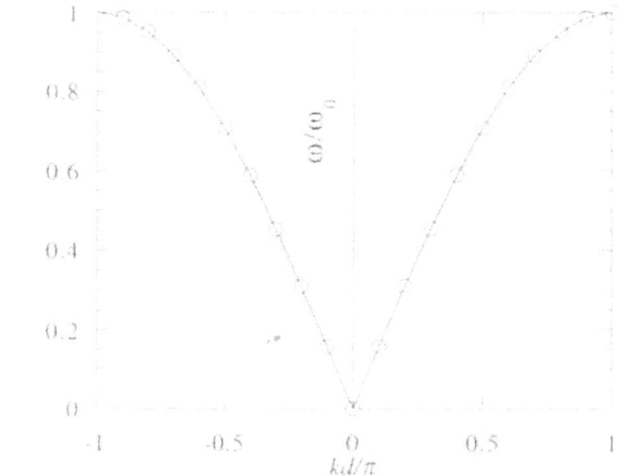

Fig. 15.2. The vibration spectrum for the one-dimensional chain of Fig. 15.1. ω_0 is the peak frequency. The points represent allowed wavenumbers for a chain of 20 atoms.

lines tangent to these curves at $q = 0$, but cut off at the same $q = \pm\pi/d$. In both cases the thermal energy approaches k_BT per mode at high temperatures.

The generalization of this calculation to three dimensions is very direct. For this problem the simple-cubic solid which we have used to illustrate solids is not the simplest case since with nearest-neighbor, central-force interactions, the structure is not stable against a shear of the lattice, so we consider the face-centered-cubic lattice (of copper and aluminum, for example) which we showed in Fig. 13.3, and redraw in Fig. 15.3. It contains a simple-cubic array of atoms, but in addition has one atom at the center of every face of every cube. We see that each atom has twelve nearest-neighbors at equal distance. The many triangles of three nearest neighbors stabilize the structure under any distortion.

The displacement of an atom initially at position $\mathbf{r}_j$ is written $\delta\mathbf{r}_j$ and the form of the displacements is written as

$$\delta\mathbf{r}_j = \frac{\mathbf{u_q}}{\sqrt{N}}\, e^{i\mathbf{q}\cdot\mathbf{r}_j} \qquad (15.4)$$

as in the one-dimensional case. We may carry out the calculation for a longitudinal mode with $\mathbf{q}$ along a cube edge, as shown in Fig. 15.3. We focus upon one neighbor to the upper right of the central atom, letting the origin of our coordinates lie at the central atom. The relative displacement of the neighbor atom, $\delta\mathbf{r}_{j+1} - \delta\mathbf{r}_j = \mathbf{u}_\mathbf{q}(e^{i\mathbf{q}\cdot\mathbf{r}_{j+1}} - e^{i\mathbf{q}\cdot\mathbf{r}_j})/\sqrt{N} = \mathbf{u}_\mathbf{q}e^{i\mathbf{q}\cdot\mathbf{r}_j}(e^{iqa/2} - 1)/\sqrt{N}$. Only the component of that relative displacement along the internuclear axis stretches the spring, so we multiply by $\cos\theta$ to obtain that component and multiply by the spring constant κ to obtain the force along the spring axis. Only the component along the x-direction survives when we add forces from all neighbors, and the component of the force is obtained by multiplying again by $\cos\theta$. We add the force for each of the other three neighbors to the right, with each contribution the same, and the four to the left. (The four in the same plane of constant x have no relative motion and give no force.) The resulting force is $8\kappa\cos^2\theta\,\mathbf{u}_\mathbf{q}e^{i\mathbf{q}\cdot\mathbf{r}_j}(\cos(qa/2) - 1)$. We set this equal to the mass of the central atom times its acceleration, $-M\omega_q^2\mathbf{u}_\mathbf{q}e^{i\mathbf{q}\cdot\mathbf{r}_j}$. The $\mathbf{u}_\mathbf{q}e^{i\mathbf{q}\cdot\mathbf{r}_j}$ cancels, so the result applies to every atom, and we obtain

$$\omega_q^2 = \frac{8\kappa}{M}\sin^2\frac{qa}{4} \quad . \tag{15.5}$$

A plot of this looks exactly like Fig. 15.2, but in this case the Brillouin Zone face comes at $2\pi/a$.

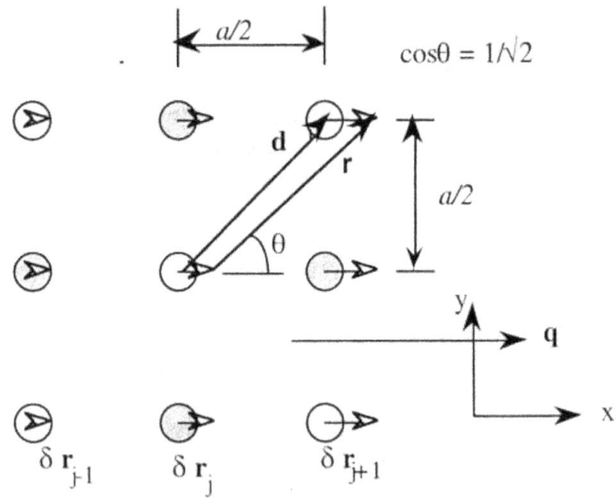

Fig. 15.3. The face-centered-cubic lattice is based upon a simple-cubic lattice with cubes of edge a , but with additional atoms at the center of each cube face. Shaded atoms lie $\pm a/2$ above or below the plane of the figure. Here a longitudinal vibrational mode has wavenumber $\mathbf{q}$ along the x-axis, parallel to a horizontal cube edge, so displacements are also along the x-direction. The force, along the x-direction, on the central atom is calculated in terms of radial springs coupling the central atom to its twelve nearest neighbors.

We redo this calculation in Problem 15.1 for a transverse mode, the same direction of $\mathbf{q}$ as in Fig. 15.3, but with the displacements in the y-direction. The calculation of the y-component of the force from the neighbor to the upper right is essentially the same, but now all four neighbors to the right do not contribute the same and different frequencies are obtained. Another transverse mode, with displacements in the z-direction, gives identical frequencies to the transverse mode with displacements in the y-direction. The calculation is formally the same for *any* wavenumber in the Brillouin Zone, but when $\mathbf{q}$ does not lie along such a symmetry direction, we do not know initially what the three directions of the three $\mathbf{u_q}$ are; we must write forces in all three directions, solve three simultaneous equations, and obtain three modes which will turn out to have $\mathbf{u_q}$ perpendicular to each other. One may be approximately longitudinal and two approximately transverse. Since there are N wavenumbers in the Brillouin Zone allowed by periodic boundary conditions, we obtain $3N$ frequencies as we expect.

In Problem 15.2 we write an expression for the total vibrational energy for a face-centered cubic crystal in thermal equilibrium in terms of such frequencies and obtain the specific heat per atom at high temperatures.

15.2 The Classical-Vibration Hamiltonian

We wrote displacements in Eq. (15.4) for a single mode in terms of a complex amplitude, with the idea that we could take the real part to obtain real displacements. It will be useful to retain such complex amplitudes and simply change variables from the $3N$ displacement components of the N atoms to $3N$ complex amplitudes, called *normal coordinates*,

$$\delta\mathbf{r}_j = \Sigma_{\mathbf{q},\lambda} \frac{\mathbf{u_q}^\lambda}{\sqrt{N}}\, e^{i\mathbf{q}\cdot\mathbf{r}_j} \tag{15.6}$$

where now the index λ has three values, representing the three modes at each wavenumber. This has introduced two independent parameters for each mode, the real and the imaginary part of $\mathbf{u_q}^\lambda$, but we now must require that $\mathbf{u_{-q}}^\lambda = \mathbf{u_q}^{\lambda*}$ (the complex conjugate) in order that the displacements be real so that there are still only $3N$ independent parameters, the mathematics will take care of any difficulties, and we may think of $\mathbf{u_q}^\lambda$ as the amplitude of the mode propagating in the direction of $\mathbf{q}$. It will be a little simpler, and easier to follow, if we proceed with the one-dimensional chain of the preceding section, and then write the result for three dimensions.

Then the displacement of the j'th atom along the line of the chain is

$$\delta x_j = \Sigma_q \frac{u_q}{\sqrt{N}} e^{iqdj} \ . \tag{15.7}$$

The total kinetic energy T, with each atom having mass M, can then be written as

$$T = {}^1\!/_2 M \Sigma_j \dot{\delta x_j}^2 = \frac{M}{2N} \Sigma_{j,q',q}\, \dot{u}_q\, e^{iqdj} \dot{u}_{q'}\, e^{iq'dj} \tag{15.8}$$

with the sums over q and q' running over the Brillouin Zone. However, the sum over j is performed first, and gives zero unless $q' = -q$, seen by proceeding exactly as for the sum in Eq. (13.11), in which case it gives N. Thus Eq. (15.8) becomes

$$T = \frac{M}{2} \Sigma_q \dot{u}_q\, \dot{u}_{-q} \tag{15.9}$$

This is the only term in the energy depending upon the $\dot{u}_q$ and therefore the only term contributing to the derivative of the Lagrangian in Eq. (3.3). Thus the momentum conjugate to the normal coordinate u_q is

$$P_q = \frac{\partial T}{\partial \dot{u}_q} = M\, \dot{u}_{-q}. \tag{15.10}$$

Note that it is the $\dot{u}_{-q}$ which enters. A second contribution has come from the term in the sum for wavenumber equal to $-q$.

The corresponding calculation of the potential energy gives a sum over ${}^1\!/_2 u_q u_{-q}$ times the effective spring constant for the mode, obtained as $8\kappa \sin^2(qa/4)$ from Eq. (15.5), which can also be written $M\omega_q{}^2$. Thus rewriting the kinetic energy in terms of the canonical momentum and adding the potential energy we obtain a Hamiltonian for the vibrational modes of

$$H = \Sigma_q \left(\frac{P_q P_{-q}}{2M} + \frac{M\omega_q{}^2 u_q u_{-q}}{2} \right). \tag{15.11}$$

This can be directly generalized to the three-dimensional case as we generalized the calculation of the spectrum. The result is exactly what we would anticipate,

$$H = \Sigma_{\mathbf{q},\lambda}\left(\frac{\mathbf{P_q P_{-q}}}{2M} + \frac{M\omega_q{}^2\mathbf{u_q u_{-q}}}{2}\right), \tag{15.12}$$

with $\mathbf{P_q} = M\,\dot{\mathbf{u}}_{\mathbf{-q}}$. The vector notation is schematic; if we have obtained the three vibrational modes at any wavenumber, we associate an amplitude $\mathbf{u_q}$ with each, which is really a scalar quantity, and each has a conjugate momentum $\mathbf{P_q}$ which is really a scalar quantity. However we shall need the direction of the displacements in constructing the electron-phonon interaction so we keep the bold-face notation for vectors. The sum over λ is a reminder that all three of the modes at each wavenumber need be added. Using this with the classical Hamilton's Equations, Eq. (3.5), gives the dynamics of the vibrations. We could also replace $\mathbf{P_q}$ by $(\hbar/i)\partial/\partial\mathbf{u_q}$ and construct a Schroedinger Equation, or an energy eigenvalue equation. We shall not do that, but in Chapter 16 shall use the properties of the momentum operator to obtain all the results we shall need.

15.3 The Electron-Phonon Interaction

There is one more classical derivation we need to perform before proceeding to the quantum-mechanical treatment of lattice vibrations (and electromagnetic waves). The presence of a lattice vibration (or a light wave) introduces changes in the Hamiltonian for electrons, and therefore coupling between electronic states. For vibrations in solids this coupling is called the *electron-phonon interaction* , though it is ultimately of classical origin. It is simplest to obtain it from pseudopotential theory, though one can also derive it for tight-binding theory (e. g., Harrison (1999)). Here we use pseudopotentials.

There are matrix elements of the pseudopotential between free-electron states only if

$$S(\mathbf{q}) = {}^1/_N\Sigma_j\, e^{-i\mathbf{q}\cdot\mathbf{r}_j} \tag{15.13}$$

differs from zero, which in the perfect lattice occurs only at lattice wavenumbers, $\mathbf{q}_0$, as we saw in Section 13.1. Now we displace each atom from its position in the perfect lattice according to Eq. (15.6). We may add these $\delta\mathbf{r}_j$ to the $\mathbf{r}_j$ of Eq. (15.13), which represent positions in the undistorted lattice, to find the new couplings which the $\delta\mathbf{r}_j$ introduce. For a vibrational mode with wavenumber which we write $\mathbf{Q}$, in the Brillouin Zone, we obtain a term in $\delta\mathbf{r}_j$ given by

$$\delta\mathbf{r}_j = \mathbf{u_Q}\, e^{i\mathbf{Q}\cdot\mathbf{r}_j}/\sqrt{N} \tag{15.14}$$

and of course the complex conjugate term, which we shall denote by c.c., must be present also. We have used a capital **Q** for the mode to avoid confusion with the **q** in Eq. (15.13). If we add this $\delta\mathbf{r}_j$ to the $\mathbf{r}_j$ in Eq. (15.13), and expand the exponential for small $\mathbf{u}_\mathbf{Q}$, we obtain

$$S(\mathbf{q}) = \frac{1}{N}\sum_j e^{-i\mathbf{q}\cdot(\mathbf{r}_j+\delta\mathbf{r}_j)} = \frac{1}{N}\sum_j e^{-i\mathbf{q}\cdot\mathbf{r}_j}\left(1 - \frac{i\mathbf{q}\cdot\mathbf{u}_\mathbf{Q}}{\sqrt{N}}\, e^{i\mathbf{Q}\cdot\mathbf{r}_j} + \text{c.c.} + \ldots\right). \quad (15.15)$$

The first term in the final form gives the result for the perfect lattice, nonzero only at lattice wavenumbers $\mathbf{q}_0$, the dots in Fig. 15.4. The second term contains a similar sum over positions in the perfect lattice, and is nonzero only if $\mathbf{q} - \mathbf{Q}$ is such a lattice wavenumber. The complex conjugate term is similarly nonzero if $\mathbf{q} + \mathbf{Q}$ is a lattice wavenumber. Thus these two terms give nonzero structure factor at "satellites" to the lattice wavenumbers, indicated by x's in Fig. 15.4. This new structure factor at $\mathbf{q} = \mathbf{q}_0 + \mathbf{Q}$ is $-i\mathbf{q}\cdot\mathbf{u}_\mathbf{Q}/\sqrt{N}$. It is multiplied by the pseudopotential form factor w_q for this system,

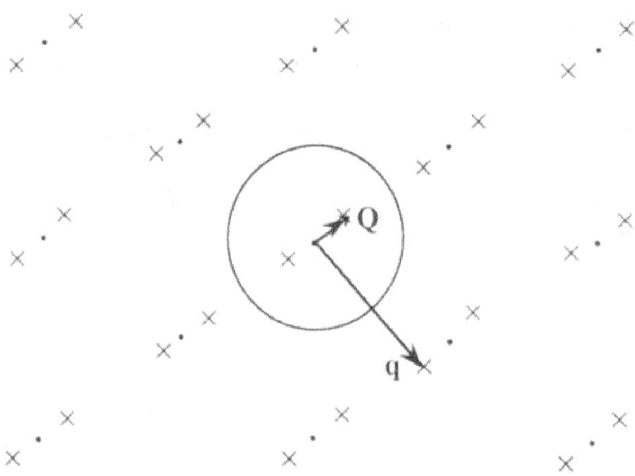

Fig. 15.4. The solid dots represent lattice wavenumbers $\mathbf{q}_0$ for which the structure factors, and therefore matrix elements of the pseudopotential, are nonzero for the undistorted crystal. In the presence of a lattice vibration of wavenumber **Q**, there arise nonzero structure factors at satellite points, $\mathbf{q} = \mathbf{q}_0 \pm \mathbf{Q}$, to each lattice wavenumber (including $\mathbf{q}_0 = 0$) indicated by x's. The circle might represent the Fermi sphere in wavenumber space for a metal. Then scattering can occur between states on the sphere (if we neglect $\hbar\omega$ in comparison to the Fermi energy ε_F) which differ in wavenumber by any such **q**.

$$<\mathbf{k}+\mathbf{q}|W|\mathbf{k}> = S(\mathbf{q})w_q = \frac{-i\mathbf{q}\cdot\mathbf{u_Q}}{\sqrt{N}}\,w_q, \qquad (15.16)$$

to obtain the matrix element between any state of wavenumber $\mathbf{k}$ and the state of wavenumber $\mathbf{k}+\mathbf{q}$. For $\mathbf{q} = \mathbf{q}_0 - \mathbf{Q}$, the complex conjugate of this expression gives the corresponding matrix element.

These matrix elements are exactly what is called the electron-phonon interaction. Regarding these lattice distortions as static (we go beyond this approximation in Chapter 16), the new structure factor can produce scattering between any two states on a surface of constant energy, such as the metal Fermi surface indicated in Fig. 15.4, which differ by the corresponding $\mathbf{q}$. We may distinguish *normal processes* as those that arise from the satellites to $\mathbf{q}_0 = 0$, for which $\mathbf{q} = \pm\mathbf{Q}$. Then the change in electron wavenumber is equal to the phonon wavenumber, corresponding to conservation of momentum with, as we shall see, the absorption or emission of one quantum of vibrational energy. We note from Eq. (15.16) that if the modes are purely longitudinal ($\mathbf{u_Q}$ parallel to $\mathbf{Q}$) or transverse ($\mathbf{u_Q}$ perpendicular to $\mathbf{Q}$), only longitudinal modes have nonzero structure factors. This is the case we shall treat in our analysis. We see, however, there are also *Umklapp processes* from satellites to nonzero $\mathbf{q}_0$. Then the change in wavenumber of the electron differs from that of the phonon by a lattice wavenumber as if the electron diffracted from the perfect lattice at the same time that it emitted or absorbed a phonon. Both longitudinal and transverse modes give Umklapp processes. They are important in solids, but in Chapter 16 we treat only the simpler case of normal processes.

If we had kept terms of second order in $\mathbf{u_Q}$ in Eq. (15.15), we would have found some terms at $\mathbf{q} = \mathbf{q}_0$, which lead to second-order terms in $S^*(\mathbf{q}_0)S(\mathbf{q}_0)$. They reduce the strength of the diffraction at each lattice wavenumber by an amount corresponding to the diffraction added, $\Sigma_{\mathbf{q}}S^*(\mathbf{q})S(\mathbf{q})$, by the satellites. The reduction factor at each $\mathbf{q}_0$ is called the *Debye-Waller factor* (discussed for example in Harrison (1970), p. 426). We will not need it here. The remarkable fact is that the thermal vibrations of the lattice do not *blur* the diffraction spot, only weaken it and add a cloud of satellite diffractions.

VI. Quantum Optics

We discussed quantum transitions between electron states due to a classical light field in Section 9.4. Similarly we could calculate electron scattering by classical vibrations of a crystal lattice using the matrix elements in terms of the vibrational amplitudes which we derived in the preceding section. However, it is often important to treat both the electrons and the fields quantum-mechanically. When the fields, or vibration amplitudes, are very large the quantum effects are not so important but they become essential when the fields are small. The treatment of electromagnetic waves and vibrational modes quantum-mechanically is called *field theory* and we now need to introduce the operators of field theory, both for these waves and modes, and for electrons. This formulation is often called *second quantization.* Their application in quantum optics is perhaps the most important for us, and the focus for this section of the book. However, it will be best to introduce the operators for electrons first, and to make the first application to phonons, quantized lattice vibrations.

In principle we already have the basic results we need. We have treated the harmonic oscillator quantum-mechanically, and have indicated that all of the findings apply to vibrational modes and to optical excitations. However, in field theory we use only a part of what we developed, the fact that the momentum operator $(\hbar/i)\partial/\partial x$ does not *commute* with the coordinate x (changing the order of the product of two "commuting operators" does not change the product). In this case, $(\hbar/i)\partial/\partial x\ (x\psi) = (\hbar/i)\psi + x(\hbar/i)\partial\psi/\partial x$ so interchanging $(\hbar/i)\partial/\partial x$ and x leaves a remainder, $\hbar/i$. This turns out to be the feature essential to defining field-theoretical annihilation and creation operators for excitations and particles. This is closely related to the point we made in Section 1.1 that it is possible to develop quantum theory without the waves which we have used throughout. Since we can obtain all of the information about the eigenstates of a harmonic oscillator, and its coupling to external fields, from the corresponding commutation relation, we did not really need the waves. Heisenberg represented the observables by matrices, rather than operators on waves. They had the same commutation relations so they led to the same results. For most contemporary physicists and engineers, the procedure with waves is much more comfortable.

Chapter 16. Operators

We begin by defining annihilation and creation operators for electrons, which is possible without introducing the field theoretical basis. This makes clear what our goals and approach should be for the harmonic oscillator, phonons, and photons. Further, we shall need both for our study of quantum optics. For electrons we simply give these as definitions for describing many-electron states, though they are derivable from a field theory in much the same way we shall derive them for oscillators.

16.1 Annihilation and Creation Operators for Electrons

We saw in Section 10.5 that when more than one electron is present, the electron state must be antisymmetric with respect to interchange of the electrons. This could be accomplished for noninteracting electrons by writing a Slater Determinant for the many electron state as we did in Eq. (10.22),

$$\Psi(\{\mathbf{r}_i\}) = \frac{1}{\sqrt{N!}}\,\mathrm{Det}\begin{pmatrix} \psi_1(\mathbf{r}_1)\,\psi_1(\mathbf{r}_2)\,\psi_1(\mathbf{r}_3) & \dots & \\ \psi_2(\mathbf{r}_1)\,\psi_2(\mathbf{r}_2)\,\psi_2(\mathbf{r}_3) & \dots & \\ \psi_3(\mathbf{r}_1) \quad \dots \quad \dots & \dots & \\ \dots & & \psi_N(\mathbf{r}_N) \end{pmatrix}. \tag{16.1}$$

The index i specifies the spin as well as the orbital for each electron and we use the capital Ψ for the many-electron state. This provides a solution of the many-electron Schroedinger Equation if the electron-electron interactions, $e^2/|\mathbf{r}_i - \mathbf{r}_j|$, are neglected. This is a very cumbersome form, and really to specify the state we only need to designate which states ψ_1, ψ_2, ...ψ_N are occupied, and the order in which we have placed them. We may introduce *creation operators* as an intuitive shortcut for specifying the many-electron state. We define the vacuum state $|0>$ and the creation operator $c_j{}^\dagger$ which places an electron in the j'th state. (The "†" represents a complex conjugate, and the operator is often called " c-j-dagger".) Then the state in Eq. (16.1) is written

$$|\Psi(\{\mathbf{r}_i\})> = c_N{}^\dagger c_{N-1}{}^\dagger ... c_2{}^\dagger c_1{}^\dagger|0>. \qquad (16.2)$$

We imagine successively creating electrons in the states ψ_1, ψ_2, etc. If we interchange two states, corresponding to interchanging $\mathbf{r}_i$ and $\mathbf{r}_j$ in Eq. (16.1), this interchanges two columns in the determinant and changes the sign of the state. This is written in terms of creation operators as the *commutation relation*,

$$c_i{}^\dagger c_j{}^\dagger + c_j{}^\dagger c_i{}^\dagger = 0. \qquad (16.3)$$

Eq. (16.3) applies to the interchange of any pair, holding the others fixed, but we shall make only nearest-neighbor interchanges.

The Pauli exclusion principle follows immediately from this commutation relation, since if any index j appears twice in the state Eq. (16.2), we could commute neighboring operators until the two were neighbors and from Eq. (16.3) $c_j{}^\dagger c_j{}^\dagger = 0$. As throughout our study, a state equal to zero is no state.

We may similarly define the complex conjugate of the many-electron state,

$$<\Psi(\{\mathbf{r}_i\})| = <0|c_1 c_2 ... c_N. \qquad (16.4)$$

The complex conjugate operators c_j clearly have the same commutation relations, $c_i c_j + c_j c_i = 0$, as the creation operators.

We wish to have normalized states, so $<0|0> = 1$ and $<0|c_j c_j{}^\dagger|0> = 1$ and we wish to have orthogonality of different states, e. g., $<0|c_j|0> = 0$. This requires commutation relations between the two kinds of operators and the choice which accomplishes the orthonormality of the many-electron states is

$$c_i c_j{}^\dagger + c_j{}^\dagger c_i = \delta_{ij}, \qquad (16.5)$$

if we also take $c_j|0> = 0$, and $<0|c_j{}^\dagger = 0$.

This completes the definition of the operators and shows that we may think of the c_j as the *annihilation operator* for the state ψ_j. For example c_j

operating on a state $c_j^\dagger c_i^\dagger|0\rangle$ containing electrons in states i and j gives, after using Eq. (16.5), $(1 - c_j^\dagger c_j)c_i^\dagger|0\rangle$. We again use Eq. (16.5) for the second term to obtain $c_j^\dagger c_i^\dagger c_j|0\rangle$ which is zero because of the $c_j|0\rangle = 0$ condition. The first term is just the initial state with the electron in the j'th state annihilated. In these terms we would say that $c_j|0\rangle$ equals zero because no state can be obtained by annihilating an electron from the vacuum.

We may confirm that each many-electron state is normalized by using Eq. (16.5) on the central pair of operators,

$$\begin{aligned}&\langle 0|c_1c_2...c_{N-1}c_Nc_N^\dagger c_{N-1}^\dagger ...\, c_2^\dagger c_1^\dagger|0\rangle \\ &\quad = \langle 0|c_1c_2...c_{N-1}(1-c_N^\dagger c_N)c_{N-1}^\dagger ...\, c_2^\dagger c_1^\dagger|0\rangle.\end{aligned} \tag{16.6}$$

Then the second term is shown to be zero by commuting the c_N to the right till we obtain $c_N|0\rangle = 0$. The first term equals the $\langle\Psi|\Psi\rangle$ for the state with the last electron removed. We may successively eliminate each electron the same way until we obtain $\langle 0|0\rangle$ which is one, proving the normalization. In Problem 16.1 we see that the same analysis shows that any two many-body states with different one-electron states occupied are orthogonal to each other.

This notation with annihilation and creation operators can also be used to express the operators for any observables for the system. We note first that the number operator is

$$n = \Sigma_j\, c_j^\dagger c_j. \tag{16.7}$$

We show this by operating with it on a state such as Eq. (16.2). For each term in the sum over j we commute the c_j and then the $c_j^\dagger$ successively past each $c_k^\dagger$ in the state. If we never come to a state with $k = j$, we come finally to $c_j^\dagger c_j|0\rangle = 0$ and no contribution to $n|\Psi\rangle$. If we do come to a state $k = j$, we obtain an extra term in commuting the c_j, which gives $c_j^\dagger(1 - c_j^\dagger c_j)$. This extra term with the 1 is just the starting state $|\Psi\rangle$, while the second vanishes as before. Thus for every state included in the list 1, 2, 3, ...N for the state $|\Psi\rangle$ we obtain a term. The N terms give us $n|\Psi\rangle = N|\Psi\rangle$, which defines n as the number operator.

In the same way, if the states ψ_j are one-electron eigenstates of the Hamiltonian with energy ε_j, the operator

$$H = \Sigma_j\, \varepsilon_j\, c_j^\dagger c_j \tag{16.8}$$

operating on the state $|\Psi>$ gives a sum of ε_j over the *occupied* states and can be considered the one-electron Hamiltonian. For free-electron states Eq. (16.8), with $\varepsilon_k = \hbar^2k^2/(2m)$, is the kinetic-energy operator. If we add a potential $V(\mathbf{r})$ seen by all of the electrons, it is represented by an operator

$$V(\mathbf{r}) \rightarrow \sum_{\mathbf{k},\mathbf{q}} <\mathbf{k+q}|V(\mathbf{r})|\mathbf{k}> c_{\mathbf{k+q}}^\dagger c_{\mathbf{k}} \qquad (16.9)$$

which couples each one-electron state $|\mathbf{k}>$ to other plane-wave states $|\mathbf{k+q}>$ (as in Eq. (13.5)). Of course $<\mathbf{k+q}|V(\mathbf{r})|\mathbf{k}> = (1/\Omega)\int e^{-i\mathbf{q}\cdot\mathbf{r}}V(\mathbf{r})\, d^3r$. The sum in Eq. (16.9) will include a sum over spin, but the matrix element $<\mathbf{k+q}|V(\mathbf{r})|\mathbf{k}>$ is nonzero only if both states have the same spin.

Finally, the electron-electron interaction can be written

$$\frac{e^2e^{-\kappa|\mathbf{r}-\mathbf{r}'|}}{|\mathbf{r}-\mathbf{r}'|} \rightarrow \frac{1}{2}\sum_{\mathbf{k},\mathbf{k}',\mathbf{q}} \frac{4\pi e^2}{(q^2+\kappa^2)\Omega} c_{\mathbf{k+q}}^\dagger c_{\mathbf{k}'-\mathbf{q}}^\dagger c_{\mathbf{k}'} c_{\mathbf{k}} \,. \qquad (16.10)$$

Again, $|\mathbf{k}>$ and $|\mathbf{k+q}>$ must correspond to the same spin and $|\mathbf{k}'>$ and $|\mathbf{k}'\mathbf{-q}>$ must correspond to the same spin. The matrix element $4\pi e^2/((q^2+\kappa^2)\Omega)$ was evaluated as in Eq. (13.9). We have included a convergence factor, $e^{-\kappa|\mathbf{r}-\mathbf{r}'|}$, as we did in Eq. (13.9). As we noted then, we could take $\kappa = 0$ at the end but we shall see in Section 20.2 that such a factor is a suitable approximation to the effect of the screening of the interaction between any particular electron pair through the motion of the other electrons present. This last operator, with or without the κ, is central to the study of the effects of electron-electron interaction on the properties of solids. Field theory provides systematic ways to approximate the effects of this term which contains four operators, though that is not our goal here

We may illustrate the subject by evaluating the energy for a pair of free electrons, with opposite spin, in the $\mathbf{k} = 0$ state, to second order in the electron-electron interaction. We write our starting, zero-order, state $|\Psi_0> = c_{0\downarrow}^\dagger c_{0\uparrow}^\dagger|0>$ and the perturbation as H_1 given in Eq. (16.10). We proceed using the perturbation theory of Section 5.4 to obtain the first- and second-order shifts in the energy,

$$\delta E = <\Psi_0|H_1|\Psi_0> + \sum_n \frac{<\Psi_0|H_1|\Psi_n><\Psi_n|H_1|\Psi_0>}{E_0 - E_n}, \qquad (16.11)$$

where $|\Psi_0>$ is the two-electron ground state with energy $E_0 = 0$ without the electron-electron interaction H_1. The $|\Psi_n>$ are excited two-electron states which "can be reached" by that interaction. We see what this means by operating upon $|\Psi_0\rangle$ with the H_1 given in Eq. (16.10). The final $c_{\mathbf{k}}$ operating on $c_{0\downarrow}^{\dagger}c_{0\uparrow}^{\dagger}|0\rangle$ will give zero unless **k** is one of the **k**= 0 states, say spin up, and then the state $|\mathbf{k+q}\rangle$ will have wavenumber **q** and the same spin up. Then $c_{\mathbf{k}'}$ will give zero except for **k'** = 0 corresponding to spin down. In this case, with **k** as spin up, $c_{\mathbf{k+q}}^{\dagger}c_{\mathbf{k'-q}}^{\dagger}c_{\mathbf{k'}}c_{\mathbf{k}}$ operating on $c_{0\downarrow}^{\dagger}c_{0\uparrow}^{\dagger}|0\rangle$ gives $-c_{\mathbf{q}\uparrow}^{\dagger}c_{-\mathbf{q}\downarrow}^{\dagger}|0\rangle$ with the matrix element $4\pi e^2/((q^2+\kappa^2)\Omega)$, where we have kept track of the sign changes in the successive use of the commutation relation, Eq. (16.5). The other term, with **k** as spin down, leads to $c_{\mathbf{q}\downarrow}^{\dagger}c_{-\mathbf{q}\uparrow}^{\dagger}|0\rangle$ with the same matrix element, but this is the same state but with the wavenumber **q** reversed. Thus we may combine them in the sum, as $c_{-\mathbf{q}\downarrow}^{\dagger}c_{\mathbf{q}\uparrow}^{\dagger}|0\rangle$ canceling the factor of $^1/_2$ in front, so that

$$
\begin{aligned}
H_1|\Psi_0\rangle &= \frac{1}{2}\sum_{\mathbf{k},\mathbf{k}',\mathbf{q}} \frac{4\pi e^2}{(q^2+\kappa^2)\Omega} c_{\mathbf{k+q}}^{\dagger}c_{\mathbf{k'-q}}^{\dagger}c_{\mathbf{k'}}c_{\mathbf{k}}c_{0\downarrow}^{\dagger}c_{0\uparrow}^{\dagger}|0\rangle \\
&= \sum_{\mathbf{q}} \frac{4\pi e^2}{(q^2+\kappa^2)\Omega}\, c_{-\mathbf{q}\downarrow}^{\dagger}c_{\mathbf{q}\uparrow}^{\dagger}|0\rangle .
\end{aligned}
\tag{16.12}
$$

The states appearing in the final sum are those which can be reached by the electron-electron interaction.

This is a very important conceptual point, and one which we shall use frequently. The interaction term in the Hamiltonian annihilates electrons in some states and creates them in others to produce intermediate states in the perturbation theory. The various combinations of interactions which enter in higher-order perturbation theory are represented by Feynman diagrams which give a visual representation of the terms which are being included, and keep track of the matrix elements which are to be used. We will not need them here. For the first-order term, the term $\langle\Psi_0|H_1|\Psi_0\rangle$ in Eq. (16.11), we note all of the terms in $H_1|\Psi_0\rangle$ are orthogonal to $|\Psi_0\rangle$, and thus give zero, except the term $\mathbf{q} = 0$. Thus, only that term contributes and gives $\langle\Psi_0|H_1|\Psi_0\rangle = 4\pi e^2/(\Omega\kappa^2)$. (It was good that we kept the κ.)

For the second-order term in Eq. (16.11) we see that each term in $H_1|\Psi_0\rangle$ produces a different intermediate state $c_{-\mathbf{q}\downarrow}^{\dagger}c_{\mathbf{q}\uparrow}^{\dagger}|0\rangle$, one with the spin-up electron excited into the state with wavenumber **q** and the spin-down electron excited into the state with wavenumber -**q**, conserving momentum. The energy of this intermediate state is $E_n = 2\hbar^2q^2/(2m)$. In completing the evaluation we do not even need to include the $|\Psi_n\rangle\langle\Psi_n|$ which appears in the expression except to note in passing that the energy denominator is $-2\hbar^2q^2/(2m)$. The $\Sigma_{\mathbf{q}}$ provides the sum over intermediate states and we simply operate again with H_1 . In this case, the only contributing terms in the new sum over **q'** will be those which "return" the system to the ground state $|\Psi_0\rangle$, the term with $\mathbf{q}' = -\mathbf{q}$, and they contribute another factor of $4\pi e^2/((q^2+\kappa^2)\Omega)$. We may complete the calculation of the energy shift by evaluating the sum as we have done before,

$$\delta E = \frac{4\pi e^2}{\Omega\kappa^2} - \Sigma_{\mathbf{q}}\left(\frac{4\pi e^2}{(q^2+\kappa^2)\Omega}\right)^2 \frac{1}{\dfrac{2\hbar^2 q^2}{2m}}$$

$$= \frac{4\pi e^2}{\Omega\kappa^2} - \frac{\Omega}{(2\pi)^3}\int 4\pi q^2 dq\left(\frac{4\pi e^2}{(q^2+\kappa^2)\Omega}\right)^2 \frac{m}{\hbar^2 q^2} \qquad (16.13)$$

$$= \frac{4\pi e^2}{\Omega\kappa^2} - \frac{8me^4}{\hbar^2\Omega}\int dq\left(\frac{1}{q^2+\kappa^2}\right)^2 = \frac{4\pi e^2}{\Omega\kappa^2} - \frac{2\pi m e^4}{\hbar^2\kappa^3\Omega}.$$

The first term represents the interaction between these two electrons, each spread out over the volume Ω. The second represents a reduction in that energy as the electrons modify their states to avoid each other. This is exactly what is called *correlation energy* for electron gases, the shift in energy due to the correlated motion of the electrons. In the chemical literature, when it is calculated for atoms or molecules, it is called configuration interaction. Eq. (16.13) would give a good value for the shift for the electron pair if κ were large enough that the second term was small compared to the first, and then presumably the terms of higher order were still smaller.

This same formulation could have been used to calculate the van-der-Waals interaction which we calculated for two coupled oscillators in Section 12.2 and for coupled atoms in Problem 12.2. The coupling $e^2/|\mathbf{r} - \mathbf{r}'|$ couples electrons on different atoms, taking them from the ground state to a state with one electron on each atom in an excited state, as seen in Problem 12.2. The resulting shift in energy is proportional to e^4 from the squared matrix element and to the reciprocal of the excitation energy as we found there.

The same formulation can be used directly with Fermi's Golden Rule to calculate scattering of electrons by each other. The treatment of matrix elements as in Eq. (16.12) is exactly the same, but instead of the energy differences appearing in the denominator they appear in a delta function for the total energy before and after. The calculations are straightforward, and just as simple as those which led to Eq. (16.13)

16.2 Stepping Operators

We return to the simple harmonic oscillator, which we treated in Section 2.5, but now seek to understand it in a form analogous to what we have just used for many-electron systems. In Section 2.5 we wrote the energy of a Harmonic oscillator with displacement coordinate x as $\frac{1}{2}M\dot{x}^2 + \frac{1}{2}\kappa x^2$. Using the procedure for obtaining a Hamiltonian we find a momentum operator, conjugate to x, of

$$P = \partial L/\partial \dot{x} = M\dot{x} \tag{16.14}$$

in terms of which the Hamiltonian is

$$H = \frac{P^2}{2M} + \frac{\kappa x^2}{2}. \tag{16.15}$$

We found the energy eigenvalues of $\varepsilon_n = \hbar\omega_0(n + \frac{1}{2})$ in terms of the classical vibrational frequency $\omega_0 = \sqrt{\kappa/M}$, and found also the eigenstates $\phi_n(x)$.

We wish here to define a *stepping operator*, which when operating on the n'th eigenstate gives the n+1st eigenstate, just as the creation operator for electrons added an electron to some state. The essential feature there was the commutation relation, Eq. (16.5), and by deriving analogous commutation relations here we shall find the corresponding stepping operators. When we generalize this to lattice vibrations we shall see that these stepping operators are the annihilation and creation operators for phonons, and when we generalize it to light modes we shall see that the stepping operators become annihilation and creation operators for photons.

The fact that the ground-state wavefunction, which we wrote in Eq. (2.40), is of the form $\psi(x) = A\ \exp(-x^2/(2L^2))$, and the first excited state is of the same form with an additional factor of x, would suggest that either the factor x or the momentum operator, $(\hbar/i)\partial/\partial x$, might serve as a stepping operator. In fact these two operators do not commute, since

$$Px = \frac{\hbar\partial}{i\partial x}x = \frac{\hbar}{i} + xP. \tag{16.16}$$

This would suggest that we define some combination of the x and P to obtain operators with commutation relations such as Eq. (16.5), which do not have the imaginary result of Eq. (16.16). Trying $o^\dagger = \omega_0 x - iP/M$ and $o = \omega_0 x + iP/M$ so that both terms have the same units gives, using Eq. (16.16), $o^\dagger o - oo^\dagger = -2\hbar\omega_0/M$. Finally then we try the definitions

$$a^\dagger \equiv \sqrt{\frac{M}{2\hbar\omega_0}}\left(\omega_0 x - \frac{iP}{M}\right)$$

$$a \equiv \sqrt{\frac{M}{2\hbar\omega_0}}\left(\omega_0 x + \frac{iP}{M}\right). \tag{16.17}$$

Again using Eq. (16.16) we obtain the commutation relations for the operators we have defined,

$$aa^\dagger - a^\dagger a = 1. \tag{16.18}$$

In Problem 16.2 we evaluate $a^\dagger a$ from Eq. (16.17), using Eq. (16.16) to obtain canceling terms, to find that the Hamiltonian

$$H = \frac{P^2}{2M} + \frac{\kappa x^2}{2} = \hbar\omega_0(a^\dagger a + 1/2) . \tag{16.19}$$

For an eigenstate of the Hamiltonian for which the energy is $\hbar\omega_0(n + 1/2)$, we have thus shown that $a^\dagger a$ *is the number operator for excitations of the oscillator.*

We can now easily show that $a^\dagger$ increases the excitation energy one step: We imagine the system in the n'th eigenstate, $a^\dagger a|n\rangle = n|n\rangle$. We then operate on the state $|n\rangle$ with $a^\dagger$ and again apply the number operator,

$$a^\dagger a\, a^\dagger|n\rangle = a^\dagger(1 + a^\dagger a)|n\rangle = (n+1)a^\dagger|n\rangle, \tag{16.20}$$

so that indeed $a^\dagger|n\rangle$ *is* in the n+1st state of excitation. We similarly find that $a^\dagger a\, a|n\rangle = (n-1)a|n\rangle$ so a lowers the excitation by one unit. It follows also that $\langle 0|a^\dagger a|0\rangle = 0$, so there is no state $a|0\rangle$. Because of these properties, $a^\dagger$ and a are also called "raising" and "lowering" operators and their counterparts will become the creation and annihilation operators for phonons and photons.

There is an odd feature of these operators which did not arise for electrons. If the state $|n\rangle$ is normalized, the state $a^\dagger|n\rangle$ is not. We see this by evaluating the normalization integral , using the commutation relation,

$$\langle n|a\, a^\dagger|n\rangle = \langle n|(1+a^\dagger a)|n\rangle = (1+n)\langle n|n\rangle. \tag{16.21}$$

We can however write a normalized state as

$$|n\rangle = \sqrt{\frac{1}{n!}}\,(a^\dagger)^n|0\rangle, \tag{16.22}$$

and it will always be appropriate to use normalized eigenstates. The more useful expressions will be

$$a^\dagger|n\rangle = \sqrt{n+1}\;|n+1\rangle, \text{ and}$$
$$a|n\rangle = \sqrt{n}\;|n-1\rangle \tag{16.23}$$

in terms of the normalized eigenstates, $|n\rangle$.

It may be helpful to make one application which illustrates the use of this formalism, and then to see how stepping operators are used for angular-momentum eigenstates, before moving to phonon operators. We imagine a harmonic oscillator to which we apply a classical oscillating force, $H_1 = -Fx\ e^{i\omega t}$ + c.c.. We may directly apply the time-dependent perturbation theory of Eq. (9.9) which becomes

$$\frac{1}{\tau} = \frac{2\pi}{\hbar}\,\Sigma_m \langle n|-Fx|m\rangle\langle m|-Fx|n\rangle[\delta(\varepsilon_n - \varepsilon_m + \hbar\omega) + \delta(\varepsilon_n - \varepsilon_m - \hbar\omega)]. \tag{16.24}$$

Proceeding as in Chapter 9, we would calculate the matrix elements from harmonic oscillator wavefunctions as $-F\int dx\psi_m(x)x\psi_n(x)$. We now have an alternative way using the definitions of $a^\dagger$ and a in Eqs. (16.17). We may add the two equations and divide both sides by $\sqrt{2M\omega_0/\hbar}$ to obtain

$$x = \sqrt{\frac{\hbar}{2M\omega_0}}\,(a^\dagger + a)\,. \tag{16.25}$$

Then we obtain

$$\langle m|-Fx|n\rangle = -F\sqrt{\frac{\hbar}{2M\omega_0}}\,\langle m|a^\dagger + a|n\rangle \left\langle \begin{array}{l} = -F\sqrt{\frac{\hbar}{2M\omega_0}}\sqrt{n+1} \text{ if } m=n+1 \\ \\ = -F\sqrt{\frac{\hbar}{2M\omega_0}}\sqrt{n} \text{ if } m = n-1, \end{array}\right. \tag{16.26}$$

and zero otherwise. Thus the product of matrix elements in Eq. (16.24) is $\langle n|-Fx|m\rangle\langle m|-Fx|n\rangle = [\hbar F^2/(2M\omega_0)](n+1)$ if the final state is $m = n+1$ and $[\hbar F^2/(2M\omega_0)]n$ if the final state is $m = n-1$. It is interesting that these matrix elements were obtained using only the commutation relations, without the wavefunctions we would have used earlier. To complete the evaluation we would need a distribution of forces, $F^2(\omega)\, d\omega$ as a function of ω, and we shall make such evaluations later. It leads to a rate $1/\tau = [\pi F^2(\omega)/(M\hbar\omega_0)]$ $(n+1)$ for raising the energy of the oscillator and a rate $1/\tau = [\pi F^2(\omega)/(M\hbar\omega_0)]n$ for lowering the energy. If we considered the case of large forces, corresponding to $F^2(\omega)$ very large, this would approach the classical limit with a raising rate greater than the lowering rate and a continual heating up of the oscillator.

16.3 Angular Momentum

Finally, we shall indicate briefly how stepping operators are used with angular-momentum eigenstates, though we shall have little occasion to use them except for the treatment of spin-orbit coupling in Section 22.5. We described angular-momentum eigenstates in terms of the spherical harmonics $Y_l^{\,m}(\theta,\phi)$ in Section 2.4. l was the total-angular-momentum quantum number and m the quantum number for the component along the z-axis, $-l \le m \le l$. As in the two-dimensional angular-momentum operator of Eq. (2.24) we may write the three-dimensional operator,

$$L = \mathbf{r}\times\mathbf{p} = (yp_z - zp_y)\hat{\mathbf{x}} + (zp_x - xp_z)\hat{\mathbf{y}} + (xp_y - yp_x)\hat{\mathbf{z}}$$

$$= \frac{\hbar}{i}\left((y\frac{\partial}{\partial z} - z\frac{\partial}{\partial y})\hat{\mathbf{x}} + (z\frac{\partial}{\partial x} - x\frac{\partial}{\partial z})\hat{\mathbf{y}} + (x\frac{\partial}{\partial y} - y\frac{\partial}{\partial x})\hat{\mathbf{z}}\right) \qquad (16.27)$$

$$= \qquad L_x\hat{\mathbf{x}} + \qquad L_y\hat{\mathbf{y}} \qquad + \qquad L_z\hat{\mathbf{z}}\,,$$

where the $\hat{\mathbf{x}}$, $\hat{\mathbf{y}}$, and $\hat{\mathbf{z}}$ are unit vectors in the three cube directions. Then the $Y_l^m(\theta,\phi)$, in terms of the coordinate system given in Fig. 2.7 shown again in Fig. 16.1, are eigenstates of the two operators,

$$L_z = \frac{\hbar}{i}\left(x\frac{\partial}{\partial y} - y\frac{\partial}{\partial x}\right) = \frac{\hbar}{i}\, r\sin\theta\frac{\partial}{r\sin\theta\,\partial\phi}, \qquad (16.28)$$

with eigenvalue $\hbar m$, and

$$L^2 = L_x^2 + L_y^2 + L_z^2 \qquad (16.29)$$

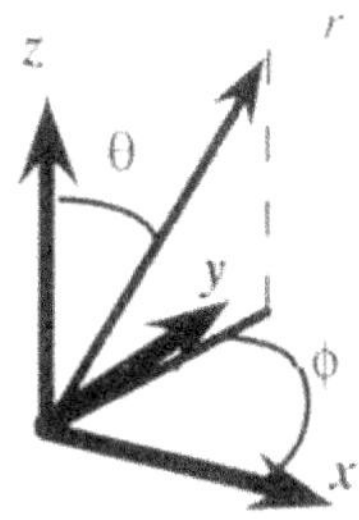

Fig. 16.1. Polar and rectangular coordinate systems.

with eigenvalue $\hbar^2 l(l+1)$.

We consider how these operators commute. We can see that the *commutator* is

$$[L^2, L_z] \equiv L^2 L_z - L_z L^2 = 0. \tag{16.30}$$

They commute, since we could expand any function of angle in spherical harmonics and operating successively with the two operators simply gives the product of the two eigenvalues for that term and the commutator cancels term by term. When two operators commute, it can be seen that states can be chosen to be eigenfunctions of both. Similarly $[L^2, L_x] = [L^2, L_y] = 0$ since we could pick polar axes along either of these axes and make the same argument. However, by writing out the terms we may see that the different components do not commute, but give

$$L_x L_y - L_y L_x = i\hbar L_z\,, \tag{16.31}$$

and the corresponding expressions with indices rotated (e. g., $x \rightarrow y$, $y \rightarrow z$, $z \rightarrow x$). These are a little like stepping operators and in fact we can construct operators which have the stepping property as

$$L_+ = L_x + iL_y,$$
$$L_- = L_x - iL_y\,. \tag{16.32}$$

We may check using Eq. (16.31), and the rotated expressions, that the commutation relations for L_z and $L_\pm$ are

$$L_z L_\pm - L_\pm L_z = \pm\, \hbar L_\pm\,. \tag{16.33}$$

With a little more operator algebra we see that

$$L_+L_- = (L_x + iL_y)(L_x - iL_y)$$
$$= L_x^2 - i(L_xL_y - L_yL_x) + L_y^2 = L_x^2 + \hbar L_z + L_y^2 = L^2 - L_z(L_z - \hbar) \quad (16.34)$$

and similarly that $L_-L_+ = L^2 - L_z(L_z + \hbar)$ so that the principal commutation relation becomes

$$L_+L_- - L_-L_+ = 2\hbar L_z . \quad (16.35)$$

From Eq. (16.33) we can see that L_- lowers the component of angular momentum along the z-axis by one unit,

$$L_z(L_-|l,m\rangle) = L_-L_z|l,m\rangle - \hbar L_-|l,m\rangle = \hbar(m-1)\,(L_-|l,m\rangle). \quad (16.36)$$

Similarly, L_+ raises the component by one unit. As with harmonic-oscillator raising and lowering operators, $L_\pm$ on a normalized state does not necessarily lead to a normalized state. We may evaluate the normalization integral for the state $L_-|l,m\rangle$ using Eq. (16.34),

$$\langle l,m|L_+L_-|l,m\rangle = \langle l,m|L^2 - L_z(L_z - \hbar)|l,m\rangle = [l(l+1) - m(m-1)]\hbar^2. \quad (16.37)$$

From this, and the counterpart for $L_+|l,m\rangle$, in terms of normalized states $|l,m\rangle$ we have

$$L_+|l,m\rangle = \hbar\sqrt{l(l+1) - m(m+1)}\;|l, m+1\rangle,$$
$$L_-|l,m\rangle = \hbar\sqrt{l(l+1) - m(m-1)}\;|l, m-1\rangle. \quad (16.38)$$

There are many relations which can be derived, relations between states with different axes, and formulae for addition of different contributions to the total angular momentum in complex systems. These can be found in almost any standard text when they are needed. We shall introduce the only one we need, that involving the addition of orbital and spin angular momentum, in Section 22.5 when we discuss spin-orbit coupling. However, here we should note the selection rules which these operators lead to, which we made use of following Eq. (9.13).

We return to the angular coordinate system of Fig. 16.1 and write states as spherical harmonics, $Y_l^m(\theta,\phi) = P_l^m(\cos\theta)e^{im\phi}$, so that $\langle l',m'|e^{i\phi}|l,m\rangle = \langle l',m'|(x+iy)/r|l,m\rangle = 1$ if $l' = l$, $m' = m+1$, and zero otherwise due to the orthogonality of the states. Similarly $x - iy$ only has nonzero matrix elements for $m' = m - 1$. It follows that perturbations proportional to x or to y, such as electric fields in the xy-plane, only couple states which differ in z-component of angular momentum by one unit, as we indicated in Section 9.4.

Chapter 17. Phonons

The generalization of these stepping operators to lattice vibrations and to light modes is quite direct. We carry it out first for lattice vibrations which may be easier to visualize. We make applications to the emission and absorption of phonons by electrons in semiconductors and then to the formation of polarons in semiconductors.

17.1 Annihilation and Creation Operators for Phonons

We begin with the Hamiltonian for the vibrational modes in Eq. (15.12) as

$$H = \Sigma_{\mathbf{q},\lambda}\left(\frac{\mathbf{P}_{\mathbf{q}}\mathbf{P}_{-\mathbf{q}}}{2M} + \frac{M\omega_{\mathbf{q}}{}^2\mathbf{u}_{\mathbf{q}}\mathbf{u}_{-\mathbf{q}}}{2}\right). \tag{17.1}$$

It can be confirmed that if we define annihilation and creation operators as in Eq. (16.17) by

$$a_{\mathbf{q}}{}^{\dagger} = \sqrt{\frac{M}{2\hbar\omega_{\mathbf{q}}}}\left(\omega_{\mathbf{q}}\mathbf{u}_{-\mathbf{q}} - \frac{i\mathbf{P}_{\mathbf{q}}}{M}\right),$$
$$a_{\mathbf{q}} = \sqrt{\frac{M}{2\hbar\omega_{\mathbf{q}}}}\left(\omega_{\mathbf{q}}\mathbf{u}_{\mathbf{q}} + \frac{i\mathbf{P}_{-\mathbf{q}}}{M}\right), \tag{17.2}$$

and evaluate

$$H = \Sigma_{\mathbf{q},\lambda} (a_{\mathbf{q}}^{\dagger} a_{\mathbf{q}} + 1/2)\hbar\omega_{\mathbf{q}} \tag{17.3}$$

we obtain Eq. (17.1). We obtain an additional term in doing this but it sums to zero in the sum over all wavenumbers (e. g., Harrison (1970) p. 408). Finally, we may replace $\mathbf{q}$ by $-\mathbf{q}$ in the first of Eq. (17.2), add it to the second and solve for $\mathbf{u_q}$ to obtain the counterpart of Eq. (16.25) as

$$\mathbf{u_q} = \sqrt{\frac{\hbar}{2M\omega_{\mathbf{q}}}} (a_{-\mathbf{q}}^{\dagger} + a_{\mathbf{q}}) . \tag{17.4}$$

The commutation relations for the operators,

$$a_{\mathbf{q}} a_{\mathbf{q}}^{\dagger} - a_{\mathbf{q}}^{\dagger} a_{\mathbf{q}} = 1 \tag{17.5}$$

carry over from Eq. (16.18) as well as the operator properties,

$$\begin{aligned} a_{\mathbf{q}}^{\dagger}|n\rangle &= \sqrt{n_{\mathbf{q}} + 1}\, |n_{\mathbf{q}} + 1\rangle, \\ a_{\mathbf{q}}|n\rangle &= \sqrt{n_{\mathbf{q}}}\, |n_{\mathbf{q}} - 1\rangle , \end{aligned} \tag{17.6}$$

from Eq. (16.23).

We saw in Eq. (16.9) that potentials $V(\mathbf{r})$ seen by the electrons are incorporated in the Hamiltonian as $\Sigma_{\mathbf{k},\mathbf{q}} \langle \mathbf{k}+\mathbf{q}|V(\mathbf{r})|\mathbf{k}\rangle c_{\mathbf{k}+\mathbf{q}}^{\dagger} c_{\mathbf{k}}$. We obtained the corresponding matrix elements between electronic states which arise from classical lattice vibrations in Eq. (15.16) and here we consider only normal process, those for which the change in electron wavenumber is equal to the vibrational wavenumber, so that these matrix elements become

$$\langle \mathbf{k}+\mathbf{q}|W|\mathbf{k}\rangle = \frac{-i\mathbf{q} \cdot \mathbf{u_q}}{\sqrt{N}} w_q . \tag{17.7}$$

We used plane waves and pseudopotentials which are appropriate for metals, but in a semiconductor the w_q in Eq. (17.7) can be replaced by a *deformation-potential constant D* (e. g., Harrison (1999)). If a dilatation of the lattice $\delta\Omega/\Omega$ shifts a band minimum by $-D\delta\Omega/\Omega$, one assumes that the displacements for a phonon $\delta\mathbf{r} = \mathbf{u_q} e^{i\mathbf{q}\cdot\mathbf{r}}/\sqrt{N}$, giving a local dilatation $\nabla\cdot\delta\mathbf{r}$, will produce a potential $V(\mathbf{r}) = -D\nabla\cdot\delta\mathbf{r}$. This leads to exactly Eq. (17.7) with w_q replaced by D. In either case, only longitudinal modes enter and $\mathbf{u_q}$ can be taken parallel to $\mathbf{q}$ so the vector notation and the sum over λ become irrelevant, $\mathbf{q}\cdot\mathbf{u_q} = qu_\mathbf{q}$. We write the term in the Hamiltonian for the interaction between electrons and classical lattice vibrations as

$$H_1 = \sum_{\mathbf{k},\mathbf{q}} \frac{-iqu_\mathbf{q}}{\sqrt{N}} w_q c_{\mathbf{k}+\mathbf{q}}^\dagger c_\mathbf{k} \ . \tag{17.8}$$

We incorporate the quantum description of the lattice vibrations by substituting from Eq. (17.4) for $\mathbf{u_q}$. This gives the electron-phonon interaction,

$$H_{e\phi} = \sum_{\mathbf{k},\mathbf{q}} -iq\, w_\mathrm{q} \sqrt{\frac{\hbar}{2M\omega_\mathbf{q}N}}\ c_{\mathbf{k}+\mathbf{q}}^\dagger c_\mathbf{k}(a_{-\mathbf{q}}^\dagger + a_\mathbf{q}) \ . \tag{17.9}$$

For convenience in using the electron-phonon interaction we may collect the factors in front as

$$V_\mathrm{q} = -iq\, w_\mathrm{q} \sqrt{\frac{\hbar}{2M\omega_\mathbf{q}}} \tag{17.10}$$

(or with w_q replaced by a deformation-potential constant D for semiconductors) so that the electron-phonon interaction becomes simply

$$H_{e\phi} = \sqrt{\frac{1}{N}}\ \sum_{\mathbf{k},\mathbf{q}} V_\mathrm{q} c_{\mathbf{k}+\mathbf{q}}^\dagger c_\mathbf{k}(a_{-\mathbf{q}}^\dagger + a_\mathbf{q}) \ . \tag{17.11}$$

Once we have obtained the form it is seen to make perfect physical sense. Electrons can be taken from a state of wavenumber $\mathbf{k}$ to one of wavenumber $\mathbf{k} + \mathbf{q}$ either by the absorption of a phonon of wavenumber $\mathbf{q}$ or by the emission of a phonon of wavenumber $-\mathbf{q}$. As always, momentum conservation is enforced by the matrix element and any energy conservation will come from the energy delta function in the Golden Rule. The form of the interaction constant, Eq. (17.10), is not so obvious except that it is proportional to the pseudopotential. In metals q will be of the order of the Fermi wavenumber so that the interaction constant V_q is of order $w_\mathrm{q}\sqrt{mE_F/M\hbar\omega_\mathrm{q}}$. We think of the form factor as being typically a tenth of the Fermi energy and $\hbar\omega_q$ being of order $\sqrt{m/M}\,E_F$ so that $V_q \approx -i\sqrt{m/M}\,w_q$ is small at some 1% of E_F, perhaps a few tenths of an electron volt. Similar values are appropriate for semiconductors. It is much better form to keep the dependence upon N explicit as in Eq. (17.11) rather than absorb it in the interaction constant of (17.10), which now does not depend upon the size of the system.

17.2 Phonon Emission and Absorption

Our first application is the emission and absorption of phonons by electrons. We begin by reducing the operators and obtain a form for the rate in terms of the occupation of electron states and phonon numbers. There is no explicit time dependence in the electron-phonon interaction so we can use the Golden Rule of Eq. (7.9) directly.

$$\frac{1}{\tau} = \frac{2\pi}{\hbar}\Sigma_f \langle i|\,H_{e\phi}|f\rangle\langle f|H_{e\phi}|i\rangle\delta(E_f - E_i), \tag{17.12}$$

with |i> the initial state and |f> the final state. We first operate on the initial state with the electron-phonon interaction,

$$H_{e\phi}|i\rangle = \frac{1}{\sqrt{N}}\,\Sigma_{\mathbf{q},\mathbf{k}}\,V_q\,(a_{\mathbf{q}} + a_{-\mathbf{q}}{}^{\dagger})c_{\mathbf{k+q}}{}^{\dagger}c_{\mathbf{k}}|i\rangle, \tag{17.13}$$

which gives two terms for each **q**, one with one less phonon than the initial state and one with one more phonon. Also, one electron has been transferred from a state **k** to a state **k** + **q** (of the same spin). Exactly as we noted for Eq. (16.12) the operation produces a final state, with some factors such as the V_q. Each such term *is* a final state, so we do not need to multiply by <f|, which would just give the constant, and then insert |f> again. The |f><f| in Eq. (17.12) is simply a reminder to notice what to insert for E_f in the delta function. We already have included a sum over the final states in the sum given in Eq. (17.13). We simply operate again by the $H_{e\phi}$ for the first matrix element. This will give us many more states, but out of those we select only the one corresponding to the initial state. That is, we select only the term in $H_{e\phi}H_{e\phi}|i\rangle$ which restores the initial phonon numbers and returns the electron to its initial state |i> which matches the <i| in Eq. (17.12). [What we have done is to ignore a Σ_f |f><f| which, mathematically, is the identity operator. Operating on any function with <f| gives the expansion coefficient for that function, for an expansion in the complete set of states |f>. Multiplying that coefficient by |f> and summing over f reexpands the function, giving the initial function.] The result is

$$\frac{1}{\tau} = \frac{2\pi}{\hbar}\frac{1}{\sqrt{N}}\Sigma_{\mathbf{q},\mathbf{k}} <i|H_{e\phi}V_q(a_{-\mathbf{q}}^\dagger + a_\mathbf{q})c_{\mathbf{k}+\mathbf{q}}^\dagger c_\mathbf{k}|i>\delta(E_f - E_i)$$

$$= \frac{2\pi}{\hbar}\frac{1}{N}\Sigma_{\mathbf{q},\mathbf{k}} V_{-q}V_q<i|(a_{-\mathbf{q}}a_{-\mathbf{q}}^\dagger + a_\mathbf{q}^\dagger a_\mathbf{q})c_\mathbf{k}^\dagger c_{\mathbf{k}+\mathbf{q}}c_{\mathbf{k}+\mathbf{q}}^\dagger c_\mathbf{k}|i>\delta(E_f - E_i). \quad (17.14)$$

Now, as always with calculations based upon annihilation and creation operators, we evaluate all expressions containing these operators. We may commute $c_\mathbf{k}^\dagger$ two steps to the right to obtain $c_\mathbf{k}^\dagger c_\mathbf{k}$, which is the number operator for the state, and we may write the result of operation on the initial state as giving a factor of $f(\mathbf{k})$, a distribution such as we introduced in Chapter 10. Using the commutation relation, Eq. (16.5), the combination $c_{\mathbf{k}+\mathbf{q}}c_{\mathbf{k}+\mathbf{q}}^\dagger$ becomes $(1-f(\mathbf{k}+\mathbf{q}))$. The operator pair $a_\mathbf{q}^\dagger a_\mathbf{q}$ is the number operator for phonons in the mode of wavenumber $\mathbf{q}$, which we replace by that number, $n_\mathbf{q}$, and $a_{-\mathbf{q}}a_{-\mathbf{q}}^\dagger$ becomes $n_{-\mathbf{q}} + 1$, using the commutation relation, Eq. (17.5).

The term with the $a_{-\mathbf{q}}a_{-\mathbf{q}}^\dagger$ represents a final state with an additional phonon, and with the electron transferred from state $\mathbf{k}$ to state $\mathbf{k}+\mathbf{q}$, so that $E_f - E_i = \hbar\omega_q + \varepsilon_{\mathbf{k}+\mathbf{q}} - \varepsilon_\mathbf{k}$. For the term with $a_\mathbf{q}^\dagger a_\mathbf{q}$ the final state has one less phonon and $E_f - E_i = -\hbar\omega_q + \varepsilon_{\mathbf{k}+\mathbf{q}} - \varepsilon_\mathbf{k}$. Thus our expression for the transition rate has become

$$\frac{1}{\tau} = \frac{2\pi}{\hbar}\frac{1}{N}\Sigma_{\mathbf{q},\mathbf{k}} V_{-q}V_q\,(1-f(\mathbf{k}+\mathbf{q}))f(\mathbf{k})(n_{-\mathbf{q}} + 1)\delta(\hbar\omega_q + \varepsilon_{\mathbf{k}+\mathbf{q}} - \varepsilon_\mathbf{k})$$

$$+ \frac{2\pi}{\hbar}\frac{1}{N}\Sigma_{\mathbf{q},\mathbf{k}} V_{-q}V_q\,(1-f(\mathbf{k}+\mathbf{q}))f(\mathbf{k})n_\mathbf{q}\,\delta(-\hbar\omega_q + \varepsilon_{\mathbf{k}+\mathbf{q}} - \varepsilon_\mathbf{k}). \quad (17.15)$$

The physics of both contributions is quite clear. The first line represents electron transitions in which a phonon of wavenumber $-\mathbf{q}$ is emitted, so to conserve momentum the electron must gain momentum $\hbar\mathbf{q}$, and must lose energy (according to the delta function) to make up for the phonon energy. The rate is of course proportional to the probability $f(\mathbf{k})$ that the initial state was occupied, and to the probability, $1 - f(\mathbf{k}+\mathbf{q})$, that the final state was previously empty since the Pauli Principle does not allow double occupation of the state. The second of the two terms in the first line, the 1 in $n_{-\mathbf{q}} + 1$, is called *spontaneous emission*, which can occur even if no phonons were initially present, if there were electrons with enough energy to create that phonon. The first term, $n_{-\mathbf{q}}$ in $n_{-\mathbf{q}} + 1$, is the additional *stimulated emission*, which can be caused by phonons already present. The second line is similarly interpreted as representing electron transitions for which a phonon is *absorbed* and the electron energy increases. They cannot occur without phonons present, and the rate is proportional to $n_{\mathbf{q}}$. It is impressive that these three distinct physical processes are all incorporated in the same analysis and described by the same parameters.

We did not yet need to specify whether we were discussing metals or semiconductors; the difference enters through the distribution functions in Eq. (17.15), the form of V_q which is used, and whether the mass is replaced by an effective mass. It may be helpful for understanding this important process to redo the analysis for a specific case, as in Problem 17.1. That problem parallels the analysis we now give of the shift in the energy of an isolated electron in a semiconductor due to the electron-phonon interaction. We shall outline the calculation for Problem 17.1 as we proceed, and plot the resulting $1/\tau$ at the end.

17.3 Polaron Self-Energy

We again use perturbation theory, but now for the shift in energy due to $H_{e\phi}$, which couples an ordinary electronic band state, with no phonons present, to higher-energy electronic states with also an emitted phonon. The coupling to higher-energy states always lowers the energy and the resulting combination of electron and lattice distortion, when it occurs in a polar crystal such as gallium arsenide, is called a *polaron*. The effect is physically analogous to the lowering in energy of a heavy ball rolling on a mattress.

We begin with a well-defined initial state, with a single electron in the state $\mathbf{k} = 0$, and no phonons present. [In Problem 17.1 $\mathbf{k}$ is not zero, but there are no phonons.] We are seeking a shift in the energy,

$$E_{pol} = \Sigma_f \frac{\langle i|H_{e\phi}|f\rangle\langle f|H_{e\phi}|i\rangle}{E_i - E_f}, \tag{17.16}$$

rather than the Golden Rule formula, Eq. (17.12) which we used before and which is used in Problem 17.1.

Only one term in the sum over $\mathbf{k}$ in the electron-phonon interaction of Eq. (17.13) contributes when we evaluate $H_{e\phi}|i\rangle$, the $c_{\mathbf{k}}$ for the single occupied state $\mathbf{k} = 0$, and only the term $a_{-\mathbf{q}}^{\dagger}$ enters if there are no phonons in the initial state. Thus we have only a sum over $\mathbf{q}$. We again need not write $\Sigma_f|f\rangle\langle f|$, but note the energy change in going from the initial to the final state, in our case $\hbar^2q^2/2m^* + \hbar\omega_q$ if the electron energies are written in terms of an effective mass as $\varepsilon_{\mathbf{k}} = \hbar^2k^2/2m^*$. In Problem 17.1 there is a difference in two kinetic energies, $\varepsilon_{\mathbf{k}}$ and $\varepsilon_{\mathbf{k+q}}$. Then with the second application of $H_{e\phi}$ we keep only the term which returns us to the initial state. Then Eq. (17.16) becomes

$$E_{\rm pol} = \frac{1}{N}\Sigma_{\bf q} \frac{V_q V_{-q} \langle i|a_{-\bf q}a_{-\bf q}^\dagger|i\rangle}{-\hbar^2 q^2/2m^* - \hbar\omega_{\rm q}} , \tag{17.17}$$

and using the commutation relation, Eq. (17.5), we have $\langle i|a_{-\bf q}a_{-\bf q}^\dagger|i\rangle = 1$. We have then eliminated all of the operators, as we must, and can complete the evaluation.

The remainder of the calculation depends upon the system we wish to treat and the approximations we wish to make. In polar crystals the strongest interaction is with a high-frequency set of modes called "optical modes", for which Eq. (17.7) is replaced by (Harrison (1999), p. 280)

$$\langle {\bf k+q}|H_{e\phi}|{\bf k}\rangle = \frac{3\sqrt{3}\,\pi e^2 e_T^* u_{\rm q}\, i}{2\sqrt{2}\varepsilon d^3 q\sqrt{N}} , \tag{17.18}$$

with e_T^* an effective charge and ε the dielectric constant. The interesting point is that the corresponding $V_{\rm q}$ is proportional to $1/q$, and the corresponding $\omega_{\rm q}$ is essentially independent of q, which makes the evaluation of Eq. (17.17) quite simple. We write $V_{\rm q} = V_0 q_0/q$ and $\omega_{\rm q} = \omega_0$ and note that the terms in Eq. (17.17) are independent of the direction of $\bf q$ so we can replace the $\Sigma_{\bf q}$ by $(\Omega/(2\pi)^3)\int 4\pi q^2 dq$ (as in Eq. (2.9)) to obtain

$$E_{\rm pol} = -\frac{\Omega}{N}\frac{V_0^2 q_0^2}{2\pi^2}\int \frac{dq}{\hbar^2 q^2/2m^* + \hbar\omega_0} . \tag{17.19}$$

Because of the form of the polar interaction in Eq. (17.18) this integral converges if we extend the integration to infinity. In such cases, in which the integral converges, it is ordinarily a good approximation to extend it to infinity. [Otherwise we might introduce a cut-off q_D which conserves the appropriate number of modes as in Eq. (10.14).] The result is

$$E_{\rm pol} = -\frac{\Omega}{N}\frac{V_0^2 q_0^2}{2\pi^2}\sqrt{\frac{2m^*}{\hbar^3\omega_0}}\int_{0,\infty} \frac{dx}{x^2+1} \tag{17.20}$$

with the final integral equal to $\pi/2$.

This lowering of the energy of the electron due to the deformation of the lattice around it turns out to be numerically rather small. Of more interest is the change in this energy with the wavenumber $\bf k$ of the electron, which could be calculated in the same way. $E_{\rm pol}$ increases in magnitude with the energy of the electron, giving an increase in the effective mass of the

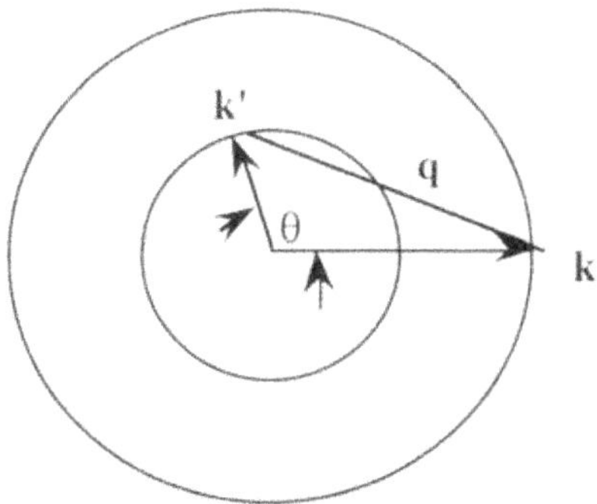

Fig. 17.1 The coordinate system for calculating phonon emission and absorption of optical phonons

resulting "polaron". It can be useful to have a means of estimating such effects.

The same approximation for the interaction $V_q = V_0 q_0/q$ and $\omega_q = \omega_0$ also makes the evaluation of the integrals for scattering analytically doable in Problem 17.1. For that problem it is convenient to sum over final-state wavenumbers $\mathbf{k}'$ rather than $\mathbf{q}$, an exactly equivalent sum. Then the sum is converted to an integral as $\Sigma_{\mathbf{q}} \to \Sigma_{\mathbf{k}'} = (\Omega/(2\pi)^3)\int 2\pi k'^2 dk' \int \sin\theta \, d\theta$ as in Fig. 17.1. Using the cosine rule we can write $q^2 = k^2 + k'^2 - 2kk' \cos\theta$.

It is usually best to utilize the delta function, $\delta(\hbar^2 k'^2/(2m^*) + \hbar\omega_0 - \hbar^2 k^2/(2m^*))$, first in cases such as Problem 17.1. This is accomplished by carrying out the integration over k', which fixes k' in terms of k, both of

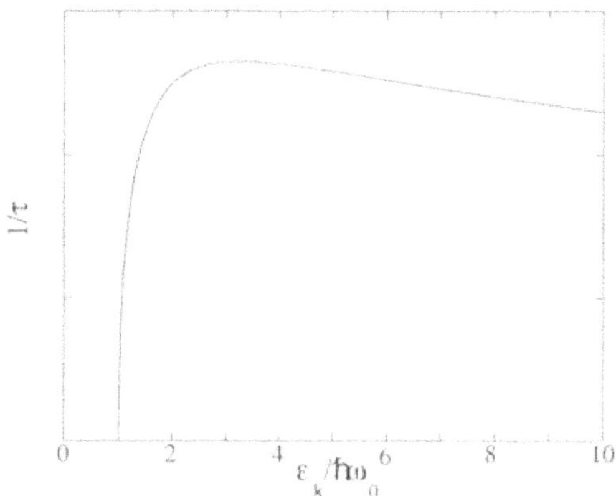

Fig. 17.2. The rate of spontaneous optical phonon (frequency ω_0) emission in a polar semiconductor as a function of electron energy ε_k, from Problem 17.1.

which still appear in the equation. The remaining integral over angle can be performed analytically. The result is plotted in Fig. 17.2.

We may note from the diagram of Fig. 17.1 that there can be no emission unless the initial energy is high enough to allow for the creation of a phonon, and that is reflected in Fig. 17.2

17.4 Electron-Electron and Nucleon-Nucleon Interactions

The electron-phonon coupling also leads to an electron-electron interaction responsible for ordinary superconductivity. Only electrons with energies very near the Fermi energy in the metal are important and their interaction is treated as a second-order coupling as in Section 9.1. The electron-phonon interaction couples the ground state of the metal to a (higher-energy) state with one electron wavenumber deflected by $\mathbf{q}$ and a phonon of wavenumber $-\mathbf{q}$ created. This state is in turn coupled to a state in which the phonon is absorbed and a second electron's wavenumber is deflected by $-\mathbf{q}$. This has the effect of a "collision" between the two electrons arising from the exchange of a *virtual* phonon, called virtual since energy is not conserved in this intermediate state. The second-order matrix element contains a $V_q{}^*V_q$ and a negative energy denominator, so it acts as an attractive interaction, which is responsible for the pairing of electrons in superconductivity.

The physical origin of this attraction can be understood in the same way we understood the polaron energy at the beginning of the preceding section. A ball rolling on a mattress has its energy lowered by deforming the mattress to lower its energy in the gravitational field. A second ball rolling on the mattress will be attracted to the depression from the first ball and they will tend to cluster, as illustrated in Fig. 17.3. We shall use that concept when we discuss the nature of the superconducting state in Section 21.4 though with these high-energy electrons and the slow-moving atoms this low-frequency concept is not as appropriate as the description given in the last paragraph. The low-frequency view can be appropriate in semiconductors and we carry it further here, first for a single electron and then for two. It

Fig. 17.3. Two polarons, arising from the electron-phonon coupling, attract each other. Similarly two balls rolling on a mattress lower their energy by deforming the mattress, and are attracted by the depression made by the other. As a consequence they tend to move closer together.

will also enable us to understand the origin of the nucleon-nucleon interaction which we introduced in Section 4.4.

We noted following Eq. (17.7) that atom displacements $\delta\mathbf{r} = \mathbf{u_q}e^{i\mathbf{q}\cdot\mathbf{r}}/\sqrt{N}$, give a local dilatation $\nabla\cdot\delta\mathbf{r}$ and in a semiconductor may produce a deformation potential $V(\mathbf{r}) = -D\nabla\cdot\delta\mathbf{r} = -i\mathbf{q}\cdot\mathbf{u_q}De^{i\mathbf{q}\cdot\mathbf{r}}/\sqrt{N}$ with $\mathbf{q}\cdot\mathbf{u_q} = qu_\mathbf{q}$ for a longitudinal mode (and zero for a transverse mode). If we think of an electron at a position $\mathbf{r}$ this gives us a perturbation to the phonon Hamiltonian and we may calculate a shift in the energy as we did for an electron in a plane-wave state in the preceding section. We are treating the electron classically at this stage of the analysis by leaving the electron position $\mathbf{r}$ in the equations. We again use Eq. (17.4) to write $u_\mathbf{q}$ in terms of annihilation and creation operators and if the zero-order state contains no phonons we obtain contributions only from $a_{-\mathbf{q}}a_{-\mathbf{q}}{}^\dagger = 1$. With an energy denominator of $\hbar\omega_\mathbf{q}$ we sum over $\mathbf{q}$ to write

$$E_{\text{pol}} = \sum\nolimits_\mathbf{q} \frac{\hbar}{2M\omega_\mathbf{q}} \frac{D^2q^2}{N} \frac{e^{-i\mathbf{q}\cdot\mathbf{r}}e^{i\mathbf{q}\cdot\mathbf{r}}}{-\hbar\omega_\mathbf{q}}, \tag{17.21}$$

the counterpart of Eq. (17.17) without the shift in electron energy in the denominator. We also kept the two phase factors which of course cancel in this case. This would be a suitable approximation if the phonon frequencies were high enough that the change in electron energy $\hbar^2q^2/(2m^*)$ were negligible. This high-frequency approximation corresponds to the physical assumption that the lattice response is effectively instantaneous.

If we do make this assumption, we may generalize Eq. (17.21) to an electron at $\mathbf{r}_1$ and a second at $\mathbf{r}_2$ so that $e^{i\mathbf{q}\cdot\mathbf{r}}$ is replaced by $e^{i\mathbf{q}\cdot\mathbf{r}_1} + e^{i\mathbf{q}\cdot\mathbf{r}_2}$ and the numerator in the final factor becomes $2 + e^{-i\mathbf{q}(\mathbf{r}_1-\mathbf{r}_2)} + e^{i\mathbf{q}\cdot(\mathbf{r}_1-\mathbf{r}_2)}$. The first term gives twice Eq. (17.21), the polaron shift for the two electrons. However, the other terms give the interaction between the two polarons. The two terms add to give the interaction in terms of the distance between the two electrons $\mathbf{r} = \mathbf{r}_2 - \mathbf{r}_1$ as

$$V_{\text{pol}}(\mathbf{r}) = -\frac{D^2}{M}\frac{1}{N}\sum\nolimits_\mathbf{q} \frac{q^2}{\omega_\mathbf{q}{}^2} e^{i\mathbf{q}\cdot\mathbf{r}}. \tag{17.22}$$

The same result could have been obtained with a classical calculation based upon the same electron-vibration interaction; note that no $\hbar$ appears in the answer.

It would be easy to make a bad approximation at this point. We might make the Debye approximation which we used for the specific heat in Section 10.23 to write $\omega_\mathbf{q} = v_sq$ and integrate to the Debye wavenumber q_D with $(4\pi q_D{}^3/3)(\Omega/(2\pi)^3) = N$. The sum multiplied by $1/N$ would then become $(3/v_s^2)(\sin q_D r/(q_D r)^3 - \cos q_D r/(q_D r)^2)$. This might be meaningful as $\mathbf{r}$ goes to 0, but the oscillations as a function of r come from the artificial cut-off at q_D. We may note a particular case for which this r-dependent result is quite wrong, the case with $q^2/\omega_q^2 = 1/v_s^2$. Then that factor may be taken out of the sum, and the sum over the exact Brillouin Zone is zero unless $\mathbf{r} = 0$. The result we obtain with the Debye integral is completely wrong for this case, and for other forms of ω_q as well.

There are different ways to proceed, including a numerical sum over the Brillouin Zone, though then one should reconsider the small-q form which we used for the coupling, $-D\nabla\cdot\delta\mathbf{r}$. One approximate way is to replace the summand by $(1/v_s^2)e^{-qa}e^{i\mathbf{q}\cdot\mathbf{r}}$ and integrate over all q, with a chosen such that the sum is N/v_s^2 with $\mathbf{r} = 0$. This yields

$$V_{pol}(\mathbf{r}) \approx -\frac{D^2}{M}\frac{1}{(1 + (2/3)^{2/3}(q_D r)^2)^2}\,. \tag{17.23}$$

This is an attractive interaction which drops off in a few interatomic distances, each of order $1/q_D$. Though it is classical, we may think of the interaction as arising from the exchange of phonons between the two electrons just as we can think of the e^2/r Coulomb interaction as arising from the exchange of photons. In fact every interaction between particles is thought of as arising from the exchange of *some* kind of particle.

When the exchanged particles are massless, the interaction has an algebraic dependence upon separation, as in Eq. (17.23). The short-range interaction between nucleons, suggested by the liquid-drop model (Section 4.4), indicated to Yukawa (1935) that the particles exchanged must have mass. We may see why by going back to the form we obtained from exchange of phonons, Eq. (17.22). We had taken the energy denominator $\hbar\omega_{\mathbf{q}}$ as the energy to create a phonon. With the phonon replaced by a particle with nonzero mass we may use the relativistic form for the energy from Problem 1.1 (including the mc^2 to create the particle). It is simplest to use it for both factors of $\hbar\omega_{\mathbf{q}}$, which is assuming the same form for the coupling as the electron-phonon coupling, and we have

$$V_{int}(\mathbf{r}) = -\frac{D^2}{M}\frac{1}{N}\sum_{\mathbf{q}}\frac{\hbar^2 q^2}{(mc^2)^2 + (\hbar c q)^2}e^{i\mathbf{q}\cdot\mathbf{r}}, \tag{17.24}$$

with m and q being the mass and wavenumber of the particle exchanged. We may again replace the sum by an integral, $(\Omega/(2\pi)^3)\int 2\pi q^2 dq \sin\theta\, d\theta$, perform the angular integral, and obtain

$$V_{int}(\mathbf{r}) = -\frac{D^2}{M}\frac{\Omega_0}{2\pi^2}\int_{0,\infty}\frac{\hbar^2 q^3 dq}{(mc^2)^2 + (\hbar cq)^2}\frac{\sin qr}{r}$$

$$= -\frac{D^2\Omega_0}{M}\frac{1}{4\pi^2 c^2 r}\int_{-\infty,\infty}\frac{e^{iqr}q^3 dq}{(mc/\hbar)^2 + q^2}. \tag{17.25}$$

This integral is easily done by contour integration (or from tables), closing the contour in the upper complex plane, where e^{iqr} makes the integral around the large semicircle zero. The integral along the axis is then equal to the residue of the pole at $q = im\,c/\hbar$, with magnitude of q equal to 2π divided by the Compton wavelength $\lambda_C = 2\pi\hbar/mc$ of a particle with mass m. (see Problem 1.1). The result is

$$V_{int}(\mathbf{r}) = -V_0\frac{e^{-\alpha r}}{\alpha r}, \tag{17.26}$$

with $\alpha = mc/\hbar$ and $V_0 = 2\pi^2 D^2\Omega_0/(Mc^2\lambda_C^3)$ for the same coupling we used for phonons. This was the form, Eq. (17.26), obtained by Yukawa for the nucleon-nucleon interaction. Knowing that the interaction had a range of the order of the nuclear size, 10^{-13} cm, which he equated to $\lambda_C/(2\pi) = \hbar/mc$, he predicted that the *meson* exchanged had a mass about a fifth of the proton mass. This was confirmed by the observation of π-mesons, *pions*, with an energy of a sixth or seventh the proton mass. Yukawa (1935) could also estimate V_0 from the known binding of the proton and neutron in the deuteron. It turned out that there were positive, negative, and neutral pions so the interaction is considerably more complicated, but there is no question that the strong interaction between nucleons arises from such mesons as proposed by Yukawa.

Chapter 18. Photons

We have completed an analysis of the quantization of lattice vibrations in solids, which came as a very natural application of our theory of harmonic oscillators to the normal modes of a solid. With our starting postulate that everything is both a particle and a wave, and all theory deriving from that one assertion, it comes as no surprise that we can directly generalize our analysis from sound waves to light waves. We shall be able to define the creation and annihilation operators exactly in parallel to those for phonons, and to write the electron-photon interaction in terms of those operators just as we did for phonons. Indeed the concept of the photon as a light-particle is more familiar than the concept of a phonon.

18.1 Photons and the Electron-Photon Interaction

Vibrations in solids were described in terms of the displacements $\delta\mathbf{r}_j(t)$ of each atom at a starting position $\mathbf{r}_j$. We saw in Section 1.3 that electromagnetic waves can be described in terms of a vector potential $\mathbf{A}(\mathbf{r},t)$ at the position $\mathbf{r}$, the position being a continuum as in the continuum treatment of sound waves in Section 1.8. This generalization is nothing new for us, corresponding for lattice vibrations to a small-q limit where ω is proportional to q, but for light it extends to arbitrarily large q. We follow our analysis of lattice vibrations, first constructing a Hamiltonian for the light.

We begin with a transformation to "normal coordinates", as in Eq. (15.6),

$$\mathbf{A}(\mathbf{r},t) = \sqrt{\frac{4\pi}{\Omega}}\,\Sigma_{\mathbf{q},\lambda}\,\mathbf{u_q}(t)^{\lambda}e^{i\mathbf{q}\cdot\mathbf{r}}. \tag{18.1}$$

The factor in front of the sum, like $\sqrt{1/N}$, is inversely proportional to the volume of the system, and the $\sqrt{4\pi}$ was included to simplify the forms which arise later. The electric field from this potential is obtained from Eq. (1.17) as

$$\mathbf{E}(\mathbf{r},t) = -\frac{1}{c}\frac{\partial \mathbf{A}}{\partial t} = -\sqrt{\frac{4\pi}{\Omega c^2}}\,\Sigma_{\mathbf{q},\lambda}\,\dot{\mathbf{u}}_{\mathbf{q}}(t)^{\lambda} e^{i\mathbf{q}\cdot\mathbf{r}} . \tag{18.2}$$

Then we may obtain the total electric-field energy as $T = (1/8\pi)\int d^3r\, E^2(\mathbf{r})$, integrated over the volume Ω. The result, in analogy with Eq. (15.9), is

$$T = \frac{1}{2c^2}\,\Sigma_{\mathbf{q},\lambda}\,\dot{\mathbf{u}}_{\mathbf{q}}{}^{\lambda}\,\dot{\mathbf{u}}_{-\mathbf{q}}{}^{\lambda} \tag{18.3}$$

with the integral over all other products of $\dot{\mathbf{u}}_{\mathbf{q}}(t)^{\lambda}e^{i\mathbf{q}\cdot\mathbf{r}}$ integrating to zero if periodic boundary conditions are satisfied on the surface of the volume Ω. Similarly, the magnetic-field energy $(1/8\pi)\int d^3r\, H^2(\mathbf{r})$ based upon the magnetic field from Eq. (1.17), $\mathbf{H} = \nabla\times\mathbf{A}$, is ${}^1/_2 q^2\Sigma_{\mathbf{q},\lambda}\mathbf{u}_{\mathbf{q}}{}^{\lambda}\mathbf{u}_{-\mathbf{q}}{}^{\lambda}$. Only the electric-field energy depends upon $\dot{\mathbf{u}}_{\mathbf{q}}{}^{\lambda}$, so we may define a momentum conjugate to $\mathbf{u}_{\mathbf{q}}$ as

$$\mathbf{P}_{\mathbf{q}}{}^{\lambda} = \partial T\ /\partial\dot{\mathbf{u}}_{\mathbf{q}}{}^{\lambda} = \frac{1}{c^2}\,\dot{\mathbf{u}}_{-\mathbf{q}}{}^{\lambda} \tag{18.4}$$

as in Eq. (15.10) and the total energy written in terms of the coordinates $\mathbf{u}_{\mathbf{q}}{}^{\lambda}$ and momenta $\mathbf{P}_{\mathbf{q}}{}^{\lambda}$ is the Hamiltonian for the light,

$$H = \Sigma_{\mathbf{q},\lambda}\left(\frac{c^2}{2}\,\mathbf{P}_{\mathbf{q}}{}^{\lambda}\mathbf{P}_{-\mathbf{q}}{}^{\lambda} + \frac{q^2}{2}\,\mathbf{u}_{\mathbf{q}}{}^{\lambda}\mathbf{u}_{-\mathbf{q}}{}^{\lambda}\right). \tag{18.5}$$

This is a sum over wavenumbers of the light and the two directions λ of polarization of the light. With no charge distributions present there is no longitudinal component. Hamilton's Equations applied to this Hamiltonian give Maxwell's Equations, or Eq. (1.20), in the absence of charges and currents. Our starting assumption of wave-particle duality (as generalized in Chapter 3) states that any system described by a Hamiltonian can be described as a particle or a wave. We could construct a Schroedinger Equation for this Hamiltonian, with a product of wavefunctions $\Pi_{\mathbf{q},\lambda}\,\psi(\mathbf{u}_{\mathbf{q}}{}^{\lambda})$ However, as for the quantum treatment of lattice vibrations in Chapter 16, we shall instead only to use the properties deriving from the commutation relations $\mathbf{P}_{\mathbf{q}}{}^{\lambda}\mathbf{u}_{\mathbf{q}}{}^{\lambda} - \mathbf{u}_{\mathbf{q}}{}^{\lambda}\mathbf{P}_{\mathbf{q}}{}^{\lambda} = \hbar/i$.

We may seek raising and lowering operators as we did for the harmonic oscillator and for lattice vibrations. The two terms must have the same units, so we may add $q\mathbf{u_q}^\lambda$ and $ic\,\mathbf{P_{-q}}^\lambda$, which by Eq. (18.5) are seen to have the same units, and scale them so that the commutator is

$$a_\mathbf{q}^\lambda a_\mathbf{q}^{\lambda\dagger} - a_\mathbf{q}^{\lambda\dagger} a_\mathbf{q}^\lambda = 1 \tag{18.6}$$

as for phonons. A definition of the annihilation operator and creation operator which achieves this is

$$\begin{aligned} a_\mathbf{q}^\lambda &= \sqrt{\frac{1}{2\hbar\omega_\mathbf{q}}}\,(q\mathbf{u_q}^\lambda + ic\mathbf{P_{-q}}), \\ a_\mathbf{q}^{\lambda\dagger} &= \sqrt{\frac{1}{2\hbar\omega_\mathbf{q}}}\,(q\mathbf{u_{-q}}^\lambda - ic\mathbf{P_q}). \end{aligned} \tag{18.7}$$

q and $\omega_\mathbf{q}$ are magnitudes. We had a choice for the phase of the definitions, and we have made the conventional choice, the same as that for the case of phonons.

With these definitions (and any choice of phase) it can be confirmed that as for phonons the total field energy from Eq. (18.5) is

$$H = \Sigma_{\mathbf{q},\lambda}\, \hbar\omega_\mathbf{q}(a_\mathbf{q}^{\lambda\dagger} a_\mathbf{q}^\lambda + {}^1/_2) \tag{18.8}$$

and using Eqs. (18.1) for the vector potential, and solving Eqs. (18.7) for $\mathbf{u_q}^\lambda$, we obtain

$$\mathbf{A} = \Sigma_{\mathbf{q},\lambda} \sqrt{\frac{2\pi\hbar\omega_\mathbf{q}}{q^2\Omega}}\,(a_\mathbf{q}^\lambda + a_{-\mathbf{q}}^{\lambda\dagger})\, e^{i\mathbf{q}\cdot\mathbf{r}}\ \hat{\mathbf{u}}_\mathbf{q}^\lambda \tag{18.9}$$

with $\hat{\mathbf{u}}_\mathbf{q}^\lambda$ a unit vector in the direction of the vector potential (and electric-field polarization) for the mode $\{\mathbf{q},\lambda\}$. Each of these steps can be readily confirmed, as can a requirement that $\hat{\mathbf{u}}_{-\mathbf{q}}^\lambda = \hat{\mathbf{u}}_\mathbf{q}^\lambda$ so that the coefficients of $e^{-i\mathbf{q}\cdot\mathbf{r}}$ can be the complex conjugate of that of $e^{i\mathbf{q}\cdot\mathbf{r}}$. In Problem 18.1 we obtain the electric field as $-(1/c)\partial\mathbf{A}/\partial t = (i\omega_\mathbf{q}/c)\mathbf{A} = iq\mathbf{A}$, square the magnitude, multiply by $1/8\pi$, and integrate over all volume, to obtain an expression for the energy quadratic in the annihilation and creation operators. Taking the expectation value for a state with $n_\mathbf{q}^\lambda$ photons in a single mode yields half the expectation value which would be obtained for the Hamiltonian H in Eq. (18.8), the other half coming from the magnetic-field energy.

In Chapter 9 we discussed the interaction between electric fields and electrons, based upon the replacement of the momentum **p** in the classical Hamiltonian by $\mathbf{p} - (-e)\mathbf{A}/c$, which gives a term in the Hamiltonian of $e\mathbf{p}\cdot\mathbf{A}/mc$. This term involving both the vector potential **A** representing the field and the electron momentum $\mathbf{p} = \hbar\nabla/i$ *is* the interaction between light and the electon. For a quantum description of the light we substitute for **A** from Eq. (18.9), and note that $cq = \omega_{\mathbf{q}}$, to obtain the electron-photon coupling in the form,

$$H_{el} = \Sigma_{\mathbf{q},\lambda} \frac{\hbar e\, \hat{\mathbf{u}}_{\mathbf{q}}{}^{\lambda}\cdot\nabla}{im\omega_{\mathbf{q}}} \sqrt{\frac{2\pi\hbar\omega_{\mathbf{q}}}{\Omega}}\, e^{i\mathbf{q}\cdot\mathbf{r}}(a_{\mathbf{q}}{}^{\lambda} + a_{-\mathbf{q}}{}^{\lambda\dagger}) \qquad (18.10)$$

with the gradient operating on the electronic wavefunction; since the unit vector $\hat{\mathbf{u}}_{\mathbf{q}}{}^{\lambda}$ is perpendicular to **q** the gradient operating on $e^{i\mathbf{q}\cdot\mathbf{r}}$ gives zero.

If we look for the effect of the resulting coupling between free electron, as we did for the electron-phonon interaction, $e^{i\mathbf{q}\cdot\mathbf{r}}$ becomes $\Sigma_{\mathbf{k}} c_{\mathbf{k+q}}{}^{\dagger} c_{\mathbf{k}}$, and the ∇ becomes $i\,\mathbf{k}$, but there is no coupling between states of the same energy, as we saw in Problem 1.1. There are only higher-order processes in which a photon is absorbed and another emitted, with the electron changing its wavenumber by a vector equal to the difference in the two photon wavenumbers. This was because we could not simultaneously conserve momentum and energy with an electron which moves at less than the speed of light, while both could be conserved for phonons if the electron moved faster than the speed of sound. However, when the electron interacts with some other system which can take up the needed momentum difference, processes involving the emission or absorption of photons become possible. Using the form, Eq. (18.10), with the gradient and $e^{i\mathbf{q}\cdot\mathbf{r}}$ allows us to treat such processes.

18.2 Excitation of Atoms

One of the most important applications of the electron-photon interaction is the transition between electronic states caused by light. In Chapter 9 we treated the ionization of atoms by light, treating the light classically as an applied alternating electric field. For transitions between atomic states within an atom, it is more appropriate to use the electron-photon interaction of Eq. (18.10) and the resulting quantum transitions. We shall treat photon absorption and emission at the same time, and this will lead us naturally into the description of lasers.

The first step will be to obtain the magnitude of the matrix element, the coefficient which accompanies the annihilation and creation operators of Eq. (18.10). To be specific, we consider an atom of beryllium, which has a ground-state configuration of $1s^22s^2$, the 1s-states being the core states. We consider transitions between this ground-state configuration and an excited configuration $1s^22s2p$ which arise from the coupling from Eq. (18.10) between the atomic 2s and 2p-states of the same spin. (States of different spin are orthogonal and since the perturbation contains no spin-dependence, the matrix element contains a factor from the spin states of $<+|-> = 0$.) The wavelength of the light with photon energies equal to $\varepsilon_{2p} - \varepsilon_{2s}$ is thousands of Angstroms, so in the matrix element between two atomic states which are only appreciable over the spread of a few Angstroms we may set $e^{i\mathbf{q}\cdot\mathbf{r}}$ equal to the value for $\mathbf{r}$ at the nucleus.

If we choose p-states with zero angular momentum along the three Cartesian axes, x, y, and z (Eq. (2.29)), the only coupling between an s-state and a p-state from the z-component of the gradient is with the p_z-state, etc. Correspondingly the matrix element of $\hat{\mathbf{u}}_\mathbf{q}^\lambda\cdot\nabla$ between such a p-state and the s-state is $<2p_z|\partial/\partial z|2s>$ times the cosine of the angle between $\hat{\mathbf{u}}_\mathbf{q}^\lambda$ and the axis of the p-state, which we write as $\cos\theta_\mathbf{q}^\lambda$. Then for each spin the coupling between the s-state and a p-state with a particular axis is written

$$<2p|H_{el}|2s> = \Sigma_{\mathbf{q},\lambda}\frac{H_\mathbf{q}^\lambda}{\sqrt{\Omega}}c_{2p}^\dagger c_{2s}(a_\mathbf{q}^\lambda + a_{-\mathbf{q}}^{\lambda\dagger}) \tag{18.11}$$

with

$$H_\mathbf{q}^\lambda = \frac{\hbar\, e\, \cos\theta_\mathbf{q}^\lambda e^{i\mathbf{q}\cdot\mathbf{r}}}{im\omega_\mathbf{q}}\sqrt{2\pi\hbar\omega_\mathbf{q}}\;<2p_z|\partial/\partial z|2s> \tag{18.12}$$

for a nucleus at $\mathbf{r}$. There is similarly a matrix element of the same form as Eq. (18.11) but with 2s and $2p_z$ interchanged. We note that $<2s|\partial/\partial z|2p_z> = -<2p_z|\partial/\partial z|2s>$. It is best for this second term to change to a sum over $-\mathbf{q}$ so that the final operators are replaced by $c_{2s}^\dagger c_{2p}(a_\mathbf{q}^{\lambda\dagger} + a_{-\mathbf{q}}^\lambda)$ and $H_\mathbf{q}^\lambda$ is replaced by $H_{-\mathbf{q}}^\lambda = H_\mathbf{q}^{\lambda *}$. Combining the two, we write the electron-photon, or electron-light, interaction coupling these two states as

$$H_{el} = \Sigma_{\mathbf{q},\lambda}\frac{1}{\sqrt{\Omega}}[H_\mathbf{q}^\lambda c_{2p}^\dagger c_{2s}(a_\mathbf{q}^\lambda + a_{-\mathbf{q}}^{\lambda\dagger}) + H_{-\mathbf{q}}^\lambda c_{2s}^\dagger c_{2p}(a_\mathbf{q}^{\lambda\dagger} + a_{-\mathbf{q}}^\lambda)]. \tag{18.13}$$

The final factor in Eq. (18.12), with units of one over length, can be evaluated using atomic wavefunctions. We shall not do that, but we carry the analysis of Eq. (18.12) a little further. Measuring angles from the z-axis as usual, and using the description of states of spherical systems from Section 2.4, $<2p_z|\partial/\partial z|2s>$ becomes

$$<2p_z|\partial/\partial z|2s> = \sqrt{\frac{3}{4\pi}}\sqrt{\frac{1}{4\pi}}\int 2\pi r^2 dr \int \sin\theta d\theta\, R_1(r)\cos\theta \frac{\partial}{\partial z} R_0(r)$$

$$= \frac{1}{\sqrt{3}}\int dr\, r^2 R_1(r)\frac{\partial R_0(r)}{\partial r} \equiv \frac{1}{r_a}. \tag{18.14}$$

In the first form we took the constant factors from the spherical harmonics Y_l^m out in front. In obtaining the second form we wrote $\partial/\partial z = \partial/\partial r\ \partial r/\partial z = \partial/\partial r \cos\theta$ and integrated over angle. The integral in the final form could be obtained numerically from tabulated wavefunctions, or from approximate hydrogen-like forms for the wavefunctions $e^{-\mu r}$ and $re^{-\mu' r}$ fit to the atomic energies (as carried out in Problem 18.2), and will inevitably be of the order of the reciprocal of the size of the atom $1/r_a$, comparable to interatomic spacings, and we proceed here by defining r_a to be the reciprocal of the matrix element as written in Eq. (18.14). For our initial study it will be sufficient to use the general form with $H_{\mathbf{q}}^{\lambda}$ as in Eq. (18.13).

We now have the parameters needed to evaluate transition rates between atomic states arising from the electron-photon interaction using the Golden Rule,

$$P_{if} = \frac{2\pi}{\hbar}\Sigma_f <i|H_{el}|f> <f|H_{el}|i>\delta(E_f - E_i). \tag{18.15}$$

The initial state $|i>$ could be the ground state (or it could be an excited state with an electron in a 2p-state but we proceed first with the electronic ground state). As for phonon absorption and emission, the various terms in the operation of H_{el} on the initial states produce various final states - times constants - from which we identify the energy change appearing in the delta function. We then select the terms from the operation of H_{el} on this final state which take it back to the initial state. Thus the $a_{\mathbf{q}}^{\lambda}$ term in Eq. (18.13), operating on the electronic ground state, reduces the number of photons in the corresponding mode, while taking the electron from the 2s-state to the 2p-state, so that for that term $E_f - E_i = \varepsilon_{2p} - \hbar\omega_{\mathbf{q}} - \varepsilon_{2s}$. The term with $a_{-\mathbf{q}}^{\lambda\dagger}$ operating on the electronic ground state *adds* a photon, while taking the electron to the 2p-state. Both steps require energy, so the delta function will never be satisfied and that term does not contribute.

Similarly, the term with $a_{\mathbf{q}}^{\lambda\dagger}$ operating on the excited electronic state adds a photon while dropping the electron from the 2p- to the 2s-state for an

$E_f - E_i = \varepsilon_{2s} + \hbar\omega_{\mathbf{q}} - \varepsilon_{2p}$, and the term with the $a_{-\mathbf{q}}{}^{\lambda}$ never satisfies the energy delta function. Only two terms from H_{el} operating on the initial state survive, and when we operate again, only one of the terms in each case will take the state back to $|i\rangle$.

For the first of these transitions, in which a photon is absorbed, the operators which appear are $c_{2s}{}^{\dagger}c_{2p}a_{\mathbf{q}}{}^{\lambda\dagger}c_{2p}{}^{\dagger}c_{2s}a_{\mathbf{q}}{}^{\lambda}$. We can commute the $a_{\mathbf{q}}{}^{\lambda\dagger}$ to the right where it becomes part of the number operator, which we write as the number $n_{\mathbf{q}}{}^{\lambda}$ of phonons in the mode. We may also commute the $c_{2s}{}^{\dagger}$ to the right to obtain the number operator for the 2s-state, which we write as the probability of its occupation, f_{2s}, and the remaining pair of operators becomes, using the commutation relation, $1 - f_{2p}$. For this class of processes involving photon absorption we obtain

$$P_{if}^{absorb} = \frac{2\pi}{\hbar} \Sigma_{\mathbf{q},\, \mathrm{spin},\lambda} \frac{H_{-\mathbf{q}}{}^{\lambda} H_{\mathbf{q}}{}^{\lambda}}{\Omega} f_{2s}(1 - f_{2p}) n_{\mathbf{q}}{}^{\lambda} \delta(\varepsilon_{2s} - \varepsilon_{2p} + \hbar\omega_{\mathbf{q}}) , \quad (18.16)$$

We make the corresponding evaluation for the second transition, in which a photon is emitted, to obtain

$$P_{if}^{emit} = \frac{2\pi}{\hbar} \Sigma_{\mathbf{q},\, \mathrm{spin},\lambda} \frac{H_{\mathbf{q}}{}^{\lambda} H_{-\mathbf{q}}{}^{\lambda}}{\Omega} f_{2p}(1 - f_{2s})(n_{\mathbf{q}}{}^{\lambda} + 1) \delta(\varepsilon_{2s} - \varepsilon_{2p} + \hbar\omega_{\mathbf{q}}). \quad (18.17)$$

As for phonons, the term in Eq. (18.17) with $n_{\mathbf{q}}{}^{\lambda}$ is stimulated emission and the term with 1 is spontaneous emission. For each of these evaluations, the sum over wavenumbers is to be replaced by an integral as in our other transition-rate calculations, $\Sigma_{\mathbf{q}} \rightarrow (\Omega/(2\pi)^3)\int d^3q$ and the energy delta-function fixes the wavenumber for each mode. Such an evaluation is carried out in Problem 18.3.

We may confirm that if the system is in equilibrium, so there is one temperature T and one Fermi energy μ, the absorption and emission rates are the same. We do this by writing the statistical factors, with each of the two energy levels written relative to the Fermi energy,

$$\text{Rate absorb} \propto \frac{1}{e^{\varepsilon_{2s}/k_BT} + 1} \, \frac{e^{\varepsilon_{2p}/k_BT}}{e^{\varepsilon_{2p}/k_BT} + 1} \, \frac{1}{e^{\hbar\omega/k_BT} - 1}$$

$$\text{Rate emit} \propto \frac{1}{e^{\varepsilon_{2p}/k_BT} + 1} \, \frac{e^{\varepsilon_{2s}/k_BT}}{e^{\varepsilon_{2s}/k_BT} + 1} \, \frac{e^{\hbar\omega/k_BT}}{e^{\hbar\omega/k_BT} - 1} \quad (18.18)$$

The delta functions in Eqs. (18.16) and (18.17) are only satisfied if $\varepsilon_{2p} = \varepsilon_{2s} + \hbar\omega$, in which case the numerators in the two Eqs. (8.18) are the same, as are the products in the denominators. It is interesting that the ± 1 terms which arose from the different statistics for Fermions and photons match with the different commutation relations in the evaluation of Eq. (18.15) making this true. It had to come out that way, as guaranteed by the detailed-balance which we discussed in Section 11.1, which always applies to equilibrium systems. In Problem 18.3 we obtain the relative occupations of the two electronic states which arise as a function of light intensity for a system in a steady-state condition, but not equilibrium.

If light coupling two levels is very intense, corresponding to very large $n_{\mathbf{q}}{}^{\lambda}$ in Eqs. (18.16) and (18.17), the additional 1 in the second of these, giving spontaneous emission, is negligible. Since in steady state the absorption rate is equal to the emission rate, the factors $f_{2s}(1 - f_{2p})$ and $f_{2p}(1 - f_{2s})$ must also be equal, corresponding to $f_{2s} = f_{2p}$. This is called *saturating* the transition. It is a feature of optical transitions which becomes important in lasers.

18.3 The Three-Level Laser

We may see how laser action arises by considering the energy levels of helium, a slightly simpler system than the beryllium discussed above. In the ground state the atomic configuration is $1s^2$ and there are excited 2s- and 2p-states. These may be obtained from Moore's (1949, 1952) tables as described in Chapter 4, in connection with Table 4.1. This gives an energy -24.6 eV for the 1s-state, -4.0 eV for the 2s-state, and -3.6 eV for the 2p-

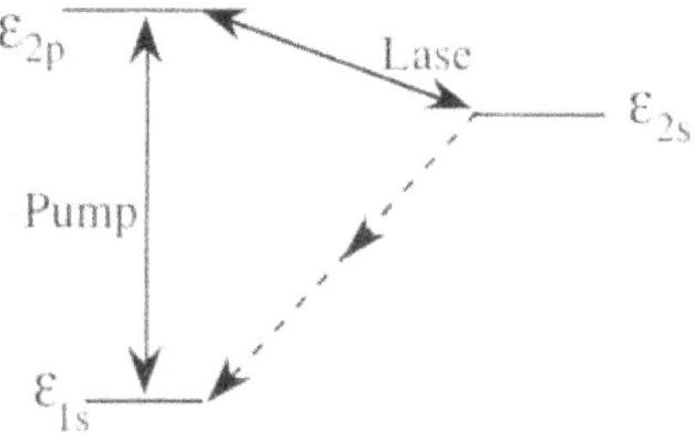

Fig. 18.1. A schematic energy-level diagram for a helium laser. In the ground state both electrons occupy the 1s-state. Light at $\hbar\omega = \varepsilon_{2p} - \varepsilon_{1s}$ pumps the electrons to the 2p state, approaching half occupation of the corresponding p-state for strong pumping. The population is then higher than that of the 2s-state, so light is emitted at $\hbar\omega = \varepsilon_{2p} - \varepsilon_{2s}$, at first by spontaneous emission, but as the intensity of the light at this frequency grows, stimulated emission is dominant. The electrons must return to the 1s-state by some other process, such as by atomic collisions.

state. These last two would both have equaled the $-e^4m/8\hbar^2 = -3.4$ eV of the hydrogen 2s- and 2p-states, except for the effect of the second proton in the nucleus and the second electron in a 1s-state which lowers the energy of the 2s-state more than the 2p-state. These numbers may not be the most convenient for practical lasers, but illustrate the mechanism of the laser. They are shown schematically in Fig. 18.1.

In equilibrium the occupation $f_0(\varepsilon_{2p}) << f_0(\varepsilon_{2s}) << f(\varepsilon_{1s})$ and, as we have seen, there are transitions up and down (as in Eq. (18.18)). If we now impose strong light at a pump frequency ω_p such that $\hbar\omega_p = \varepsilon_{2p} - \varepsilon_{1s}$, electrons will be pumped to the higher level, approaching equal concentration for the p-state which is coupled by light of this polarization and for the ground state. At the same time, the occupation of the 2s-level remains low and we have an inverted population of the 2p-state relative to the 2s-state. With the excess number of electrons in the 2p-state, there will be spontaneous emission into modes with frequency ω_l with $\hbar\omega_l = \varepsilon_{2p} - \varepsilon_{2s}$. Then if there are parallel mirrors which capture some of that light as a standing wave, the intensity in one mode will build up; with more electrons in the 2p-state than in the 2s-state, the stimulated emission will also be greater than the absorption, and the energy in that mode will grow to a very high level. This is the laser action, and with one mirror with less than 100% reflectivity intense light will emerge from exactly this mode. There must be a way for the electrons which are thus transferred to the 2s-state to return to the 1s-state if the process is to continue. There is no direct optical transition between such s-states, but the transfer can occur by collisions between helium atoms or with the wall. In other systems a fourth level between the two lower levels can provide a path of allowed transitions (i. e., if it was a p-state). There may be more than one mode in the mirror system (more than one wavelength of light) which could support lasing action, but ordinarily one will dominate. Helium, as described here, would be a gas laser. Similar systems can be constructed in solids using impurity states in an insulator to provide the counterpart of the levels of Fig. 18.1. More importantly, lasers can be made using the band states. In order to discuss them we must first consider interband transitions.

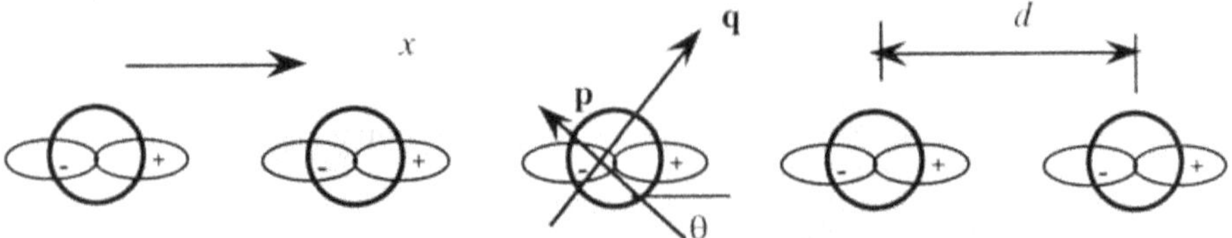

Fig. 18.2. A chain of atoms, as in Section 6.1, but with a p-state as well as an s-state on each atom. Light of wavenumber **q** and polarization **p** couples the states on each atom.

18.4. Interband Transitions

We treat only the simplest meaningful case, again electrons moving in a chain of atoms, and generalize the result to three-dimensional systems. We shall need two bands, so a second state for each atom is introduced as indicated in Fig. 18.2. We form s-bands

$$\varepsilon_{k'}{}^{c} = \varepsilon_s + 2V_{ss\sigma}\cos k'd \tag{18.19}$$

which we think of as empty conduction bands in Fig. 18.3, and p-bands,

$$\varepsilon_k{}^{v} = \varepsilon_p + 2V_{pp\sigma}\cos kd\,. \tag{18.20}$$

at a lower energy, which we think of as filled valence bands. We do not include for the present any effects of the coupling $V_{sp\sigma}$.

We consider the coupling between a valence-band state $|k^v\rangle = (1/\sqrt{N})\Sigma_i|p_i\rangle e^{ikdi}$ and a conduction band state $|k'^c\rangle = (1/\sqrt{N})\Sigma_j|s_j\rangle e^{ik'dj}$ due to the electron-light coupling from Eq. (18.13), which is

$$\langle k'^c|H_{el}|k^v\rangle = \frac{1}{N}\,\Sigma_{i,j}\,\Sigma_{\mathbf{q},\lambda}\frac{H_{-\mathbf{q}}{}^{\lambda}}{\sqrt{\Omega}}\,\langle s_j|c_{sj}{}^{\dagger}c_{pi}(a_{\mathbf{q}}{}^{\lambda\dagger} + a_{-\mathbf{q}}{}^{\lambda})|p_i\rangle e^{-i(k'dj-kdi)}. \tag{18.21}$$

If we include only the coupling between orbitals on the same atom, discussed in Section 18.2, then only terms with $i = j$ will enter, each with the phase of the light on that atom, $e^{-iq_{||}dj}$ from $H_{-\mathbf{q}}{}^{\lambda}$, with $q_{||}$ the component of $\mathbf{q}$ along the chain. Then the sum over j will give zero unless $k' = k + q_{||}$, in which case it will give N. We designate those terms in the sum by $q_{||} = k' - k$. Only the mode polarized in the plane of the figure contributes so we drop the sum over λ and $\theta_{\mathbf{q}}{}^{\lambda}$ becomes the θ in Fig. 18.2 . This gives then

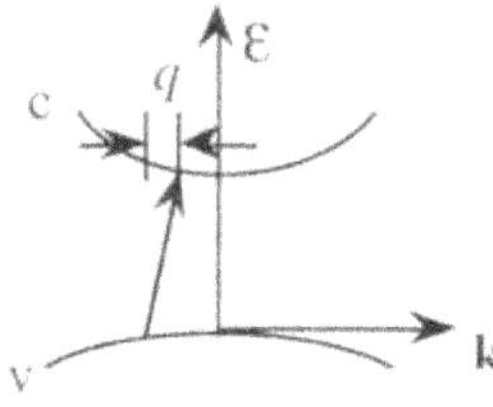

Fig. 18.3. Bands for the one-dimensional chain of Fig. 18.2. The upper band is s-like, the lower, p-like. The photon wavenumber q is ordinarily very small so a transition, indicated by the upward arrow, is almost vertical.

$$<k'^{c}|H_{el}|k^{v}> = \Sigma_{\mathbf{q}(q\| = k' - k)} \frac{H_{-\mathbf{q}}}{\sqrt{\Omega}} (a_{\mathbf{q}}^{\dagger} + a_{-\mathbf{q}}). \tag{18.22}$$

with

$$H_{-\mathbf{q}} = \frac{-\hbar e \cos\theta}{im\omega_{\mathbf{q}} r_a} \sqrt{2\pi\hbar\omega_{\mathbf{q}}} \tag{18.23}$$

from Eq. (18.12) and (18.14), analogous to the $H_{\mathbf{q}}{}^{\lambda}$ of Eq. (18.12), but without the factor $e^{i\mathbf{q}\cdot\mathbf{r}}$. There are of course also complex conjugate matrix elements $<k^{v}|H_{el}|k'^{c}>$. In semiconductors the interatomic matrix elements, $i \neq j$, actually dominate, which follows already from the fact that the dielectric susceptibilities of semiconductors are much higher than would be obtained as a sum of atomic polarizabilities. The needed matrix elements of the gradient in Eq. (18.10) can be obtained approximately from the corresponding tight-binding matrix elements (Harrison (1999), p. 219) as

$$< l,m \mid \nabla \mid l',m > = -\frac{m\mathbf{d}V_{ll'm}}{\hbar^2} = -\eta_{ll'm}\frac{\mathbf{d}}{d^2}, \tag{18.24}$$

with the $\eta_{ll'm}$ the same coefficients as given in Eq. (6.6). However, it will be simpler to proceed here with the intraatomic form.

We may generalize to matrix elements between band states in three-dimensional crystals. We must then retain the sum over polarization λ , the $\cos\theta$ again is $\cos\theta_{\mathbf{q}}{}^{\lambda}$, and $\mathbf{k}'$ is $\mathbf{k} + \mathbf{q}$. In semiconductor systems one is often interested in transitions between states in quantum slabs, discussed in Section 2.3. Then the formulation has similarity both to the one-dimensional and three-dimensional cases discussed here. It will be adequate for the limited discussion we make to leave the matrix element as $H_{\mathbf{q}}$,which we obtained with intra-atomic coupling for the chain in Eq. (18.23).

We consider first a bulk crystal, or a region in a crystal which can be regarded as bulk. Of most interest will be the transitions near the threshold energy, equal to the band gap, as illustrated in Fig. 18.3. The photon wavenumber can be obtained from $q = \omega/c$ to find that it is very small, or we can note that the photon energy required to cross the gap is in the optical range, with thousands of Angstroms wavelength, so q will be of the order of one thousandth of the Zone dimensions $2\pi/d$. The transitions are almost exactly vertical on a diagram such as Fig. 18.3. In fact, q is smaller by a factor of one hundred than shown there.

If we now imagine a semiconductor as shown in Fig. 18.3 in thermal equilibrium, there will be a low density of electrons in the conduction band and (if there is no doping) an equal concentration of holes in the valence band as we saw in Section 10.5, and Problem 10.3 in particular. Occasionally an electron can drop into an empty state of very nearly the same wavenumber in the valence band, emitting a photon. Exactly as often the reverse transition is made in which the same hole is remade and initial conduction band state occupied through the absorption of a photon of the same wavenumber. This is guaranteed for equilibrium systems by detailed balance (Section 11.1) and shown in detail for two discrete levels in Eq. (18.18). Such a detailed treatment would be more intricate for this case, but the result is guaranteed.

If we now increase the electron and hole density, without increasing the number of photons - this is taking the system out of equilibrium - photons will be created at a greater rate than they are absorbed. (This is the reverse of what was done in Problem 18.3, where the light intensity was increased over equilibrium values, leading to steady-state absorption.) This increased emission will be in proportion to both increases in carrier density. Exactly this effect is accomplished in a *light-emitting diode* as illustrated in Fig. 18.4. A confinement region, with reduced gap to hold the carriers, is surrounded by an n-type region (with excess electrons due to doping) on one side and a p-type region (with excess holes due to doping) on the other side. In equilibrium, carriers will redistribute and bands shift relative to an overall Fermi energy for the system such that again detailed balance will prevail. However, if a voltage is applied raising the potential of the electrons on the left relative to the right, the flow shown will enhance the electron density in the center. The same potential will drive holes from the right, enhancing the hole density in the central region and producing excess light at a frequency corresponding to $\hbar\omega$ slightly greater than the gap. The greater the applied voltage, the greater the intensity of the light produced.

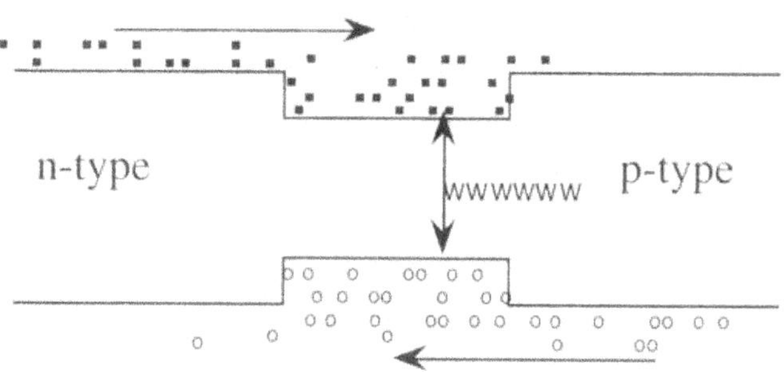

Fig. 18.4. A light emitting diode. Applying a voltage which drives the carriers in the direction given by the arrows will increase the concentration of both types of carriers in the confinement region (the region with a narrower gap) above thermal equilibrium and produce excess light at the energy of the gap, as indicated.

It is often true that the distribution of the electrons in the confinement region remains a Fermi distribution at some temperature, such as the temperature of the thermal phonon distribution (Section 10.2), but with a Fermi energy μ_c (called a *quasi-Fermi level* since there is no overall equilibrium) which is high, in accord with the high carrier concentration. Similarly the Fermi energy μ_v associated with the holes is low in accord with their high concentration. A convenient way to calculate the output of a light-emitting diode is to evaluate each contributing rate, or the factors corresponding to Eq. (18.18) which determine that rate, *for equilibrium*, but use the relevant Fermi energy which can be shifted out of equilibrium in the end.

For any pair of coupled levels, the absorption is proportional to the probability $f_v \approx 1$ that the valence-band state is occupied, times the probability $1 - f_c \approx 1$ that the conduction-band state is empty times the $\langle n_q \rangle$ for the photon. For the light-emitting diode we imagine an equilibrium distribution of photons, so the absorption is proportional to the number of thermal photons in the modes involved, $\langle n_q \rangle = 1/(e^{\beta\hbar\omega_q} - 1) \approx e^{-E_g/k_BT}$ according to Eq. (10.10), with E_g the gap. We write the total absorption rate per unit volume of material as

$$R_{absorb} = R_T\, e^{-E_g/k_BT}, \tag{18.25}$$

where R_T could be estimated as in Problem 18.3.

Calculation of the corresponding emission rate is simplest if the carrier densities are low enough that we can approximate the Fermi distribution by a Boltzmann distribution. Then, the emission is proportional to the probability $f_v \approx e^{-(\varepsilon_c - \mu_c)/k_BT}$ that the conduction-band state is occupied, times the probability $1 - f_c \approx e^{-(\mu_v - \varepsilon_v)}$ that the valence-band state is empty times $\langle n_q \rangle$ +1 for the photon. However again $\langle n_q \rangle \approx e^{-E_g/k_BT} \ll 1$ so spontaneous emission, the second term in $n_q + 1$, dominates and the rate is proportional to $e^{-(\varepsilon_c - \mu_c)/k_BT} e^{-(\mu_v - \varepsilon_v)/k_BT} = e^{-E_g/k_BT} e^{(\mu_c - \mu_v)/k_BT}$. In thermal equilibrium $\mu_c = \mu_v$ and this rate must equal the absorption rate so the proportionality constant must be the same as in Eq. (18.25). Thus the emission rate per unit volume, whether or not $\mu_c = \mu_v$, is given by

$$R_{emit} = R_T\, e^{-E_g/k_BT}\, e^{(\mu_c - \mu_v)/k_BT}. \tag{18.26}$$

By introducing a current as in Fig. 18.4 we increase μ_c relative to μ_v and increase the rate exponentially above the thermal rate, $R_T\, e^{-E_g/k_BT}$. We might mention that there are "Sommerfeld corrections" to the emission (and absorption) from Coulomb attraction between electrons and holes. They are apparently large, but not often included.

In the light-emitting diode we envisage excess carriers in comparison to thermal equilibrium. If we can actually achieve an inverted population, higher electron occupation probability f_C for states at the conduction-band edge than the corresponding f_V at the valence-band edge as illustrated in Fig. 18.5a, we may have lasing action just as in the three-level laser of Section 18.3. This would be a solid-state laser. It is accomplished by adding mirrors (often Bragg mirrors consisting of alternate layers of different refractive indices which can reflect light of a particular wavelength). Then a particular light mode can grow in intensity until stimulated emission is dominant, just as described for the three-level laser.

Accomplishing this is made difficult by the decreasing joint density of states for $\Delta E(k)$ near the band gap, as indicated in Fig. 18.5b. Various techniques may be used to alleviate this difficulty. One is to insert quantum wells as illustrated in Fig. 18.6a, so that subbands are formed in the quantum wells, with a density of states which is constant, as was seen for electrons moving in two dimensions in Problem 2.2. Multiple wells can be introduced. Such wells considerably enhance the behavior as seen in Fig. 18.6. An additional difficulty arises in real systems such as gallium arsenide, which has bands analogous to those shown to the right in Fig. 13.9. The conduction band mass is small, as shown, so that the electrons are concentrated near $k = 0$ and so also is the light-hole mass small. However, most holes will be in the heavy-hole band, with heavier mass and therefore higher density of states. These range to larger k and thus to wavenumbers which do not contribute to the lasing. One of these heavy-hole bands is

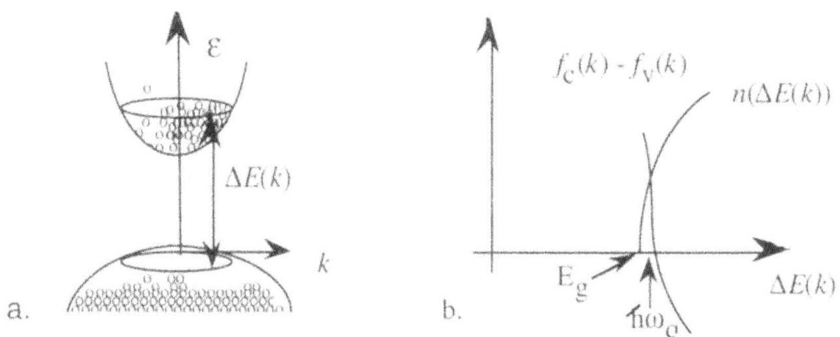

Fig. 18.5. In Part a, at small k, the occupation of conduction-band states is higher than that of valence-band states, positive $f_C(k) - f_V(k)$ as shown in Part b, and lasing can occur between such states. However, the joint density of states $n(\Delta(k)) = n_c(\varepsilon(k))n_V(\varepsilon(k))$ is small in that region. Mirrors are chosen to select a $\Delta E(k) = \hbar\omega_q$ which will optimize the emission.

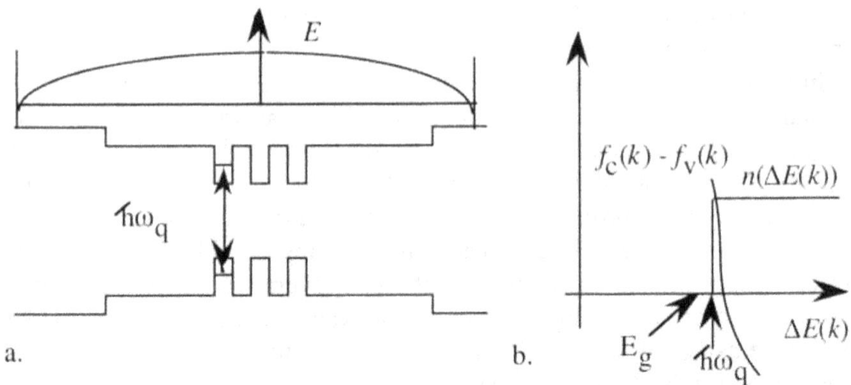

Fig. 18.6. Quantum wells are introduced in the system of Fig. 18.4 for a solid-state laser. Then the density of states for the resulting subbands rises abruptly at the minimum energy, as shown in Part b. It also concentrates the carriers where the electric fields for the lasing mode, shown as the curve above in Part a, are largest.

pushed deeper in energy by spin-orbit coupling as we shall see in Section 22.5, and the other can be shifted with a "strain-layer superlattice" system in which the layers serving as quantum wells are compressed, or made thinner (and expanded parallel to the planes) so that the p-states which are oriented perpendicular to the layers are shifted down in energy as illustrated in Fig. 18.7. (It was seen in Section 13.5 that the top of each valence band corresponds to one p-state.) These are the p-states which form the heavy-hole subbands for motion in the plane of the figure so only the light-hole subband is present at the top of the valence band, an important improvement.

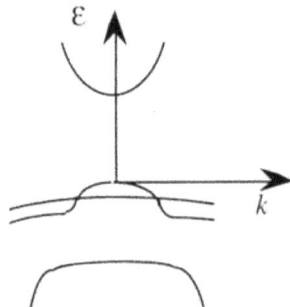

Fig. 18.7 A strain-layer superlattice can shift the heavy-hole band downward so that almost all electrons and holes are concentrated near k = 0 and contribute to laser action.

Chapter 19. Coherent States

We saw after Eq. (2.46) that a harmonic oscillator wave packet oscillated back and forth like a classical oscillator. Such a packet state of the oscillator is called a *coherent state*. The light which emerges from a laser is also such a coherent state and the concept is very important in quantum theory, though it is essentially a classical effect. It is interesting to consider it further now that we have quantized the phonon and photon fields. We begin with the quantum description of the simple harmonic oscillator, but return to a classical description of the driven oscillator before continuing the quantum description. We finally relate the results to coherent light.

19.1 Coherence in a Harmonic Oscillator

Any wavefunction $\phi(x)$ for a harmonic oscillator can be expanded in eigenstates $\Sigma_n A_n\phi_n(x)$ and we may use the time-dependence of each eigenstate, with $\varepsilon_n = \hbar\omega_0(n + {}^1\!/_2)$ in terms of the classical frequency ω_0, to obtain the time-dependence of ϕ as

$$\phi(x,t) = \Sigma_n A_n\phi_n(x)e^{-i\omega_0(n+1/2)t} . \qquad (19.1)$$

We noted in Section 2.5 that it is a consequence of this equal spacing of levels that any $\phi(x)$ must return to the same function every period, $2\pi/\omega_0$. It also follows that half-way between these returns the wavefunction will be $\phi(-x)$, the classical behavior of an oscillator. In order to have this classical behavior, we *must* have a linear combination of different states of excitation; a single eigenstate has a probability distribution symmetric in x at all times. If it is a combination of many states of excitation, we must also have coherent phases for the coefficients of the states of adjacent quantum numbers. We may see this be evaluating the expectation value of x using Eqs. (16.23) and (16.25)

$$\langle\phi|x|\phi\rangle = \sqrt{\frac{\hbar}{2M\omega_0}}\, \Sigma_{n,m} A_m{}^* A_n \langle m|(a^\dagger + a)|n\rangle\, e^{-i\omega_0(n-m)t}$$
$$= \sqrt{\frac{\hbar}{2M\omega_0}}\, \Sigma_n \left[\sqrt{n+1}A_{n+1}{}^* A_n e^{i\omega_0 t} + \sqrt{n}A_{n-1}{}^* A_n\, e^{-i\omega_0 t}\right]. \tag{19.2}$$

If the A_n , and in particular the phase, varies randomly from A_n to A_{n+1} the various terms will average to a small value. If they are *coherent*, varying for example from one n to another as $e^{i\alpha n}$, then the terms will have the same phase and lead to a large value. Coherence refers to such relations between the A_n which can make the expectation value for the coordinate large. If only one A_n were nonzero ($\phi(x)$ would be an energy eigenstate) $\langle\phi|x|\phi\rangle$ would be zero as we noted before. If two adjacent A_n were nonzero at $1/\sqrt{2}$, the sum of $A_{n+1}A_n$ would be $^1/_2$. If the magnitude of A_n is peaked at n_0 but varies slowly with n (and varies in phase as $e^{i\alpha n}$), we may replace the sum by an integral $\Sigma_n A_{n+1}{}^* A_n = \int dn\, A_n{}^* A_n\, e^{i\alpha} \approx e^{i\alpha}$, the maximum magnitude. Then Eq. (19.2) becomes

$$\langle\phi(x,t)|x|\phi(x,t)\rangle \approx \sqrt{\frac{2\hbar n_0}{M\omega_0}}\cos(\omega_0 t + \alpha). \tag{19.3}$$

This is exactly the classical limit for a harmonic oscillator with amplitude x_0 such that the energy is $^1/_2 M\omega_0{}^2 x_0{}^2 = n_0\hbar\omega_0$, with α the phase at time $t = 0$. We see that the coherent state is the classical state. A classical oscillator has a value of x and of P at time $t = 0$, and its future position and momentum are determined for all times in terms of these values. Thus a classical oscillator cannot be *in*coherent. Incoherence is a quantum effect, in this case arising because an oscillator may be in a single eigenstate, with completely uncertain phase, or may be in a combination of many eigenstates, again with uncertain phase unless the relative phases of each of the combinations are specially related.

This is a subtle point, and physicists often think of coherence - as opposed to incoherence - as a quantum effect. That is not correct from our point of view. We will attempt to keep this clear when we discuss coherence in terms of light modes in Section 19.4, but it may be desirable to treat the coherent state of the classical harmonic oscillator slightly further.

19.2 A Driven Classical Oscillator

We study the behavior of a classical oscillator in order to understand the quantum system more clearly. We will in fact be able to identify the classical response with that we have calculated quantum-mechanically using time-dependent perturbation theory, and to identify the energy loss with the transitions which we have calculated quantum-mechanically using the Golden Rule.

Let x be the classical coordinate of a mass M which feels a spring force $F = -\kappa x$, corresponding to spring constant κ, and has a normal-mode frequency $\omega_0 = \sqrt{\kappa/M}$. It will be convenient to write the driving force as $-eE\,e^{-i\omega t}$ (as for a negative charge $-e$), or the real part of that. In fact the equations of motion are linear so that we may proceed with a complex force and complex displacement and take real parts at the end if we wish. If $-eE$ is complex, the phase may be different from zero at $t = 0$. The classical equation of motion is

$$M\ddot{x} = -\kappa x - \mu\dot{x} - eEe^{-i\omega t}, \tag{19.4}$$

where we have included also a small viscous term $-\mu\dot{x}$ which would damp out any initial vibrations, leaving only the driven motion proportional to $e^{-i\omega t}$. Substituting that form, the time derivatives become factors of $-i\omega$ and we may solve for x as

$$x = \frac{-eEe^{-i\omega t}}{M\,(\omega_0{}^2 - \omega^2 - i\omega\mu/M)}\,. \tag{19.5}$$

The real part of $1/[M\,(\omega_0{}^2 - \omega^2 - i\omega\mu/M)]$ gives displacements exactly in phase with the applied force, $-eE\cos\omega t$ if $-eE$ is real. However, there is also an imaginary part of $1/[M\,(\omega_0{}^2 - \omega^2 - i\omega\mu/M)]$ which gives displacements out of phase with the force, proportional to $-eE\sin\omega t$ if $-eE$ is real. This is also just like a complex impedance in an ac electrical circuit in which there are in-phase and out-of-phase components of the current response to an applied ac voltage.

Thus we may define a complex frequency-dependent polarizability $\alpha(\omega)$ for this system by associating a dipole $p = -ex$ with the oscillator, and

$$p = \alpha(\omega)E\,e^{-i\omega t}. \tag{19.6}$$

Then if we write the small viscous term $\omega\mu/M$ as δ this polarizability is

$$\alpha(\omega) = \frac{e^2}{M} \frac{1}{\omega_0^2 - \omega^2 - i\,\delta}$$
$$= \frac{e^2}{M} \frac{\omega_0^2 - \omega^2}{(\omega_0^2 - \omega^2)^2 + \delta^2} + i\,\frac{e^2}{M} \frac{\delta}{(\omega_0^2 - \omega^2)^2 + \delta^2}. \qquad (19.7)$$

The real and imaginary parts are plotted in Fig. 19.1.

The real part, which is all that is left if δ is very small, is in tune with our physical experience. If we drive a classical oscillator at low frequency, it moves in phase with our push. If we drive it well above its natural frequency it is furthest from us, and accelerating toward us, as we pull hardest, a reverse in the displacement relative to the force. And, if we drive it near the resonant frequency the response becomes very large. The imaginary part only becomes large near the resonant frequency. It corresponds to a velocity in phase with the force rather than displacement in phase with the force, and work is done by this force in phase with the velocity. The energy is dissipated by the viscous term. Such a classical solution is very complete, and can therefore be preferable to a quantum solution when a classical approximation is appropriate.

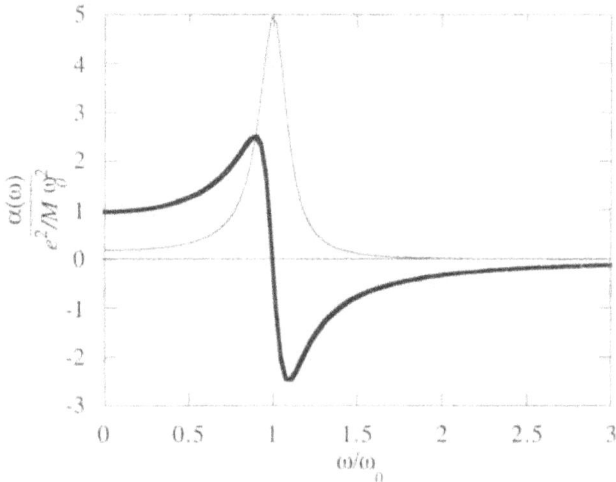

Fig. 19.1. A plot of the real part (heavy line) and the imaginary part (light line) of the polarizability based upon a harmonic dipole oscillator, from Eq. (19.7) with δ taken as $0.2\omega_0^2$.

19.3 A Driven Quantum Oscillator

It is of interest to compare this classical result with the two limiting cases which we have treated quantum-mechanically. We calculated a polarizability for a static applied field using perturbation theory already in Problems 5.2 and 5.3. We can redo this for a time-dependent field by generalizing Eq. (5.28) to time-dependent fields. Using this first-order theory for the state will retain equations of first order in the field, as in the preceding section. We carry it out first generally, and then apply it to the harmonic oscillator. We again expand the state of the system in eigenstates $|j\rangle$ of the unperturbed Hamiltonian, as in Eq. (5.8), but now we let the coefficients depend upon time and include a specific time-dependent factor for the zero-order energy,

$$|\psi\rangle = \sum_j u_j(t) e^{-i\varepsilon_j t/\hbar} |j\rangle. \tag{19.8}$$

We then seek corrections to a state $|j\rangle$ due to a time-dependent perturbation which has matrix elements $\langle i|H^1|j\rangle e^{-i\omega t}$, with $\langle i|H^1|j\rangle$ independent of time. Thus we substitute Eq. (19.8) in the Schroedinger Equation, Eq. (1.16), and multiply on the left by $\langle i|$ to obtain

$$-\frac{\hbar}{i}\frac{u_i(t)}{\partial t}e^{-i\varepsilon_i t/\hbar} + \varepsilon_i u_i(t)e^{-i\varepsilon_i t/\hbar} = \sum_j u_j(t)\langle i|H^1|j\rangle e^{-i\omega t}e^{-i\varepsilon_j t/\hbar} + \varepsilon_i u_i(t)e^{-i\varepsilon_i t/\hbar}. \tag{19.9}$$

The second terms on both sides cancel. To first order in the perturbation, $u_j = 1$ for the unperturbed starting state $|j\rangle$ and the others can be neglected in the sum on the right which already has a first-order matrix element. We see that u_i varies as $e^{i(\varepsilon_i - \varepsilon_j - \hbar\omega)t/\hbar}$ so we can differentiate and solve for u_i as

$$u_i(t) = \langle i|H^1|j\rangle \frac{e^{i(\varepsilon_i - \varepsilon_j - \hbar\omega)t/\hbar}}{\varepsilon_j + \hbar\omega - \varepsilon_i}. \tag{19.10}$$

This is the time-dependent counterpart of Eq. (5.28) which we shall use again in Section 19.5. If we add a perturbing term proportional to $e^{i\omega t}$ it gives the corresponding expression with ω replaced by $-\omega$. Combined with Eq. (19.8) this gives a *dressed state* $|j\rangle$ containing corrections to the unperturbed state.

We now apply this result to a driven harmonic oscillator. The force $-eEe^{-i\omega t}$ which we have introduced corresponds to a term in the Hamiltonian of $H^1 = eExe^{-i\omega t}$ and we can write x in terms of raising and lowering operators for the harmonic-oscillator states as in Eq. (16.25). Using these we obtain

$$H^1|n\rangle = eE\sqrt{\frac{\hbar}{2M\omega_0}}\,[\sqrt{n+1}\,|n+1\rangle + \sqrt{n}\,|n-1\rangle]e^{-i\omega t}. \tag{19.11}$$

Thus there are two corrections to a zero-order state $|n\rangle$, the first is higher in energy by $\varepsilon_i - \varepsilon_j = \hbar\omega_0$ and the second lower by the same amount. The first-order state becomes

$$|\psi\rangle = |n\rangle + eE\sqrt{\frac{\hbar}{2M\omega_0}}\left\{\frac{\sqrt{n+1}\,e^{-i(\omega-\omega_0)t}}{\hbar(\omega-\omega_0)}\,e^{-i\omega_0 t}|n+1\rangle + \frac{\sqrt{n}\,e^{-i(\omega+\omega_0)t}}{\hbar(\omega+\omega_0)}\,e^{+i\omega_0 t}|n-1\rangle\right\}. \tag{19.12}$$

The factors $e^{\pm i\omega_0 t}$ preceding $|n\pm1\rangle$ in the final two terms are factors appearing in Eq. (19.8) if energies are measured relative to that of $|n\rangle$, so no factor appears with the first term, $|n\rangle$. To obtain the polarizability from this we evaluate $\langle\psi|-ex|\psi\rangle$ using again Eq. (16.25). All time-dependent phase factors $e^{\pm i\omega_0 t}$ cancel in these first-order terms and we obtain

$$\langle\psi|-ex|\psi\rangle = -\frac{e^2Ee^{-i\omega t}}{M\omega_0(\omega-\omega_0)} - \frac{e^2\omega n\,Ee^{-i\omega t}}{M\omega_0(\omega^2-\omega_0^2)}\,. \tag{19.13}$$

There is an important lesson from this result. We have used a complex perturbation $eEe^{-i\omega t}$ and should add its complex conjugate, to correspond to a real perturbation. This will give a term equal to Eq. (19.13) with ω replaced by $-\omega$. However, the sum for the two second terms is purely imaginary, proportional to $i\sin\omega t$. If we proceed more carefully, as with the small viscous term which we introduced with the classical equations, this term will lead to the absorptive term which we shall treat separately in a moment. Similarly, the sum of the two first terms will lead to one term which is purely imaginary and a second which leads to a real $2e^2E\cos\omega t/[M(\omega_0^2 - \omega^2)]$. This corresponds exactly to the real part of the classical result, Eq. (19.7), if δ is taken equal to zero. Perturbation theory which led to Eq. (19.10) correctly gives the dressing of the state, and the polarization of the state, as long as the frequency ω is far enough from resonance that absorption is not occurring and the denominator $\omega_0^2 - \omega^2$ is not becoming so small that the corrections are large. It is the direct generalization of the perturbation theory of Eq. (5.28) to time-dependent fields.

For the absorptive term alone, the imaginary part, we can return to the simple time-dependent perturbation theory of Eq. (9.9), using the matrix elements obtained from Eq. (19.11)

$$\frac{1}{\tau} = \frac{2\pi}{\hbar} \sum_j \sum_\omega H_{0j}(\omega) H_{j0}(\omega)\, \delta(\varepsilon_0 - \varepsilon_j + \hbar\omega)$$
$$= \frac{\pi e^2}{M\omega_0} \sum_\omega E_\omega{}^2[(n+1)\delta(\hbar\omega - \hbar\omega_0) + n\delta(\hbar\omega + \hbar\omega_0)]. \tag{19.14}$$

The first term represents excitation of the oscillator and the second de-excitation (with negative ω), with n the initial quantum number of the oscillator. This is still treating the alternating field as a classical field, rather than as the quantized field of Chapter 16.

To compare with the classical result we would multiply the first term by $\hbar\omega_0$ and the second by $-\hbar\omega_0$ to obtain the net quantum rate of energy absorption,

$$R_{ab}{}^{quant} = \frac{\pi e^2 \hbar}{M} \sum_\omega E_\omega{}^2 \delta(\hbar\omega - \hbar\omega_0)\,. \tag{19.15}$$

We can now identify this absorption with the imaginary part of the classical polarizability in Eq. (19.7). We must allow a distribution of light fields E_ω with frequencies in the range of $\omega = \omega_0$ in Eq. (19.7). Then for each there is a dipole induced equal to $\alpha(\omega)E_\omega$ and work done by the field equal to the field times the rate of change of the dipole, $-i\omega\alpha(\omega)E_\omega{}^2$. We should sum this over all frequencies to obtain a classical rate of energy absorption,

$$R_{ab}{}^{class} = \frac{e^2}{M} \sum_\omega E_\omega{}^2 \frac{\omega\delta}{(\omega_0{}^2 - \omega^2)^2 + \delta^2}\ . \tag{19.16}$$

For small δ^2 this function of frequency is seen to be strongly peaked at $\omega = \omega_0$ such that an integral over ω (from $-\infty$ to ∞) of the final factor is $\pi/2$. Thus the final factor in Eq.(19.16) can be written $\pi\delta(\omega-\omega_0)/2 = \pi\hbar\delta(\hbar\omega - \hbar\omega_0))/2$ and if we add an equal contribution for the complex conjugate fields we obtain exactly the quantum result, Eq. (19.15). An interesting aspect of this comparison is that the energy delta function which appears in the Golden Rule is present also in the classical result. It is not original with quantum theory.

To treat both the real and imaginary part correctly at the same time, as we did for the classical case, we might introduce the quantum-mechanical density matrix, with a small imaginary term as in the classical treatment. There are also Green's-function formulations which accomplish this, but both go beyond the scope of this text.

It may be useful to be reminded of these classical results, which are just the counterpart of the quantum effects we have been treating. All we have added to classical physics is our starting assertion of wave-like, as well as particle-like, characteristics of everything. When we work in a large-quantum-number limit, where the wave aspects become unimportant, we must obtain the classical results. This is called the *correspondence principle.*

19.4 Coherent Light

In Chapter 18 we introduced normal coordinates for the photon field and noted that each mode, wavenumber and polarization, behaved as a harmonic oscillator. When we talk of coherent states of light we are talking about the states of a single mode. We are constructing states of the system which are sums of different excitation levels (or numbers of photons) and coherent amplitudes of the different excitation levels *in that single mode.* We should not get this concept mixed with a wave packet which is a combination of amplitudes in different modes.

For the mode in question, with a particular wavenumber $\mathbf{q}$, we construct combinations of states with different excitation levels,

$$|\psi\rangle = \Sigma_{n_q} A_{n_q} e^{-in_q\omega_q t}\, |n_q\rangle\ , \tag{19.17}$$

again with a time-dependent factor for the energy of each harmonic-oscillator state. If we let the magnitude of A_{n_q} be slowly varying with n_q, but peaked at some large value n_0, and we let the phase vary as $e^{i\alpha n_q}$ then we obtain coherent light. Most results from the preceding section carry over. We wrote the vector potential in terms of annihilation and creation operators in Eq. (18.9) as

$$\mathbf{A} = \Sigma_{\mathbf{q},\lambda} \sqrt{\frac{2\pi\hbar\omega_\mathbf{q}}{q^2\Omega}}\,(a_\mathbf{q}^{\lambda} + a_{-\mathbf{q}}^{\lambda\dagger})\, e^{i\mathbf{q}\cdot\mathbf{r}}\ \hat{\mathbf{u}}_\mathbf{q}^{\lambda}\ . \tag{19.18}$$

We obtain a vector potential $<\psi|\mathbf{A}|\psi>$ varying as $\sqrt{2\pi n_0\hbar\omega_q/q^2\Omega}\exp[i(\mathbf{q}\cdot\mathbf{r} - \omega_0 t - \alpha)]$ from the first term, $a_{\mathbf{q}}^{\lambda}$, in analogy with Eq. (19.3). We obtain a complex-conjugate expression from the second term, $a_{-\mathbf{q}}^{\lambda\dagger}$, for the opposite $\mathbf{q}$. This is a coherent state in which the vector potential and the electric field vary in a well-prescribed way as a function of position and time, $\cos(\mathbf{q}\cdot\mathbf{r} - \omega_q t - \alpha)$. This is simply a classical light wave propagating with a wavenumber $\mathbf{q}$, exactly as the coherent state of the harmonic oscillator in Section 19.1 oscillated in a prescribed way.

It should be no surprise that lasers, in which the light is produced from stimulated emission, produce coherent light. However, it does not seem simple to show it from our equations. The inverted population of electrons in the atoms apparently behaves as a highly excited charged harmonic oscillator. Once there is light present, the light stimulates emission in phase with itself and this oscillating dipole radiates according to Maxwell's equations, producing coherent radiation with a dependence upon position and time corresponding to the phase of the starting radiation. In just this way a radio transmitter produces coherent radiation, with well-determined variation of the field corresponding to the current in the antena.

Again, we have been discussing only the coherence between the phases of different excitation levels *for a single optical mode.* If we wish to discuss pulses of light, light packets in real space, we must match the phase of *neighboring* modes relative to each other. This is a different kind of coherence, but requires that the state within each of the modes is also coherent between excitation levels. Otherwise we cannot associate a phase with that mode, needed to construct the packet.

19.5 Electromagnetically-Induced Transparency

We discuss one further aspect of coherent light, closely related to our treatment of the harmonic oscillator in Section 19.1. It has been discussed much more completely by Harris (1997) and we follow a part of his analysis. We treat the simplest case first, the hydrogen atom with an applied static, or dc, field. We shall then return to the more interesting case with this dc field replaced by a light field. In neither case is the quantization of the light essential so we can proceed with coherent classical light.

We have sketched the hydrogen levels in Fig. 19.2, with the states $|2s>$ and $|2p>$ having the same energy, as we saw in Section 4.1, and the state $|1s>$ at lower energy. The dc field E_c is called the "coupling field" and it splits the energies of the 2s- and 2p-states in hydrogen. Such splitting of degenerate levels by an electric field is called the *Stark Effect,* and the splitting is usually of second-order in the field, but for 2s- and 2p-states we can see that it is first order. The coupling can be written $eE_c<2p|x|2s>$,

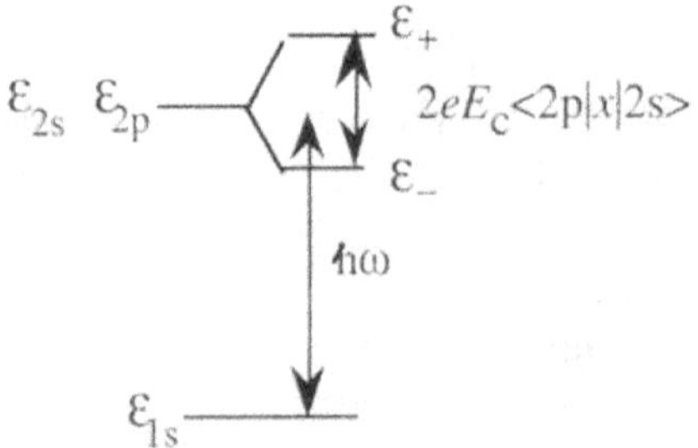

Fig. 19.2. The energy levels of the hydrogen atom, with an electric field E_c applied which produces a Stark splitting of the 2s- and 2p-levels into levels at ε_+ and ε_-. The system then responds to a probe beam of frequency ω according to a polarizability as shown in Fig. 19.3

which turns out for these states to be $3eE_c(\hbar^2/(me^2))$. Two such coupled degenerate states form "bonding" and "antibonding" combinations $|\psi_\pm\rangle = (|2s\rangle \pm |2p\rangle)/\sqrt{2}$ with energies

$$\varepsilon_\pm = \varepsilon_{2s} \pm eE_c\langle 2p|x|2s\rangle , \qquad (19.19)$$

as we saw already in Section 5.1. We now apply a "probe field", an optical field $E_p e^{-i\omega t}$ (we could take the real part afterward) also indicated in Fig. 19.2, which couples the 1s to the 2p-orbital, and therefore to the $|\psi_\pm\rangle$. We may use first-order perturbation theory to correct the occupied 1s-state to $|1s\rangle^{(1)}$. For time-dependent perturbations we may write the first-order state using Eqs. (19.8) and (19.10). We measure energies from ε_{1s} in writing the phase factors $e^{-i\varepsilon_\pm t/\hbar}$. We obtain

$$|1s\rangle^{(1)} = |1s\rangle + \frac{|\psi_-\rangle\langle\psi_-|exE_p|1s\rangle e^{-i\omega t}}{\varepsilon_{1s} + \hbar\omega - \varepsilon_-} + \frac{|\psi_+\rangle\langle\psi_+|exE_p|1s\rangle e^{-i\omega t}}{\varepsilon_{1s} + \hbar\omega - \varepsilon_+}$$

$$(19.20)$$

$$= |1s\rangle + \left(\frac{|\psi_-\rangle}{\varepsilon_{1s} + \hbar\omega - \varepsilon_-} - \frac{|\psi_+\rangle}{\varepsilon_{1s} + \hbar\omega - \varepsilon_+}\right)\frac{\langle 2p|x|1s\rangle eE_p}{\sqrt{2}}\, e^{-i\omega t} .$$

In the last step we noted that the 2s-state is not coupled to the 1s-state and the p-states in the two terms enter with opposite sign. We may proceed in the same way to obtain the dipole associated with this state, $\langle 1s|^{(1)}\text{-}ex|1s\rangle^{(1)}$, and equate it to $\alpha(\omega)eE_p e^{-i\omega t}$ to first order in the field to obtain the real part of the polarizability $\alpha(\omega)$ as

$$\alpha(\omega) = \left(\frac{1}{\varepsilon_{1s} - \varepsilon_- + \hbar\omega} + \frac{1}{\varepsilon_{1s} - \varepsilon_+ + \hbar\omega}\right)\langle 1s|x|2p\rangle\langle 2p|x|1s\rangle e^2 . \qquad (19.21)$$

This is plotted in Fig. 19.3. We see that midway between the two levels the polarizability vanishes. One can examine the state at that frequency to see that only the 2s-state is contained in the first-order state, so there is no dipole. Correspondingly a medium made of such systems will have no refraction, and there is no loss in the system because there is no coupling between the 1s- and 2s-states. This is called an *electromagnetically induced transparency* at this intermediate frequency.

For any atom but hydrogen the two states ε_{2s} and ε_{2p} would not be degenerate. We might guess that electromagnetically-induced transparency could be produced if instead of applying a dc electric field to the two states of the same energy we applied a coupling field $E_c e^{-i\omega_c t}$ (or the real part of this) with $\hbar\omega_c = \varepsilon_{2p} - \varepsilon_{2s}$ much as we coupled harmonic oscillator states with an ac field in Section 19.3. This speculation turns out to be correct, and it can be understood by treating the coupling field exactly as we treated the driving field in Section 19.3. We expand our state now in only two states in Eq. (19.8), which we label $i = 2p$ and $j = 2s$ Then Eq. (19.9) becomes

$$-\frac{\hbar}{i}\frac{u_{2p}(t)}{\partial t}) = u_{2s}(t)\langle 2p|H^1|2s\rangle e^{-i(\varepsilon_{2s}-\varepsilon_{2p}-\hbar\omega_c)t/\hbar} \qquad (19.22)$$

and the corresponding equation with 2s and 2p interchanged. In both cases the phase factor on the right becomes one when the coupling field is tuned

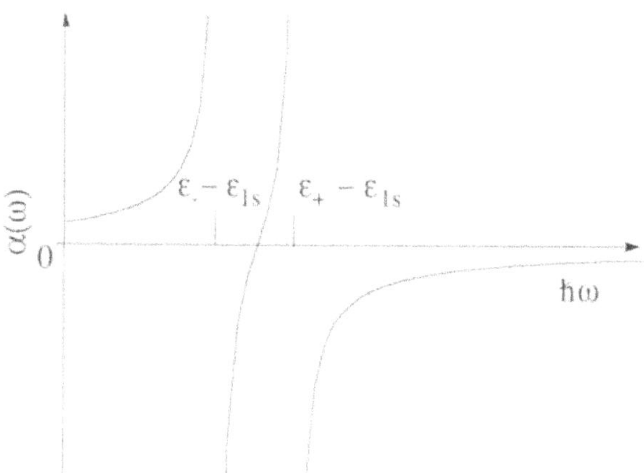

Fig. 19.3. The polarizability of the Stark-split hydrogen atom, Fig. 19.2, as given in Eq. (19.21).

appropriately. The two equations can be combined as $-\hbar^2\partial^2 u_{2s}/\partial t^2 = \langle 2s|H^1|2p\rangle\langle 2p|H^1|2s\rangle u_{2s}$ with solutions $u_{2p} = u_{2s} = e^{-i\Omega_R t}$ and $u_{2p} = -u_{2s} = e^{\,i\Omega_R t}$. Here Ω_R called the *Rabi frequency*,

$$\Omega_R{}^2 = \frac{\langle 2s|H^1|2p\rangle\langle 2p|H^1|2s\rangle}{\hbar^2}\,. \qquad (19.23)$$

Just as the perturbation-theory correction to the states could be generalized to time-dependent potentials in Eq. (19.10), two nondegenerate states can be coupled by an appropriate time-dependent field to form bonding and antibonding *dynamic states*. These two dynamic states play exactly the role played by the Stark-split 2s and 2p states in hydrogen. They are peculiar states, and not eigenstates as we have discussed extensively, in that the two terms in the wavefunction, $|\psi\rangle = (u_{2s}e^{-i\varepsilon_s t/\hbar}|2s\rangle \pm u_{2p}e^{-i\varepsilon_p t/\hbar}|2p\rangle)e^{\pm i\Omega_R t}$, do not change phase at the same rate. However, they respond to a probe field much as did the Stark-split hydrogen states.

If we now introduce a probe light wave at a frequency ω_p near $(\varepsilon_{2p} - \varepsilon_{1s})/\hbar$ (note that this is the energy difference between the *1s-state* and the 2p-state) it will couple an occupied electronic 1s-state to an upper dynamic state at $\varepsilon_{2p} + \hbar\Omega_R$ and a lower dynamic state at $\varepsilon_{2p} - \hbar\Omega_R$, just as the probe in the degenerate case coupled the 1s-state to states split by $\pm \langle 2p|exE_c|2s\rangle$. For probe frequencies between these dynamic levels raised and lowered by the Rabi frequency, the two modes enter the response with opposite signs, just as they did for the degenerate case in Eq. (19.21). With vanishing polarizability at the midpoint between the states, the system is completely transparent. There is no absorption nor refraction. Again this is accomplished with a state with only 1s- and 2s-occupation. Absorption rises with the square of the frequency difference from the crossing point.

The range of quantum-optic effects, of which this is an example, is enormous. Our object here is only to introduce the concepts and methods by which they are understood, not to explore the many possibilities.

VII. Many-Body Effects

The step which made electronic structure understandable was the one-electron approximation, which we introduced in Section 4.2. In looking at the state of one electron, the effects of other electrons were included in an average way by including an averaged potential from those electrons. This one-electron picture provided us with states in terms of which we could discuss transitions and tunneling and optical absorption and emission. They also proved the basis for statistical analysis when many particles were present and could be used to estimate total-energy changes when atoms were rearranged or moved. We turn finally to some cases in which this one-particle picture is inadequate and see how we can proceed to understand such systems.

Chapter 20. Coulomb Effects

The principal interaction between electrons is the Coulomb interaction, and it is the basis of most of the effects we shall discuss. When we discuss superconductivity in Chapter 23, the important interaction between electrons will arise indirectly through the phonons. In particle and nuclear physics the interactions come from fundamentally different sources as we saw in Section 17.4. There are many qualitatively different effects arising from the interaction between particles. If we understand the physical nature of any effect, we can ordinarily frame the problem in terms of that understanding, much as we took variational wavefunctions to correspond to bonding states in molecules, or propagating states in solids. Including many-body effects is not a straight-forward addition of another term to the one-particle Hamiltonian; it is an asking of new questions. We begin with a discussion of Coulomb shifts, which arise because the charge on an individual electron is not infinitesimal.

20.1 Coulomb Shifts

We made a one-electron approximation in constructing electronic states in atoms in Chapter 4. This was a seeking of approximate many-electron states in the form of a product wavefunction, or an antisymmetric combination of product wavefunctions, of the form $\psi_1(\mathbf{r}_1)\psi_2(\mathbf{r}_2)...\psi_N(\mathbf{r}_N)$ for the N electrons present. This led to a one-electron eigenvalue equation with a potential based upon which states were occupied, and the solution of that equation gave a set of energy eigenstates ε_i, the lowest of which were occupied in the ground state, corresponding for example to a $1s^22s^22p^2$ configuration for carbon. We indicated that these eigenvalues were approximately equal to the removal energy of an electron from the corresponding state. It is also true that the energy required to transfer an electron in the atom to an excited state of the atom is given approximately by the difference in the eigenvalues corresponding to the states between which the electron is transferred. For example, changing from a $1s^22s^22p^2$ to a $1s^22s2p^3$ configuration for carbon requires approximately ε_{2p} - ε_{2s}. The new configuration actually corresponds to a slightly different charge distribution and potential which should be used to obtain new eigenvalues, but we have neglected such small corrections, which are many-body effects.

However, if we were to remove a second electron from an atom, going from $1s^22s^22p$ to $1s^22s^2$ for carbon, it is clear that much more energy would be required to remove that second electron than the ε_{2p} which was required for the first, It would be working against the extra $-e^2/r$ from the doubly-charged atom as it was removed. Similarly, adding an electron to a neutral carbon atom, going to a $1s^22s^22p^3$ configuration would not gain the energy ε_{2p}. An electron returning to the ionized atom to make it neutral gains ε_{2p}, but that coming to a neutral atom gains less by an energy equal to the Coulomb interaction U between two p-electrons. It is of order seven electron volts for silicon and the heavier elements but over eleven eV for carbon (estimates are given for all the elements in Harrison (1999), p. 9). This corresponds to e^2/r with r of the order of 2Å as expected for charge distributions of atomic size. This is all in accord with the familiar fact that the *electron affinity* of an atom, the binding energy of the additional electron in a negatively charged atom, is much smaller than the ionization energy of the neutral atom. The difference is this Coulomb interaction U which is also approximately equal to the difference in the first and second ionization energies of the same atom (assuming both removals are from the same state, e. g., a 2p-state).

One might have thought that this Coulomb effect would spoil the prediction of cohesive energy of an alkali halide which we made in Section 6.3. We took the energy gained in forming the solid as the energy gained in adding an electron to the halogen atom, minus the energy required to remove it from the alkali atom. Here we would say that an energy U should be added to the free-atom term value we used for the halogen. That is true, but the energy of that added electron is also lowered by the presence of the six positive alkali *ions* surrounding it, raised by the twelve nearest halogen ions, etc. The sum over neighbors is called the Madelung energy, equal to $-1.8e^2/d$ (see, for example, Harrison (1999) 326ff), and approximately cancels the Coulomb U. The cancellation is no accident. The atoms in the ionic crystal select a spacing such that the transfer of electrons between states on different atoms does not greatly change their proximity to the nuclei. Such cancellations have made many of the simplified one-electron estimates meaningful in spite of real Coulomb shifts.

One might also have thought that such Coulomb shifts did not apply to the transfer of an electron from the valence band in a semiconductor to the conduction band in the semiconductor since we think of both states as spread throughout the entire crystal. This would be misguided since the crystal is in fact made of atoms and an atomic description is also meaningful for the crystal. Thus we may think of the transfer of an electron to the conduction band as a transfer from a bonding state (for which the energy eigenvalue applies) to an antibonding state *in a site where the bond levels are both occupied.* Thus we might expect the eigenvalues - the results of a band calculation - to underestimate the gap by an energy of the order of the U for the constituent atoms. That is true, but we may also see that this enhancement of the gap is reduced by a factor of $1/\varepsilon$, with ε the dielectric constant equal to 12 for silicon,

$$\delta E_{\text{gap}} = \frac{U}{\varepsilon}. \tag{20.1}$$

This is not because the dielectric medium intervenes between the interacting electrons, but because an extra electron in a bond polarizes the *surrounding* medium so that the potential is $+e^2/(\varepsilon r)$ and reduced by a factor of $1/\varepsilon$ at the surface of the atom or bond. This enhancement of the gap, of order 7 eV/12 ≈ 0.5 eV for silicon, relative to band calculations, is seen experimentally. It can be calculated more completely by the mathematical methods of many-body theory, as by Hybertsen and Louie (1985), but it is given rather well by Eq. (20.1) for all semiconductors and insulators (Harrison (1999), 207ff).

Adolph, Gavrilenko, Tenelsen, Bechstedt and Del Sol (1996) have made a rather complete study of the effect of this enhancement on various properties. It is found that the enhancements tend to be rather independent of wavenumber in the bands, as suggested by the Eq. (20.1), so that the correction is approximately a displacement of the entire band, without changing the dependence upon wavenumber. This is sometimes called a "scissors" operation, like cutting a page on which the calculated bands are plotted, and shifting the conduction bands upward in energy by U/ε in order to describe the real excitations to conducting states. For calculating energy shifts by perturbation theory, as for the dielectric response in Problem 5.3, or band curvatures in the so-called $\mathbf{k}\cdot\mathbf{p}$ method, one should again use the enhanced band gap, including the contribution from Eq. (20.1), in the energy denominators. This is not completely obvious, and would not be the case if the perturbation-theory shifts corresponded essentially to excitation within each bond site, as would be the case if the excitons discussed briefly in Section 14.3 were localized to a bond site and had binding energy (relative to a separated electron and hole) given by Eq. (20.1). Excitons are in fact spread over many bonds in semiconductors and much more weakly bound relative to a separated electron and hole.

We note finally that in a metal, for which we think of $\varepsilon = \infty$, no enhancement is expected and the metallic conductivity associated with a finite density of excited states per unit energy at the Fermi energy is retained. As we go from a semiconducting state to a metallic state by shifting the conduction bands downward, the dielectric constant ε increases and the real excitation energies become closer to those from the band calculation until they coincide exactly as the gap becomes zero.

20.2 Screening

Asserting an infinite dielectric constant for a metal is an oversimplification which really applies only to fields constant in time and constant in space. In this section we consider the effects of space-dependent and time-dependent applied fields which redistribute the electronic charge and modify the applied potential, an effect called *screening*. The problem requires a *self-consistent solution* : in order to calculate the potential which is present in the system we need to know the charge distribution, and to calculate the charge distribution we need to know the potential. There are two ways in which such self-consistent solutions are often obtained. For numerical solution one guesses the potential, perhaps as a superposition of free-atom potentials, and then calculates the wavefunctions and charge distribution. From this charge distribution one recalculates the potential, and repeats - or iterates - the process until both potential and charge distribution have settled down at the self-consistent solution. The second, which we shall use here, is to linearize the response to the potential, allowing an expansion in independent components. Then the response equation and Poisson's Equation can be solved together self-consistently.

The simplest basic formulation is the Fermi-Thomas method, a semiclassical theory which we discuss here. We shall then outline the quantum treatment of the same effects and give the results. The Fermi-Thomas approximation envisages a net potential $V(\mathbf{r})$ (including any modifications from charge redistribution) which varies slowly over distances of the order of the electron wavelength, a few Angstroms in metals. Then we imagine the system of electrons in equilibrium, with a single Fermi energy as described in Section 10.5. Since the potential is varying slowly in space, we may consider a region at $\mathbf{r}$ where the potential is essentially constant and the electronic energies given by $\hbar^2k^2/(2m) + V(\mathbf{r})$. The states will be filled to the Fermi energy, an energy $\hbar^2k_F{}^2/(2m) + V(\mathbf{r})$ with a Fermi wavenumber related to the electron density in that region $n(\mathbf{r})$, given in Eq. (2.10) as $k_F{}^3(\mathbf{r}) = 3\pi^2n(\mathbf{r})$. However, the Fermi energy to which we fill is the statistical Fermi energy μ which we introduced in Section 10.5. and is a constant of the system. Therefore,

$$\frac{\hbar^2 k_F{}^2(\mathbf{r})}{2m} + V(\mathbf{r}) = \frac{\hbar^2(3\pi^2 n(\mathbf{r}))^{2/3}}{2m} + V(\mathbf{r}) = \mu \tag{20.2}$$

at all $\mathbf{r}$. We see that if the net potential, $V(\mathbf{r})$, varies slowly with position, so also must the electron density. Through Poisson's Equation we know that there must be a contribution $V_s(\mathbf{r})$, called the *screening potential,* to this net potential $V(\mathbf{r})$ satisfying $-\nabla^2 V_s(\mathbf{r}) = 4\pi e^2 n(\mathbf{r})$. The net potential $V(\mathbf{r})$ also contains other applied contributions $V_0(\mathbf{r})$, such as the potential from the nuclei, which we ordinarily know at the outset. The Fermi-Thomas method solves these two equations together. It can even be applied to an atom with Z electrons and an applied potential $V_0(\mathbf{r}) = -Ze^2/r$, solving for $n(r)$ rather than for the wavefunction as in the more complete quantum calculation. It is called *semiclassical* because it retains the Pauli principle but not the full wave mechanics. The method has not proven very useful for such systems for which the real electron density varies as rapidly with position as in atoms and molecules.

Of much greater interest is the application to metals for which the electron density is rather uniform. It is then appropriate to linearize Eq. (20.2) and the equations can be solved analytically. We in fact see from Eq. (20.2) that the change in electron density due to a small change in net potential is

$$\delta n(\mathbf{r}) = -\frac{3mn^{1/3}}{(3\pi^2)^{2/3}\hbar^2} V(\mathbf{r}) = -n(\varepsilon)\delta V(\mathbf{r}), \tag{20.3}$$

with the density of states $n(\varepsilon)$ per unit energy and per unit volume given in Eq. (2.11) and evaluated at the Fermi energy. We may understand the final form by noting that as the potential fluctuates from point to point the filling varies much as the depth in a swimming pool varies as the floor fluctuates up and down. This is illustrated in Fig. 20.1 for the electron gas.

Once we have linearized the equations it becomes appropriate to Fourier transform any applied potential as $V^0(\mathbf{r}) = \Sigma_{\mathbf{q}} V_{\mathbf{q}}{}^0 e^{i\mathbf{q}\cdot\mathbf{r}}$ and treat each Fourier component separately to obtain a *dielectric function* which describes the modification of each term in the potential by the redistribution of the electron gas. The screening potential for each component will have the same dependence $e^{i\mathbf{q}\cdot\mathbf{r}}$ and the coefficient is written $V_{\mathbf{q}}{}^s$, so the coefficient for the net potential is

$$V_{\mathbf{q}} = V_{\mathbf{q}}{}^0 + V_{\mathbf{q}}{}^s. \tag{20.4}$$

In Poisson's Equation, given above, the $\nabla^2 V_s(\mathbf{r})$ becomes $-q^2 V_{\mathbf{q}}{}^s$ so

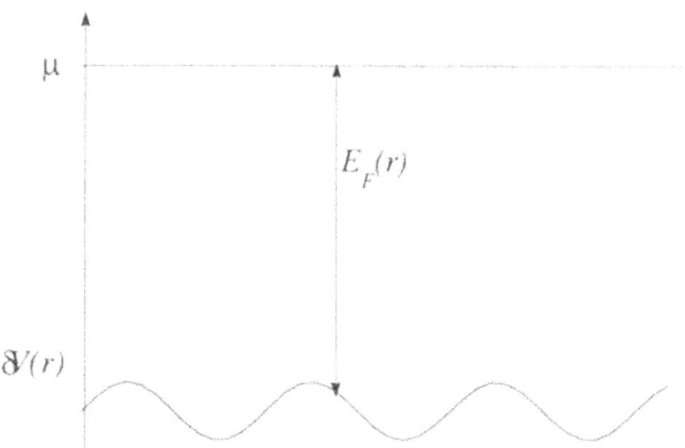

Fig. 20.1. In the linearized Fermi-Thomas approximation the statistical Fermi energy μ is a constant of the system, but if the potential fluctuates, the kinetic energy level to which states are filled, E_F, must fluctuate, giving an electron-density fluctuation equal to $-\delta V(\mathbf{r})$ times the density of states $n(\varepsilon_F)$ per unit energy per unit volume near the Fermi energy.

$$V_{\mathbf{q}}{}^{s} = \frac{4\pi e^2}{q^2}\, n_{\mathbf{q}}{}^{s}. \tag{20.5}$$

Finally, in terms of Fourier components Eq. (20.3) becomes

$$n_{\mathbf{q}}{}^{s} = -n(\varepsilon_F) V_{\mathbf{q}}. \tag{20.6}$$

We may solve these three equations together by adding $V_{\mathbf{q}}{}^{0}$ to both sides of Eq. (20.5) so the left side becomes $V_{\mathbf{q}}$. Then $n_{\mathbf{q}}{}^{s}$ on the right side of Eq. (20.5) is written in terms of $V_{\mathbf{q}}$ using Eq. (20.6) and the result solved for $V_{\mathbf{q}}$ as

$$V_{\mathbf{q}} = \frac{V_{\mathbf{q}}{}^{0}}{\varepsilon(q)} \tag{20.7}$$

with

$$\boxed{\varepsilon(q) = 1 + \frac{4\pi e^2 n(\varepsilon_F)}{q^2} \equiv 1 + \frac{\kappa^2}{q^2},} \tag{20.8}$$

called the wavenumber-dependent dielectric function because any component of the applied potential, or applied electric field, of wavenumber q is reduced by a factor of $1/\varepsilon(q)$. The Fermi-Thomas screening parameter κ is given by

$$\kappa^2 = 4\pi e^2 n(\varepsilon_F) = \frac{4me^2k_F}{\pi\hbar^2}, \tag{20.9}$$

obtained using Eq. (2.11) for $n(\varepsilon_F)$.

Recall that we have linearized the equations so it is really valid only for small perturbations and we have used the Fermi-Thomas approximation which assumes perturbations which vary slowly with position, meaning that q is small compared to k_F. In spite of these limitations in principle, the theory often works well quantitatively even when we go beyond those limitations. This may be because the self-consistent solution prevents large unrealistic deviations from the correct solution.

We already made use of this screening in obtaining the matrix elements of an empty-core pseudopotential in Section 13.1. Because the problem was linearized, we could calculate the screening of each atomic pseudopotential in a metal separately. This is a case where the applied potential, $w^0(r) = -Ze^2/r$ for $r > r_c$ and zero otherwise, changes abruptly with position, but the result is a useful one. Our first step was to obtain the Fourier expansion of this applied potential in Eq. (13.9). [Actually we sought $\langle\mathbf{k} + \mathbf{q}|w^0(r)|\mathbf{k}\rangle = (1/\Omega_0)\int d^3r\, w^0(r)e^{i\mathbf{q}\cdot\mathbf{r}}$, which is the same thing.] The integration required a convergence factor, $e^{-\kappa r}$, but we could take κ equal to zero afterward to obtain

$$\langle\mathbf{k} + \mathbf{q}|w^0(r)|\mathbf{k}\rangle = -\frac{4\pi Ze^2}{\Omega_0 q^2}\cos qr_c. \tag{20.10}$$

We are to divide this Fourier component by the dielectric function of Eq. (20.8) to obtain $-4\pi Ze^2\cos qr_c/[\Omega_0(q^2 + \kappa^2)]$, the form which we used.

We noted further at that point that this form, with a $q^2 + \kappa^2$ in the denominator, was exactly what was obtain in the integral $\langle\mathbf{k} + \mathbf{q}|w^0(r)|\mathbf{k}\rangle$ with the convergence factor, so we know the inverse Fourier transform. For the simple Coulomb potential with $r_c = 0$ the *screened Coulomb potential* becomes

$$-\frac{Ze^2}{r} \rightarrow -\frac{Ze^2}{r}e^{-\kappa r}. \tag{20.11}$$

[It is best to use the $r_c = 0$ form here because of small terms, proportional to κ, dropped in the integral.] The effect of the screening is very clear in Eq. (20.11). It simply eliminates in a smooth way the large-distance tale of the potential, leaving the potential very much the same at small distances.

This makes an extraordinary simplification of the theory of metals (see, for example, Harrison (1999), Chapters 12-14). In this Fermi-Thomas theory the interaction energy between metallic atoms must also take this screened Coulomb form (actually the interaction between two metal atoms, with valences Z_1 and Z_2 and core radii r_{c1} and r_{c2} becomes $Z_1Z_2e^2\cosh\kappa r_{c1}\cosh\kappa r_{c2}e^{-\kappa r}/r$). Much of the dynamics and statics of interacting metal atoms becomes describable in terms of simple, two-body, central-force interactions, with additional volume-dependent terms in the energy. In this case it has been possible to incorporate these many-body terms arising from the interaction between the electrons in the metal in a simple self-consistent theory.

The assumption of potentials slowly varying with position, which was intrinsic to Fermi-Thomas theory, can be eliminated by a full quantum theory, while still retaining the linearization which is the most essential aspect. To first order in the potential $\delta V(\mathbf{r})$, or its Fourier components $\langle\mathbf{k} + \mathbf{q}|\delta V(\mathbf{r})|\mathbf{k}\rangle$ or $\langle\mathbf{k} + \mathbf{q}|w(\mathbf{r})|\mathbf{k}\rangle$, we may calculate the modified free-electron states in first-order perturbation theory as

$$|\mathbf{k}\rangle^{(1)} = |\mathbf{k}\rangle + \frac{|\mathbf{k}+\mathbf{q}\rangle\langle\mathbf{k}+\mathbf{q}|\,\delta V(\mathbf{r})|\mathbf{k}\rangle}{\frac{\hbar^2}{2m}(k^2 - |\mathbf{k}+\mathbf{q}|^2)} . \tag{20.12}$$

We may square this, keep terms linear in $\delta V(\mathbf{r})$, and sum over occupied states $k < k_F$ to obtain the $\delta n(\mathbf{r})$ in terms of $\delta V(\mathbf{r})$, which we obtained in the Fermi-Thomas approximation in Eq. (20.3). The rest of the analysis proceeds exactly as above, leading to a more complicated dielectric function given by (e. g., Harrison (1970)),

$$\varepsilon(q) = 1 + \frac{\kappa^2}{2q^2}\left(\frac{1 - q^2/(2k_F)^2}{q/k_F}\ln\left|\frac{2k_F + q}{2k_F - q}\right| + 1\right), \tag{20.13}$$

This is called the *Hartree dielectric function*, since it is based upon the Hartree approximation discussed in Section 4.2, or the *Lindhard dielectric function* after the first person to derive it. It replaces the less-accurate Eq. (20.8) and is almost as easy to use.

This quantum dielectric function is plotted in Fig. 20.2, and compared with the Fermi-Thomas approximation to it. As expected, they approach each other at small q where the assumptions of Fermi-Thomas theory apply.

They also both approach one at large q so that indeed the Fermi-Thomas theory has done quite well. However, if we look closely at the region near $q/k_F = 2$ we notice a subtle fluctuation in the quantum dielectric function. Examination of Eq. (20.13) near this point shows that it takes the form $(q-2k_F)\ln|q - 2k_F|$ and has a negatively infinite $d\varepsilon/dq$ at that point. In spite of the subtlety of the singularity, it has significant consequences. If we use this dielectric function to screen a spherically-symmetric potential, as in Eq. (20.11), we find (e. g., Harrison (1970)) a term which varies at large distances as $\cos(2k_Fr)/(k_Fr)^3$, rather than exponentially as in Eq. (20.11). These large-distance fluctuations, called *Friedel oscillations*, are real and interesting, but have turned out to have surprisingly few consequences. At large q one may also see that the quantum dielectric function approaches one as $1/q^4$ rather than as the $1/q^2$ in the dielectric function of Eq. (20.8).

We finally consider time-dependent screening. There are many aspects which can be described in terms of transport theory, as in Chapter 11, with the addition of a potential which then depends self-consistently upon the distribution function $f(\mathbf{p}\,\mathbf{r}t)$. One of the most important many-body effects, plasma oscillations, can be understood this way and in fact in the even simpler approximation described at the end of Section 11.3. It is basically the same calculation which we made for the speed of sound in Section 1.8 but now for a charged electron gas. We characterized the compressional wave by a displacement of the medium in the z-direction given by $u(z, t) =$

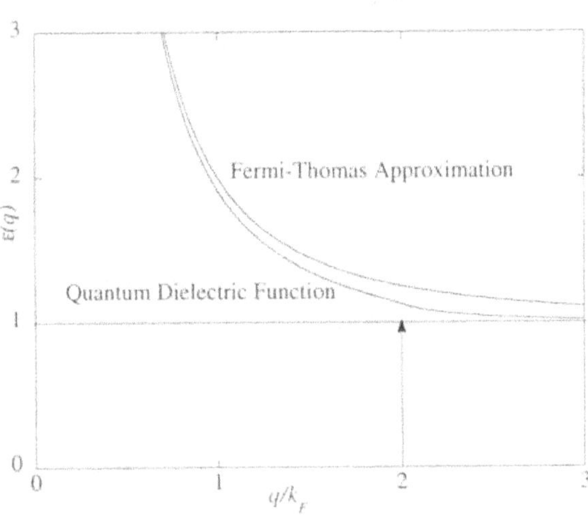

Fig. 20.2. The quantum dielectric function of Eq. (20.13), with κ/k_F taken equal to one and the Fermi-Thomas dielectric function of Eq. (20.8) with the same parameters.

$u_0e^{i(qz-\omega t)}$ (or the real part) which in an electron gas of density n_0 gives rise to a density fluctuation $-n_0\nabla u(z, t) = -iqn_0u_0e^{i(qz-\omega t)}$. This gives rise to a potential fluctuation, given by Poisson's Equation, Eq. (20.5), as $V_qe^{i(qz-\omega t)} = -iq(4\pi e^2/q^2)n_0u_0e^{i(qz-\omega t)}$. The corresponding force on an electron is $-\nabla V(z) = -4\pi e^2n_0u_0e^{i(qz-\omega t)}$, which we may equate to the electron mass m times the acceleration of the gas, $-\omega^2u_0e^{i(qz-\omega t)}$. Equating these leads to a frequency ω_p, called the *plasma frequency*, given by

$$\omega_p{}^2 = \frac{4\pi e^2n_0}{m} . \tag{20.14}$$

In contrast to sound waves, it has turned out to be independent of wavenumber because of the long-range nature of the electron-electron interaction. They are a many-body effect, a direct consequence of the electron-electron interaction. They show up in any complete treatment, classical or quantum-mechanical, of the dynamics of an electron gas. Once they are suggested, they can be understood by the simple argument we have given. For n_0 corresponding to metallic densities these turn out to have energies of order $\hbar\omega_p \approx 10$ eV. They are observed in the energy-loss spectrum of high-energy particles passing through metal foils.

It may be interesting that the same argument could have been made for a collection of metallic ions, each of mass M and charge Z, if we ignored the important effect of the electrons present. This gives the ion plasma frequency as $\omega_{IP}{}^2 = 4\pi Z^2e^2/(M\Omega_0)$, with Ω_0 the volume per ion. The effect of the real compensating electron gas is to screen the interactions, reducing the force, the acceleration, and therefore the frequency-squared, by a factor of $1/\varepsilon(q)$. If we use the Fermi-Thomas dielectric constant, the result at long wavelengths, small q, is

$$\omega^2 = \frac{4\pi Z^2e^2}{\kappa^2M\Omega_0}q^2. \tag{20.15}$$

With the interactions screened, as we have discussed, the frequency becomes proportional to the wavenumber as for a sound wave in a metal, which is what this compressional wave is. This prediction of the speed of longitudinal sound in a metal, called the *Bohm-Staver speed of sound*, is in reasonable accord (within 10 or 20%) with experiment for the simple metals. If we substitute Eq. (20.9) for κ^2 we see that the speed of sound is given by $\sqrt{Zm/(3M)}$ times the Fermi velocity $\hbar k_F/m$ of the electrons.

Chapter 21. Cooperative Phenomena

There are systems for which many-body effects make dramatic changes in the behavior. These ordinarily arise from the cooperative effect of many particles, like the condensation of vapor into a liquid. Hear we discuss such cooperative phenomena in the context of quantum theory. One of the most familiar such cooperative phenomena is ferromagnetism, but we postpone that discussion till the next chapter after we discuss other magnetic effects.

21.1 Localization and Symmetry Breaking

Another cooperative phenomenon is associated with localization of electronic states and can be understood already in the simple Li_2 molecule which we treated in Section 5.1. We found a ground state, in our one-electron approximation, with both valence electrons in a bonding orbital, and consequently with a 50% chance at any moment of being on the same atom. If we imagine pulling these atoms slowly apart, $V_{ss\sigma}$ decreases and eventually becomes unimportant, but we are retaining a state with a 50% chance of both electrons on the same atom with a corresponding Coulomb interaction U. Clearly the energy will be lower if we change to a state with one electron on each atom. This corresponds to a *correlated* motion of the electrons since they tend to avoid each other, rather than each forming a one-electron state, independent of the other, as we assumed. The correlated state is also often called a *localized* state. We proceed to see how the system changes from a bond-like to a localized state.

We describe each electron in terms of four states, the 2s-state on each atom with the spin σ either up ($\uparrow$) or down ($\downarrow$). The Hamiltonian, in the operator notation of Section 16.1, would be written

$$H_0 = \Sigma_{i,\sigma}\, \varepsilon_s\, c_{i\sigma}{}^{\dagger} c_{i\sigma} + V_{ss\sigma} \Sigma_\sigma \left(c_{1\sigma}{}^{\dagger} c_{2\sigma} + c_{2\sigma}{}^{\dagger} c_{1\sigma}\right). \tag{21.1}$$

In the context of this formulation, we can add the electron-electron interaction which describes the increase in Coulomb energy U if two electrons are on the same atom as opposed to separate atoms, and they would need to have opposite spin if on the same atom. That electron-electron contribution to the Hamiltonian is

$$H_1 = U \Sigma_i\, c_{i\uparrow}{}^{\dagger} c_{i\uparrow} c_{i\downarrow}{}^{\dagger} c_{i\downarrow}\,. \tag{21.2}$$

As we have indicated, adding the interaction with four operators tremendously complicates any problem. However, this problem started out so simple that we can in fact solve it exactly, in the context of this formulation. This has become a two-electron problem, but there are only six two-electron states. One basis state is with a spin-up electron on atom one and a spin-down electron on atom two, which we write $c_{1\uparrow}{}^{\dagger} c_{2\downarrow}{}^{\dagger}|0\rangle$. Another is $c_{1\downarrow}{}^{\dagger} c_{2\uparrow}{}^{\dagger}|0\rangle$. There are also two basis states with both electrons on the same atom, $c_{1\uparrow}{}^{\dagger} c_{1\downarrow}{}^{\dagger}|0\rangle$ and $c_{2\uparrow}{}^{\dagger} c_{2\downarrow}{}^{\dagger}|0\rangle$ and two basis states with parallel spin, $c_{1\uparrow}{}^{\dagger} c_{2\uparrow}{}^{\dagger}|0\rangle$ and $c_{1\downarrow}{}^{\dagger} c_{2\downarrow}{}^{\dagger}|0\rangle$. These six basis states have energies $2\varepsilon_s$, $2\varepsilon_s$, $2\varepsilon_s + U$, $2\varepsilon_s + U$, $2\varepsilon_s$, and $2\varepsilon_s$, respectively, before introducing the $V_{ss\sigma}$. In addition, the first basis state is coupled to the third and fourth by $V_{ss\sigma}$ (which couples individual electron states of the same spin on the two atoms) and so also is the second basis state couple to the third and fourth by $V_{ss\sigma}$. All other states are uncoupled. The corresponding six-by-six Hamiltonian matrix can be solved easily. The last two basis states, with parallel spin, are eigenstates with energy $2\varepsilon_s$, uncoupled to the other basis states or each other. We could think of these states as having one electron on each atom or one in a bonding state and the other in an antibonding state of the same spin. (When the state is written out, terms with both electrons on the same atom cancel as the Pauli Principle tells us they must.)

The remaining four eigenstates are even and odd combinations of the remaining four basis states. Of most interest is the ground state, which will be even. One even combination is: $[c_{1\uparrow}{}^{\dagger} c_{1\downarrow}{}^{\dagger}|0\rangle + c_{2\uparrow}{}^{\dagger} c_{2\downarrow}{}^{\dagger}|0\rangle]/\sqrt{2}$, with energy of $2\varepsilon_s + U$. The other is: $[c_{1\uparrow}{}^{\dagger} c_{2\downarrow}{}^{\dagger}|0\rangle + c_{1\downarrow}{}^{\dagger} c_{2\uparrow}{}^{\dagger}|0\rangle]/\sqrt{2}$ with energy $2\varepsilon_s$. We may verify that they are coupled by $2V_{ss\sigma}$ and solve the quadratic equation for the two even eigenstates as

$$\varepsilon_{\pm} = 2\varepsilon_s + \frac{U}{2} \pm \sqrt{\left(\frac{U}{2}\right)^2 + 4V_{ss\sigma}{}^2}\,. \tag{21.3}$$

Use of creation operators for the states assures the appropriate antisymmetry, which is of no consequence in this simple example. With only one orbital per atom there are no exchange terms in the model. The minus sign in Eq. (21.3) gives the ground state. Note that if U is neglected it gives $2(\varepsilon_s + V_{ss\sigma})$, with $V_{ss\sigma}$ negative, the solution we obtained in Chapter 5. If on the other hand, $V_{ss\sigma}$ becomes very small, the energy approaches $2\varepsilon_s$. The electrons indeed separate onto different atoms. Furthermore, the energy and the state vary smoothly between the two limits as the atoms are separated from each other. Because of the smooth variation, over the entire range of $V_{ss\sigma}/U$, all states are correlated to some extent but the correlations only become important when U is of order or larger than $V_{ss\sigma}$. [In passing we note that in addition to the ground state there is a high-energy state obtained with the plus in Eq. (21.3). There is also one odd state with energy $2\varepsilon_s$ which, with the two parallel-spin states mentioned before, form a triplet, three states of the same energy, and corresponding, it turns out, to parallel spins of $1/2$ units each totaling one unit of spin with three orientations. There is also an odd state with energy $2\varepsilon_s + U$. We are only interested here in the ground state.] This system is frequently discussed in terms of the exchange interaction which we introduced in Section 4.2, but we regard that as misleading and confusing. There is only one orbital per atom and exchange can only enter if we introduce artificial self-exchange and self-direct interactions as we discussed in Section 4.2. For the understanding of these systems there is considerable advantage in not introducing these artificial effects and in retaining only the real electron-electron interaction.

It is interesting to compare the energy we obtain by an exact solution, Eq. (21.3), with the one-electron solutions which we have used throughout the book. Evaluating the expectation value for the Hamiltonian, Eqs. (21.1) and (21.2) with respect to these bonding states yields $U/2 - 2|V_{ss\sigma}|$, compared as the curve "HF" (for Hartree-Fock) with Eq. (21.3) plotted as "Exact" in Fig. 21.1. We imagine this as a plot of energy versus spacing, since the abscissa, $U/|V_{ss\sigma}|$, increases with increasing spacing. The exact energy is lower than the approximate solution (this follows from the variational argument of Section 4.2), but they become rather close at small spacings where the real molecules are rather well described by the one-electron approximation. The large error of $U/2$ at large spacings, mentioned above, is apparent.

It is also possible to improve upon the one-electron solution by allowing the spin-up solution to be of the form $\sin\eta|1\uparrow\rangle + \cos\eta|2\uparrow\rangle$ and the spin-down solution to be of the form $\cos\eta|1\downarrow\rangle + \sin\eta|2\downarrow\rangle$, with the coefficients chosen to retain normalization but allow segregation of the electrons by

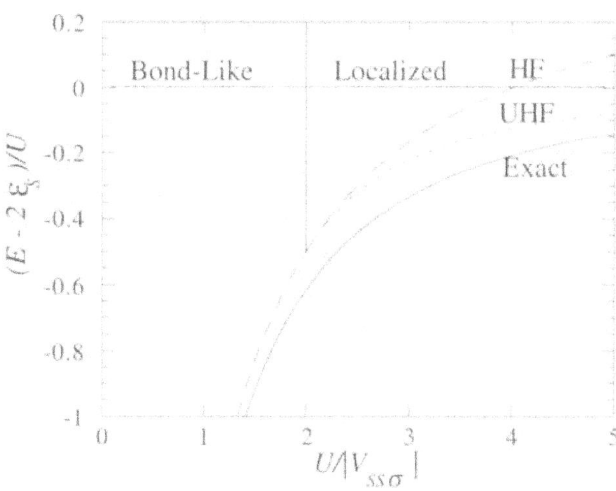

Fig. 21.1. A plot of the exact total energy, Eq. (21.3), for the two-level model of the Li_2 molecule, compared with the one-electron approximation (HF) used in Chapter 5. Also shown is a third solution, UHF, a one-electron approximation in which the spin-up and spin-down electrons are allowed to break symmetry and localize, dividing the region into bond-like and localized regions.

varying η , and the total energy expectation value of the Hamiltonian is minimized with respect to η. This is another example of selecting a variational solution which encompasses the physical concept which we think is important. This particular choice is called *Unrestricted Hartree-Fock* (e. g., Harrison (1999), 595ff), and the result is plotted as UHF in Fig. 21.1. Indeed it eliminates the $U/2$ discrepancy at large spacing, but it retains too small a binding by a factor of two and in fact for $U < 2|V_{ss\sigma}|$ the minimum energy comes at the symmetric state, giving the energy of the HF solution, as seen.

This Unrestricted Hartree-Fock approach does incorporate an appropriate spin segregation on the two atoms, but it is misleading in doing it in a discontinuous way (the second derivative of the energy with respect to the abscissa is discontinuous at $U = 2|V_{ss\sigma}|$). Frequently that is not a serious drawback. When a condensed-matter system has two qualitatively different states, such as the localized and the bond-like solutions, the energy of the localized solution is ordinarily minimum at larger spacing. The bond-like solution may have a minimum energy at a smaller spacing. This is sketched in Fig. 21.2, giving the energy of a system as a function of the volume of that system. If that minimum for the bond-like state is higher in energy, as shown there, the stable state is the localized one. However, if a pressure P is applied, there is an additional term in the energy equal to $P(\Omega - \Omega_0)$, also sketched in Fig. 21.2. Adding it to the E_{tot} shifts the minimum to a slightly lower volume and brings the bond-like energy minimum down. The common tangent, also drawn in Fig. 21.2, is also rotated counterclockwise. If sufficient pressure is applied that this common tangent has positive slope, the energy will be lower in the bond-like state and the system will make a *first-order transition* (a transition with a discontinuous change in volume) to the bond-like state. If the approximate descriptions of these two states are good in the region of their minimum energy, the prediction can be accurate, and it does not matter than neither description is very good at intermediate volume. That intermediate volume is not accessed by the experiments. [See Harrison (1999), Chapter 16, for studies of this aspect for the rare earths and actinide metals.] There may be cases where the entire range of states between two limits is accessed, and then a more complete description such as Eq. (21.3) may be needed.

The most important results from this section are, first, seeing that the electron-electron interaction can fundamentally change the approximations which are appropriate for discussing the systems, and, second, that it is not always necessary to study the most difficult intermediate case. Often transitions are made between states of condensed matter which are

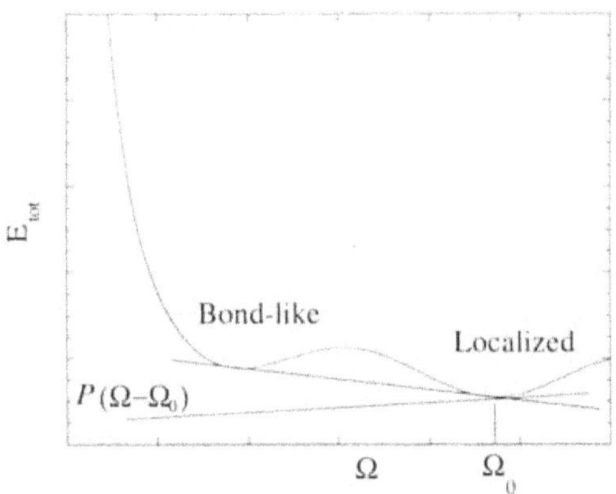

Fig. 21.2. A schematic plot of total energy versus volume for a system which has lowest energy with the electrons localized to their atoms at a large volume Ω_0. If a pressure P is applied, an additional term is added, a straight line with a slope equal to P. If the slope of that pressure line exceeds the negative slope of the common tangent shown, the system will transform to the bond-like state at the smaller volume.

fundamentally different, but both may be simply understandable.

21.2 The Hubbard Hamiltonian

The Hubbard Hamiltonian is the direct extension of the Hamiltonian given in Eqs. (21.1) and (21.2) to a long chain of N atoms such as we discussed in Section 6.1,

$$H = \Sigma_{i,\sigma}\, \varepsilon_s\, c_{i\sigma}{}^{\dagger} c_{i\sigma} + V_{ss\sigma}\Sigma_{i,\sigma}\, (c_{i\sigma}{}^{\dagger} c_{i+1\sigma} + c_{i+1\sigma}{}^{\dagger} c_{i\sigma})$$
$$+\; U\,\Sigma_i\, c_{i\uparrow}{}^{\dagger} c_{i\uparrow} c_{i\downarrow}{}^{\dagger} c_{i\downarrow}\,, \qquad (21.4)$$

usually with one electron per atom and sometimes extended to a square lattice or simple-cubic lattice. It introduces the essential feature of the Coulomb interact but eliminates any unnecessary complications, as it did for the two-atom case discussed in the preceding section. It cannot be solved analytically, as was the two-atom problem, because the basis contains so many N-electron states but the approximations introduced there can give insight into the behavior of such a system and important solid-state systems which share these features. We discuss here symmetry-breaking and antiferromagnetic insulators.

With only the first line in Eq. (21.4) we constructed one-electron states and obtained the simple energy band $\varepsilon_k = \varepsilon_s + 2V_{ss\sigma}\cos kd$. We can similarly construct two-electron states

$$|\psi(2)\rangle = \left(\frac{1}{\sqrt{N}}\Sigma_i\, e^{ikdi} c_{i\sigma}{}^{\dagger}\right)\left(\frac{1}{\sqrt{N}}\Sigma_j\, e^{ik'dj} c_{j\sigma'}{}^{\dagger}\right)|0\rangle$$
$$= \frac{1}{N}\;\Sigma_{i,j}\, e^{ikdi} e^{ik'dj} c_{i\sigma}{}^{\dagger} c_{j\sigma'}{}^{\dagger}|0\rangle\,. \qquad (21.5)$$

This approach is used for $N = 2$ in Problem 21.1. These states are normalized (except if $k' = k$ and $\sigma' = \sigma$, in which case the state is zero). Once we have seen how the normalization has worked out, we can see that the expectation value of the first sum in Eq. (21.4) with respect to this state is $2\varepsilon_s$ per atom and for the second is $2V_{ss\sigma}(\cos kd + \cos k'd)$. This clearly generalizes to many-electron product states, which are also antisymmetrized by our use of creation operators.

Of course it is the second line in Eq. (21.4) which is interesting. For each atom in a sum over i', $\langle\psi(2)|U\Sigma_{i'} c_{i'\uparrow}{}^{\dagger} c_{i'\uparrow} c_{i'\downarrow}{}^{\dagger} c_{i'\downarrow}|\psi(2)\rangle$, the only contribution from the states of Eq. (21.5) will come when both i and j are equal to i'. If the spins are parallel, the $c_{i'\uparrow}{}^{\dagger}c_{i'\uparrow}{}^{\dagger}$ for the state is zero and there is no contribution. If the spins are antiparallel we obtain U/N^2 and there are N such terms, so the Coulomb energy is U/N, as we obtained in the last section for $N = 2$. This also generalizes to the N-electron state, with the lowest half of the band filled with $N/2$ electrons of each spin, in that there is no contribution from states of parallel spin, and the expected contribution for no correlated motion of the electrons of antiparallel spin. Each spin-up electron on a given site will see on average a half an electron of opposite spin, and no electrons of the same spin. We can add this up for all spin-up electrons to obtain $NU/4$ and all interactions have been included once. (We mentioned the double counting of interactions if we add shifts for *all* electrons in Section 4.2.)

As in the two-atom case, there is reason to correlate the motion. We can in fact see an *instability* of the many-electron generalization of Eq. (21.5) using an Unrestricted Hartree-Fock state with lower symmetry as we did for the two-atom problem, and see that a lower energy can be obtained. We do this for one electron per atom and proceed by constructing Bloch sums as in the first form in Eq. (21.5), but with different coefficients on odd- and even-numbered atoms, shifting spin-up electrons to one set and spin-down electrons to the other. This will shift the energy, through the U-term in Eq. (21.14), for each electron differently upon the different atoms. If we knew the result, a net average fraction $\sin^2\eta$ of up-spin electrons on even atoms and $\cos^2\eta$ on odd atoms, and thus a shift from the final term of $(U/2)\sin^2\eta$ upward for up-spin on even atoms and down-spin on odd atoms, and a shift of $(U/2)\cos^2\eta$ upward for down-spin on even atoms and up-spin on odd atoms, we could proceed with the one-electron calculation for each set. However, we do not know η so we must proceed self-consistently to guess η, do a band calculation, and then use the resulting states to estimate the shift and thus η, as we did for the screening calculation.

The important results can be gotten rather easily. Given a value of η we have a simple band calculation with a Bloch sum of spin-up states on even atoms $\sqrt{2/N}\,\Sigma_{i\ (\mathrm{even})}e^{ikdi}c_{i\uparrow}{}^{\dagger}|0\rangle$ with energy $\varepsilon_s + (U/2)\sin^2\eta$. It contains no nearest-neighbor atoms so $V_{ss\sigma}$ does not enter the expectation value of its energy. It is, however, coupled by $2V_{ss\sigma}\cos kd$ to a Bloch sum of spin-up states on odd atoms, $\sqrt{2/N}\,\Sigma_{i\ (\mathrm{odd})}e^{ikdi}c_{i\uparrow}{}^{\dagger}|0\rangle$, and is not coupled to any other Bloch sum. The variational solution, or band calculation, is carried out with a state given by $u_1(k)$ times the Bloch sum on odd atoms plus $u_2(k)$ times the Bloch sum on even atoms. The solution is obtained from the solution of a quadratic equation as

$$\varepsilon_k = \varepsilon_s + \frac{U}{2}(\sin^2\eta + \cos^2\eta) \pm \sqrt{\left(\frac{U}{2}\frac{\cos^2\eta - \sin^2\eta}{2}\right)^2 + (2V_{ss\sigma}\cos kd)^2}$$

$$(21.6)$$

$$= \varepsilon_s + \frac{U}{2} \pm \sqrt{U^2\cos^2 2\eta/16 + 4V_{ss\sigma}{}^2\cos^2 kd} \ .$$

These bands are plotted in Fig. 21.3 choosing the - for $kd < \pi/2$ and the + for $kd > \pi/2$. We really have doubled the cell size by taking alternate atoms different and should plot both in a Brillouin Zone with $kd < \pi/2$ as done in the dotted curve to the left, but the scheme we use makes clearer the effect of breaking the symmetry. We see that it has opened up a gap at $\pi/2$, just where the Fermi wavenumber comes. Energy is lowered by populating only states below the gap.

These were the bands for spin-up electrons. Of course the bands for spin-down electrons are exactly the same, but they correspond to electrons shifted to the other set of atoms. Spin alternation between atoms is called a *spin-density wave*, or an *antiferromagnetic state*, and in this case with a gap opened up at the Fermi energy, it is an antiferromagnetic insulator.

To complete the calculation we must calculate, given a particular η in Eq. (21.6), the charge distribution and the resulting shifts which should equal $(U/2)\cos^2\eta$ and $(U/2)\sin^2\eta$ for a self-consistent choice. The solution of two simultaneous equations leading to Eq. (21.6) is exactly parallel to that

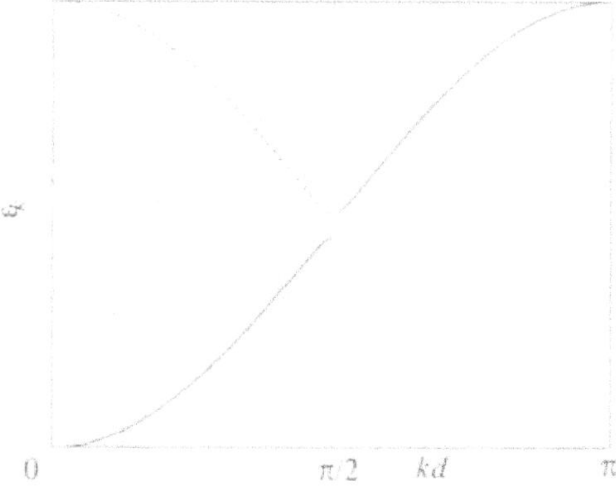

Fig. 21.3. A plot of the bands with broken symmetry. With alternate atoms different, the primitive cell is larger and the Brillouin Zone is reduced to $0 < kd < \pi/2$ with two bands as shown. They may also be understood as opening a gap in the original bands for $0 < kd < \pi$.

leading to Eq. (5.15) for molecular orbitals, and the states take the same form. If we define a $V_2 = 2V_{ss\sigma}\cos kd$ and a $V_3 = U\cos 2\eta/4$, then the coefficients of the two Bloch sums become $\sqrt{(1 \pm \alpha_p)/2}$ with $\alpha_p = V_3/\sqrt{V_2{}^2 + V_3{}^2}$. Then the fraction of probability density on an even site for a Bloch state of wavenumber k is $(1 \pm V_3/\sqrt{V_2{}^2 + V_3{}^2}\,)/2$. We may sum over the states of one spin, and divide by half the number of atoms to obtain the charge from that spin on a particular atom as

$$Z_\pm = \frac{2}{N}\frac{Nd}{2\pi}\int_{-\pi/2d,\pi/2d}\frac{dk}{2}\left(1 \pm \frac{U\cos 2\eta/4}{\sqrt{U^2\cos^2 2\eta/16 + 4V_{ss\sigma}{}^2\cos^2 kd}}\right)$$

$$= \frac{1}{2} \pm \frac{a}{2\pi}\int_{-\pi/2,\pi/2} dkd\,\frac{1}{\sqrt{a^2 + \cos^2 kd}}\,, \tag{21.7}$$

with

$$a = \frac{U\cos 2\eta}{8|V_{ss\sigma}|}\,.$$

(The integral is for the half-filled band.) But this $Z_\pm$ is what we have written $\cos^2\eta$ or $\sin^2\eta$ or $(1 \pm \cos 2\eta)/2$, so our self-consistency condition is

$$\frac{8|V_{ss\sigma}|}{U} = \frac{1}{\pi}\int_{-\pi/2,\pi/2} dkd\,\frac{1}{\sqrt{a^2 + \cos^2 kd}} \approx \frac{2}{\pi}\ln\left|\frac{4}{a}\right| \tag{21.8}$$

We related the integral to an elliptic integral, and wrote the final form for small a. Correction terms are of order $a^2 \ln|a|$ but this form is reasonably accurate for $0 < a < 2$. There is always a solution and a gap since, no matter how small U is, a $\cos 2\eta$ can be chosen small enough that $\log|4/a|$ will be large enough to match the left side. The gap is given by $E_g = {}^1/{}_2 U\cos 2\eta = 4|V_{ss\sigma}|a$. We may exponentiate the first and last forms in Eq. (21.8) to obtain

$$E_g = 16|V_{ss\sigma}|\, e^{-4\pi|V_{ss\sigma}|/U}. \tag{21.9}$$

There is always an instability against such a transition in a one-dimensional case at zero temperature, but these qualifications are important. At finite temperature there will be some electrons excited across the gap which contribute to the final term in Eq. (21.7) with opposite sign. We may in fact repeat the evaluation for finite temperature using a Fermi distribution with the Fermi energy midgap, $\varepsilon_m = \varepsilon_s + U/2$ as seen from Eq. (21.6). Then the integrand in Eq. (2.18) is multiplied by $1/(e^{(\varepsilon_k - \varepsilon_m)/k_BT} + 1)$ and from Eq. (21.6) we see that $\varepsilon_k - \varepsilon_m = 2V_{ss\sigma}\sqrt{a^2 + \cos^2 kd}$. We then subtract a term for electrons excited above the gap by subtracting a factor $1/(e^{-(\varepsilon_k - \varepsilon_m)/k_BT} + 1)$ so that the integrand in Eq. (21.8) is multiplied by $\tanh(2V_{ss\sigma}\sqrt{a^2 + \cos^2 kd}/k_BT)$ and the limits on the integral remain the same. At low temperature, this tanh approaches a step function at the $\pi/2$ limits and the result is unchanged. However, at higher temperatures it becomes proportional to $2V_{ss\sigma}\sqrt{a^2 + \cos^2 kd}$, cancels the factor in the integrand which is diverging at small a and eliminates the $\log|4/a|$ which always allowed a solution. The conclusion is that when the zero-temperature gap becomes small, of the order of kT , the transition cannot occur. In a similar way, if the system were two- or three-dimensional, the singularity is lost and the transition is not required. [This can be seen by replacing the chain by many chains, weakly-coupled to each other, so that the Fermi wavenumber along the chain varies slightly with transverse wavenumber and provides the smearing of the cut-off, which temperature provided above.]

We treated the case of the exactly half-filled band, with the net spin alternating from atom to atom. However, a similar argument can be made for a one-dimensional system with k_F different from half filling. We simply introduce a spin-density fluctuation varying as $\cos(2k_Fd)$ and band splitting, just as in Eq. (21.6) is produced at the Fermi surface. This is called an *incommensurate spin-density wave* since the period of the fluctuation is no longer locked to multiples of the lattice spacing.

Of course our analysis was based upon Unrestricted Hartree-Fock, not an exact solution, but the conclusions are believed to apply to real systems: one-dimensional metals are regarded as unstable with respect to formation of a spin-density wave, forming a gap and an insulating state. There are complications concerning phase transitions in lower-dimensional systems, arising from very large statistical fluctuations. The nature of the transitions depends upon the order in which limits are taken; for example, the size of the system becoming infinite or the coupling between an array of one-dimensional systems becoming small. These questions are too mathematical to discuss in detail here, but we may note that if such an insulating state is formed in a three-dimensional system, and the gap is small, the gap will decrease and finally disappear as the temperature is raised. This occurs sharply at a *transition temperature*, but the gap and the energy are a continuous function of temperature, with a discontinuity only of the second derivative of the energy with respect to temperature. Such a transition is called a *second-order transition*. The disappearance of ferromagnetism at the Curie temperature and of superconductivity at the critical temperature are other examples of second-order phase transitions.

The two-dimensional analog of the transition to an antiferromagnetic insulating state which we have discussed in detail for one dimension is important because it is closely related to high-temperature superconductivity. It concerns a square lattice with a half-filled band. Then the energy bands generalize to $\varepsilon_{\mathbf{k}} = \varepsilon_s + 2V_{ss\sigma}(\cos k_x d + \cos k_y d)$ and a square Brillouin Zone. When half-filled, the Fermi surface is at $\varepsilon_{\mathbf{k}} = \varepsilon_s$ or at $k_y = \pm(\pi/d - |k_x|)$, as illustrated in Fig. 21.4. Then in the Unrestricted Hartree-Fock Approximation an antiferromagnetic state with alternate atoms polarized in opposite direction opens up a gap over the entire Fermi surface exactly as in one dimension and the analysis which we gave for that case applies. If a large gap is formed, then adding small second-neighbor terms which deform the original Fermi surface are of little consequence. This explains the antiferromagnetic insulating state of the copper-oxide compounds, which however become metallic if they are doped away from half-filling and form the high-temperature superconductors.

One way of treating such systems, estimating total energies and spin densities, is to approximate the integrals over a band by a *special point* $\mathbf{k}^*$, a wavenumber which seeks to represent an average of the band. (This is also called a Baldereschi point (Baldereschi (1973)) and discussed in Harrison (1999) 348ff.) For the one-dimensional band it would be the wavenumber half-way to the Zone edge, $k^* = \pi/2d$, where the leading Fourier component of the band, $\cos kd$, is zero. For a square lattice, the two sets of leading Fourier components, $\cos k_x d + \cos k_y d$ and $\cos (k_x + k_y)d + \cos (k_x - k_y)d$, vanish at the point $\mathbf{k}^* = [\pi/(2d), \pi/(2d)]$ half-way from the origin to the corner of the BZ. In the particular case, shown in Fig. 21.4, the introduction

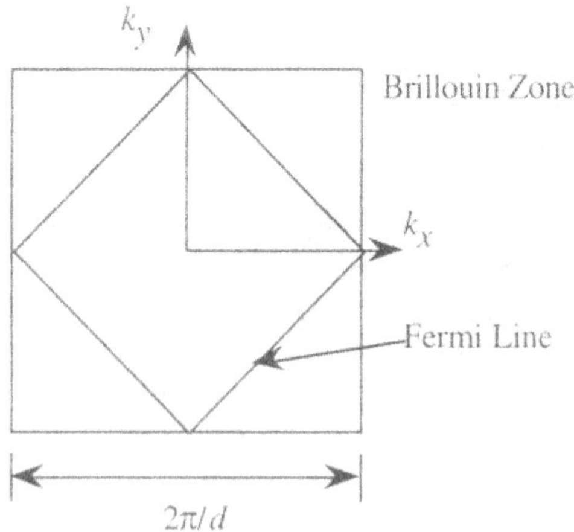

Fig. 21.4. A two-dimensional square lattice with spacing d gives a square Brillouin Zone with edge $2\pi/d$. For nearest-neighbor tight-binding bands, the Fermi line (Fermi surface in two dimensions) is the square shown with all states occupied inside.

of the antiferromagnetic order reduces the Brillouin Zone to the small rotated square, the Fermi line, and we wish to integrate over that portion of wavenumber space. Then the special point becomes $[0,\pi/(2d)]$. The bands are evaluated at that point and the problem reduces to the two-level problem we treated in Section 21.1. The magnetism of the cuprates was treated in exactly this way in Harrison (1987).

21.3 Peierls Distortions

Having found an instability of a system does not mean that it will make a transition along that path. Sometimes there are other instabilities and a system will tend to follow the strongest one. In the particular case of a half-filled band in a one-dimensional metal, a Peierls distortion (see, for example, Peierls (1979)) always provides an instability. For this distortion we imagine starting with the undistorted metallic chain and allow alternate atoms to be displaced to the right and the left by some distance u. This will cost some elastic energy, proportional to u^2 if the system was initially at the equilibrium spacing. However, it will also open up a gap at $k = \pi/2d$ through a matrix element proportional to u, as we have seen for the electron-phonon interaction, whether we were using pseudopotentials or tight-binding. This produces bands as we found in Eq. (21.6) with $U\cos 2\eta/4$ replaced by a term proportional to u and a gap opened up as in Fig. 21.3. The analysis proceeds exactly as for the antiferromagnetic instability and there is always a solution with some finite distortion u.

There are some important features to notice about this result. First, the alternate displacement has paired the atoms up, so that if this is a lithium chain we may think of it having formed Li_2 molecules with bonding and antibonding states. There is a residual coupling between them, so they form bonding and antibonding bands, but this is a completely natural result and could have been anticipated without the analysis. We may further note, that there will be electron pairs in each bond site, every $2d$ along the chain, and less charge in the sites between. We have produced a *charge-density wave* by this distortion, very much like the spin-density wave we produced in the preceding section. It is a different instability, and which can lower the energy most will depend upon U, $V_{ss\sigma}$, the electron-phonon coupling, and the elastic rigidity.

Another feature may be very helpful for understanding cooperative phenomena in general and superconductivity in particular. The state which we have found corresponds to a finite amplitude of a phonon mode for the wavenumber $q = \pi/d$. If we wished to discuss this in terms of phonons we must make a mixture of excitations of different numbers of phonons and these must have coherent phases as we showed in Section 19.1. This order in the system is sometimes called *off-diagonal long-range order*, referring to order in a density matrix which we have not discussed. The coherent state we have generated is essentially a classical state, as we saw in Chapter 19, and requires that the number of phonons present be poorly defined. We shall see that the construction of a superconducting state requires, in a very similar way, that the number of electrons present be poorly defined, a much more difficult situation to imagine.

There is a second feature which both the spin-density wave and the charge-density wave have in common with superconductivity, and that is the dependence of the gap on the coupling which caused it. In the case of the spin-density wave, caused by U, it was given in Eq. (21.9) as $e^{-4\pi|V_{ss\sigma}|/U}$, with U appearing in the denominator of the exponent. Similarly, the gap in the charge-density wave contains the electron-phonon interaction in the denominator of an exponent. The total energy depends upon the same factor and we note that $e^{-4\pi|V_{ss\sigma}|/U}$ cannot be expanded in a series $\Sigma_n a_n U^n$. That means also that it would never be possible to obtain these states proceeding from the normal state and treating the coupling in perturbation theory, *even if we carried it to all orders in U*. It was only possible to obtain the state by proposing a variational solution which reflected the instability in question. In hindsight, one can see that this was the essence of the theory of superconductivity by Bardeen, Cooper, and Schrieffer (1957).

21.4 Superconductivity

Superconductivity is a cooperative state in a metal which arises from an attraction between electrons. The origin of that attraction is the electron-phonon interaction, which we saw in Section 17.3 causes the lattice to deform and lower the energy of an electron, as a polaron. A second electron which happened to be at the same position would also have its energy lowered by that distortion, which means that there is an attraction between the two electron arising through the electron-phonon interaction, much as two heavy balls rolling on a mattress which we illustrated in Fig. 17.3. There is also a Coulomb repulsion between the electrons, but the superconducting state manages to take advantage of a net attraction from the phonons, at least in conventional superconductors predating the cuprates.

This attraction leads to an instability which was pointed out by Cooper (1956). He addressed the state of a three-dimensional metal, with all states filled to a Fermi sphere of radius k_F and in which there was some attractive interaction $V(\mathbf{r}_i - \mathbf{r}_j)$ between pairs of electrons. He then sought a state $\psi(\mathbf{r}_1,\mathbf{r}_2)$ for two electrons which had energy lower than simply placing them at the Fermi surface; this would be an instability. This state could be written in terms of the center-of-mass **R** and a relative coordinate **r**, and in fact factored into functions of those two coordinates separately since there are no terms in the Hamiltonian coupling them. The center-of-mass factor will be $e^{i\mathbf{K}\cdot\mathbf{R}}$ and the lowest-energy state will have $\mathbf{K} = 0$. The remaining factor is written $\psi(\mathbf{r}_1 - \mathbf{r}_2)$.

This two-electron state was taken to consist of two electron of opposite spin and must be orthogonal to all of the occupied states for $k < k_F$, so he sought to expand it in terms of the plane waves outside that surface,

$$\psi(\mathbf{r}_1 - \mathbf{r}_2) = \Sigma_{\mathbf{k}>k_F} a_{\mathbf{k}} \frac{e^{i\mathbf{k}\cdot(\mathbf{r}_1 - \mathbf{r}_2)}}{\sqrt{\Omega}} \tag{21.10}$$

This is often referred to as pairing of two electrons moving in opposite directions but it is just pairing without any drift momentum $\hbar\mathbf{K}$. One may expect that a spherically-symmetric state is of lowest energy, so the expansion can be made instead in spherical waves $\sqrt{1/(2\pi R)}\ \sin(k'r)/r$, in a large sphere of radius R, (these are orthonormal eigenstates of $-\hbar^2\nabla^2/m$ based upon the reduced mass, $m/2$, with pair energy $2\varepsilon_{k'} = \hbar^2 k'^2/m$) as

$$\psi(r) = \sum\nolimits_{k'>k_F} a_{k'} \sqrt{\frac{1}{2\pi R}}\ \frac{\sin k'r}{r}, \tag{21.11}$$

with r the distance between electrons. This is a variational solution for the Hamiltonian, which we write $H = -\hbar^2\nabla^2/m + V(r)$, and the resulting eigenvalue equation $H\psi = \varepsilon\psi$ can be rewritten as

$$\left(\frac{-\hbar^2\nabla^2}{m} - \varepsilon\right)\sum\nolimits_{k'>k_F} a_{k'} \sqrt{\frac{1}{2\pi R}}\ \frac{\sin k'r}{r} = -V(r)\sum\nolimits_{k'>k_F} a_{k'} \sqrt{\frac{1}{2\pi R}}\ \frac{\sin k'r}{r}. \tag{21.12}$$

We multiply by $\sqrt{\frac{1}{2\pi R}}\ \frac{\sin kr}{r}$ and integrate over the volume to obtain

$$(2\varepsilon_k - \varepsilon)a_k = -\sum\nolimits_{k'>k_F} a_{k'} \frac{2}{R}\int dr\ \sin k'r\ \sin kr\ V(r). \tag{21.13}$$

The matrix element $(2/R)\int dr\, \sin k'r\, \sin kr V(r)$ is taken to be negative. We neglect its variation with k and k' over a small range $\Delta\varepsilon$ of states near the pair Fermi energy where we let the a_k be nonzero, and write that matrix element $-V$, for an attractive potential.

We may now obtain a solution, and the method is different from any we have used before. Again, for V a constant, the right side of Eq. (21.13) is independent of k so we may define it to be a constant C and solve for $a_k = C/(2\varepsilon_k - \varepsilon)$. Then substituting for a_k on both sides in Eq. (21.13), and canceling the C from both sides, we have

$$1 = V \sum_{k'>kF} \frac{1}{2\varepsilon_{k'} - \varepsilon} . \tag{21.14}$$

To see the solutions, we plot the right side as a function of ε as in Fig. 21.5. The sum diverges wherever ε is equal to one of the $2\varepsilon_{k'}$ in the range over which we have taken the $a_{k'}$ nonzero, as seen in the figure. The solutions of Eq. (21.14) come where this sum is equal to 1, also sketched in the figure. We see that we have a solution just to the right of each $2\varepsilon_{k'}$, a state of slightly larger energy. Much more importantly, we have one solution at very much lower energy, an energy well below the Fermi energy. This is indeed the instability Cooper sought. Had the potential not been attractive, $V(r) > 0$, the singular solution would have been of higher energy, far to the right in Fig. 21.5.

We may finally solve for the energy of the Cooper pair from Eq. (21.14). We see from Fig. 21.5 that ε for the Cooper-pair state is well removed from all of the $2\varepsilon_{k'}$ over which we sum so the summand is smooth and we may replace the sum by an integral, $\Sigma_{k'} \rightarrow \int d\varepsilon_{k'}\, n(\varepsilon_{k'})/2$ with $n(\varepsilon_{k'})/2$ the number of one-electron states per unit energy (and per spin; e. g., electron-one with spin up) in the system, taken as independent of energy over the small one-electron energy range $\Delta\varepsilon$. Then the integral may be performed to obtain

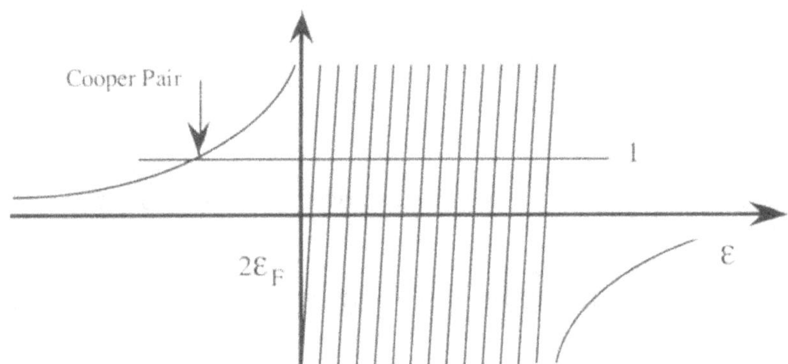

Fig. 21.5. A plot of $V \Sigma_{k'} 1/(2\varepsilon_{k'} - \varepsilon)$, summed over a set of pair states just above the Fermi energy, as a function of ε. Where that sum equals 1, there is a solution of Eq. (21.14) and a state.

$$1 = \frac{Vn(\varepsilon_F)}{4} \ln\left|\frac{2\Delta\varepsilon - \varepsilon}{-\varepsilon}\right| \tag{21.15}$$

with ε measured from the Fermi energy. We divide through by $Vn(\varepsilon_F)/4$ and exponentiate both sides. The energy of the Cooper-pair state ε is small compared to the energy range $2\Delta\varepsilon$ so that the result is

$$\varepsilon = -2\Delta\varepsilon\, e^{-4/Vn(\varepsilon_F)}. \tag{21.16}$$

Note the resemblance to the energy of the antiferromagnetic state given in Eq. (21.9) in the appearance of the interaction in the denominator of the exponent. This exponential will ordinarily be very small, as we assumed in taking ε small compared to $\Delta\varepsilon$.

The next task undertaken by Bardeen, Cooper and Schrieffer was to seek the ground state. In constructing the single Cooper pair, we used terms from a range of states and we cannot simply repeat the process for additional states because of the Pauli Principle. A variational solution was tried in which the number of pairs of electrons was ill-defined, as was the number of phonons in the Peierls state. That is, the BCS state was taken to be of the form,

$$|\Psi_{BCS}\rangle = \prod_{\mathbf{k}} (u_{\mathbf{k}} + v_{\mathbf{k}} c_{-\mathbf{k}\downarrow}^{\dagger} c_{\mathbf{k}\uparrow}^{\dagger})|0\rangle\,. \tag{21.17}$$

Each factor in this extended product contains one term ($u_{\mathbf{k}}$) with no electrons and one ($v_{\mathbf{k}} c_{-\mathbf{k}\downarrow}^{\dagger} c_{\mathbf{k}\uparrow}^{\dagger}$) with two electrons. This was exactly the key point. A Hamiltonian was taken with electron kinetic energies and with an electron-electron coupling which could scatter the electron pairs from one state to another, $\Sigma_{\mathbf{k}'\mathbf{k}} V_{\mathbf{k}'\mathbf{k}}\, c_{\mathbf{k}'\uparrow}^{\dagger} c_{-\mathbf{k}'\downarrow}^{\dagger} c_{-\mathbf{k}\downarrow} c_{\mathbf{k}\uparrow}$, as in Eq. (16.10). Given the variational state, Eq. (21.17), the calculation is straightforward and followed the calculation of the Cooper pairs given above. They evaluated $\langle\Psi_{BCS}|H|\Psi_{BCS}\rangle$ and used Lagrange multipliers to fix the *expectation value* of the total number of electrons and the normalization conditions $u_k^* u_{\mathbf{k}} + v_{\mathbf{k}}^* v_{\mathbf{k}} = 1$. This was minimized with respect to all $u_{\mathbf{k}}$ and $v_{\mathbf{k}}$. In place of the constant C an *energy-gap parameter* $\Delta_{\mathbf{k}}$ was defined by

$$\Delta_{\mathbf{k}} = -\sum_{\mathbf{k}'} u_{\mathbf{k}'}^* v_{\mathbf{k}'} V_{\mathbf{k}'\mathbf{k}} \tag{21.18}$$

and taken to be Δ, independent of $\mathbf{k}$, by taking $V_{\mathbf{k}'\mathbf{k}}$ equal to a constant, $-V$, as we did for the Cooper-pair calculation. Note that to contribute to Δ both $u_{\mathbf{k}'}$ and $v_{\mathbf{k}'}$ must be nonzero for each $\mathbf{k}'$ and for Δ to be large the phase of

$u_{\mathbf{k}'}{}^* v_{\mathbf{k}'}$ must be coherent in the sense we discussed in Chapter 19 from one $\mathbf{k}'$ to another. Thus the state is quite analogous to the classical Peierls distortion discussed in Section 21.3. The counterpart of Eq. (21.15) is called the energy-gap equation and solved using an energy range $\Delta\varepsilon$ within the Debye energy $\hbar\omega_D = \hbar v q_D$ of the Fermi energy. This Debye energy is the range of phonon energies as defined for Eq. (10.14), and believed to be the range of energy over which that electron-electron interaction is attractive. The resulting formula for the energy-gap parameter was close to Eq. (21.16) for the Cooper-pair energy,

$$\Delta = 2\hbar\omega_D \, e^{-1/Vn(\varepsilon_F)}. \tag{21.19}$$

with $n(\varepsilon_F)$ as always the number of electron states (including the factor of two for spin) per unit volume and per unit energy.

The energy gain in forming the superconducting state was found to be $^1/_4 n(\varepsilon_F)\Delta^2$. A gap of 2Δ was opened in the excitation spectrum, analogous to that shown in Fig. 21.3. It is possible to construct a drifting superconducting state, simply by transforming to a moving coordinate system, equivalent to shifting the entire superconducting state, and each electronic wavefunction, by some wavenumber $\mathbf{q}$. This can be seen from Eq. (21.17) and (21.18) to have the effect of multiplying the energy-gap parameter by $e^{2i\mathbf{q}\cdot\mathbf{r}}$. Then $\Delta(\mathbf{r})$ becomes essentially a superconducting wavefunction, such as had been introduced earlier (without the factor 2 in the exponent) by Landau and Ginsburg (1950). This Landau-Ginsburg theory is a quantum theory of superconductivity, based upon the single superconducting wavefunction describing the many-electron superconducting state.

In a superconducting ring $\Delta(\mathbf{r})$ may increase in phase by some integral number of 2π's, associated with a particular value of the current, and that number cannot change without forcing $\Delta(\mathbf{r})$ to go through zero somewhere along the ring, requiring a macroscopic energy. This is the origin of persistent current: it will not decrease at all over extraordinarily long times. This is also the origin of quantized flux. Each increase in the number of 2π's produces an additional magnetic flux quantum of $\pi\hbar c/e$. The consequences of this theory by Bardeen, Cooper and Schrieffer are immense, and very many of them were obtained in the original paper, possibly the most extraordinary physics paper of this century.

Chapter 22. Magnetism

We return to a number of aspects of magnetism which have not been discussed, ending with a discussion of ferromagnetism. Magnetic-field strengths are most frequently given in units of the gauss, and to evaluate expressions we will need to go beyond the values for $\hbar^2/m$, and e^2 in eV and Å which we gave in Eq. (1.10). When working with magnetic fields, in gauss, we can substitute e in electrostatic units (esu) and all other values in centimeter-gram-second (cgs) units,

$e = 4.8\times10^{-10}$ esu.
$\hbar = 1.054\times10^{-27}$ erg-sec. (equivalent to 6.6×10^{-16} eV-sec.)
$m_e = 9.1\times10^{-28}$ g.
$c = 3\times10^{10}$ cm/sec.

and our results will be in cgs. In this chapter, as in Section 2.4, we write the electron mass as m_e to avoid confusion with the quantum number m for the z-component of angular momentum.

22.1 Free Electrons in a Magnetic Field

We found in Section 3.3 that the effects of a magnetic field **H** on the dynamics of an electron could be included using a vector potential **A** from which the field could be derived as $\mathbf{H} = \nabla\times\mathbf{A}$. Then in the kinetic energy in terms of the momentum, **p** is to be replaced by $\mathbf{p} - (-e/c)\mathbf{A}$. Thus for a free electron, the Schroedinger Equation becomes

$$\frac{1}{2m_e}\left(\frac{\hbar\nabla}{i} - \frac{-e}{c}\,\mathbf{A}(\mathbf{r},t)\right)^2 \psi(\mathbf{r},t) = i\hbar\,\frac{\partial\psi(\mathbf{r},t)}{\partial t}. \qquad (22.1)$$

Different choices may be made for the vector potential which will give the same constant magnetic field **H** in the z-direction,

$$H_z = \frac{\partial A_y}{\partial x} - \frac{\partial A_x}{\partial y}, \tag{22.2}$$

and they will yield different forms of energy eigenstates. The different choices are called different *gauges*, and although the wavefunctions are different for different gauges they are all equivalent and will give the same properties. For constructing free-electron states the Landau gauge, $A_y = Hx$, is the most convenient. Then the eigenvalue equation from Eq. (22.1) becomes

$$\frac{1}{2m_e}\left(-\hbar^2\frac{\partial^2}{\partial x^2} + \left(\frac{\hbar}{i}\frac{\partial}{\partial y} - \frac{-e}{c}Hx\right)^2 - \hbar^2\frac{\partial^2}{\partial z^2}\right)\psi(\mathbf{r}) = \varepsilon\psi(\mathbf{r}). \tag{22.3}$$

We try a solution of the form

$$\psi(\mathbf{r}) = \phi(x)\, e^{ik_y y}\, e^{ik_z z} \tag{22.4}$$

and substitute it in Eq. (22.3) to obtain

$$\frac{1}{2m_e}\left(-\hbar^2\frac{\partial^2}{\partial x^2} + \left(\hbar k_y - \frac{-e}{c}Hx\right)^2 + \hbar^2 k_z^2\right)\phi(x)\, e^{ik_y y}\, e^{ik_z z} = \varepsilon\phi(x)\, e^{ik_y y}\, e^{ik_z z}$$

or (22.5)

$$-\frac{\hbar^2}{2m_e}\frac{\partial^2}{\partial x^2}\phi(x) + \frac{1}{2m_e}\left(\hbar k_y - \frac{-e}{c}Hx\right)^2\phi(x) + \frac{\hbar^2 k_z^2}{2m_e}\phi(x) = \varepsilon\phi(x).$$

This equation may come as a complete surprise! We have found that $\phi(x)$ satisfies the harmonic oscillator equation, for an oscillator centered at

$$x_0 = -\frac{\hbar k_y c}{eH} \tag{22.6}$$

and spring constant $\kappa = e^2H^2/(m_e c^2)$. This corresponds to a frequency $\omega = \sqrt{\kappa/m_e}$ or

$$\omega_c = \frac{eH}{m_e c}, \tag{22.7}$$

the classical cyclotron frequency, or Larmour frequency. (Note that it may be obtained in radians per second using the constants from the beginning of the chapter.) Then the result becomes sensible, though unexpected. A cyclotron orbit, viewed from along the y-direction, has a motion $x = r_0 \cos\omega_c t$, as does a harmonic oscillator and we may think of the state, Eq. (22.4), as a tight-binding sum $\Sigma_n e^{ik_y d_y n}$ of circular orbits centered at points $(x_0, d_y n)$ as illustrated in Part a of Fig. 2.1. The energy associated with the harmonic-oscillator state $\phi_n(x)$ is $\hbar\omega_c(n + 1/2)$ so from Eq. (22.5) we have eigenstate energies of

$$\varepsilon = \hbar\omega_c(n + 1/2) + \frac{\hbar^2 k_z^2}{2m_e} . \tag{22.8}$$

The energy is independent of the wavenumber k_y which determines the position along x of the orbit through Eq. (22.6). The electron may propagate freely along the field, as for a classical orbit, contributing the energy $\hbar^2 k_z^2/(2m_e)$. We may think of the quantization as coming from the circular orbits [this point is tricky because the states depend upon the gauge chosen for the vector potential] giving states spaced equally in energy for motion in the xy-plane, $\hbar^2 k^2/(2m_e) = \hbar\omega_c(n + 1/2)$, and therefore with equal spacing in cross-sectional area in the xy-plane of wavenumber space. This is shown schematically in Part b of Fig. 22.1. There we construct cylinders in wavenumbers space, with axes parallel to the z-axis, and with cross-sectional areas differing by $2\pi m_e\omega_c/\hbar$ (from $\Delta(\hbar^2 k^2/2m_e) = \hbar\omega_c$) from one to the next.

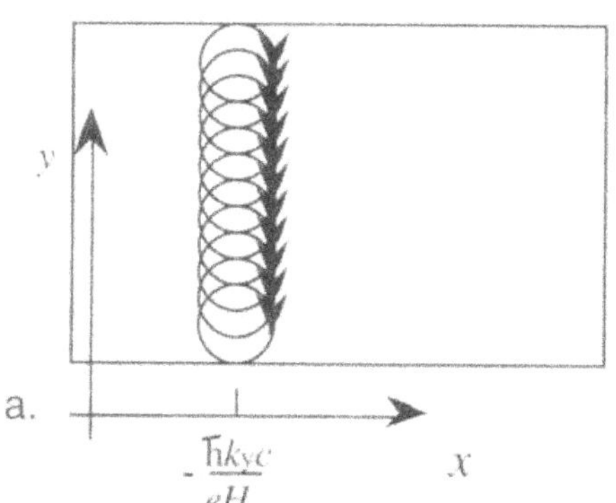

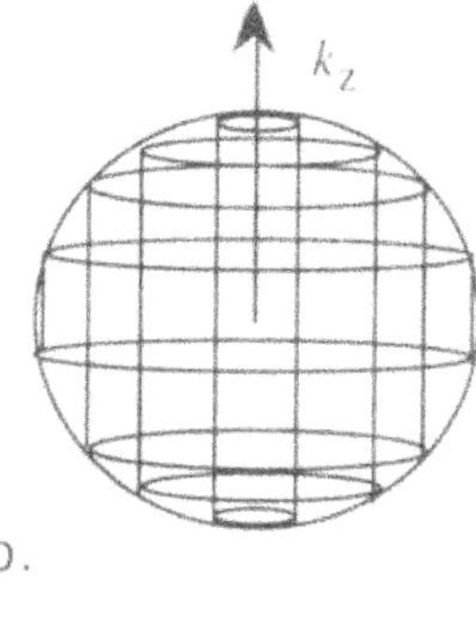

Fig. 22.1. In Part a is a schematic sketch of a Landau level for a free electron in a magnetic field. Part b illustrates the quantization of such levels in wavenumber space with fixed orbit areas in the xy-plane, and free propagation along the field in the z-direction. The surrounding sphere might represent the Fermi surface, containing the occupied states.

Each cylinder represents an allowed xy-motion, with varying propagation along the z-axis.

As the field is increased, ω_c increases and the separation of the cylinders increases. If we imagine a system, as in Part a of Fig. 22.1, with dimensions L_x, L_y, and L_z, we may apply periodic boundary conditions for the state, Eq. (22.4), in the y- and the z-directions. This gives a spacing of the levels along k_y of $2\pi/L_y$ and the range of wavenumbers k_y is limited by Eq. (22.6): if x_0 is restricted to a range L_x, then k_y is restricted to a range $eHL_x/\hbar c$. Thus the number of allowed k_y-values is $eHL_xL_y/(2\pi\hbar c)$, proportional to the cross-sectional area of the system, as it should be. Further, the number of allowed k_y also increases in proportion to the magnetic field so that the number of states up to a certain energy, such as that for the Fermi sphere shown in Part b of Fig. 22.1, remains approximately constant. As the field increases the cylinders expand but the number of states accommodated on the cylinders within the sphere remains nearly fixed.

The number of states within a thin shell of energy at that surface *does* vary with field. It is proportional to the area of cylinder within the shell and just as a cylinder becomes tangent to the sphere the area becomes much larger. Thus the density of states at the Fermi energy fluctuates as the field is increased. This fluctuation shows up in the diamagnetic susceptibility an effect called the *de Haas-van Alphen Effect.* The period of the fluctuation in $1/H$ gives a direct measure of the cross-sectional area of the Fermi surface and proved a powerful tool in the study of Fermi surfaces of simple metals, mentioned in Section 14.1. Generally the number of electrons is kept exactly fixed, so the Fermi sphere must fluctuate very slightly as the density of states varies. In addition to these fluctuations there is a small smooth increase in the total energy as the field increases, describable by a diamagnetic susceptibility χ_d, negative, with the energy increase given by $-{}^1/_2\chi_d H^2$. That susceptibility is not so easy to derive, but it is given by (e. g., Seitz (1940) 583ff)

$$\chi_d = -\frac{Ne^2}{4k_F{}^2 m_e c^2} \tag{22.9}$$

for a free-electron gas of N electrons per unit volume. This is an important consequence of quantum theory because one can rigorously show that a charged *classical* gas has vanishing diamagnetic susceptibility. The field simply deflects classical electrons without changing their energy. Only in quantum theory can there be a magnetic susceptibility. For a quantum gas the susceptibility can be understood physically in terms of currents due to edge states, for example harmonic-oscillator wavefunctions $\phi_n(x - x_0)$ with x_0 adjusted such that the nodes of that wavefunction come at a surface of constant x where vanishing boundary conditions are applied. These provide current in the y-direction at this surface, which is not completely canceled by states with reversed velocities in this region.

Another important feature of Eq. (22.9) is the appearance in the denominator of m_ec^2, the rest energy of the electron of a half-million electron volts. The other factors inevitably lead to an energy (since χ is dimensionless) which is of the order of electron volts. Thus the magnetic susceptibilities tend to be very small, of order 10^{-6}, while electric susceptibilities are of order one. This appearance of rest energy in the denominator, and also the smallness of the values, signifies that magnetism is a relativistic correction to classical behavior, as well as a quantum effect. This is closely associated with the fact, which we shall indicate in the following section, that when quantum theory was made relativistic by Dirac, the electron spin, its associated magnetic moment, and the resulting contributions to the susceptibility seemed to come automatically. A consequence is that it can be dangerous to discuss magnetic phenomena as we are doing without including other relativistic effects. There seems not to be any serious error for the properties we discuss here. We return to further discussion of magnetic susceptibilities in Section 22.3, finding a canceling paramagnetic contribution.

The motion of electrons in a *two-dimensional* electron gas, with the magnetic field perpendicular to the plane, is especially interesting. The analysis given above is applicable, but there is only a single state associated with motion in the z-direction, such as $\sqrt{2/d}\,\sin(\pi z/d)$ for a slab of thickness d. Then the cylinders shown in Fig. 22.1 are reduced to circles, each accommodating some number of electrons. As the field is increased, electrons leave the outer-most occupied circle as it expands until it is completely empty, and then electrons begin to leave the next circle in. One might expect some peculiar behavior just at the field where one circle is completely occupied and the next empty state is $\hbar\omega_c$ above it. von Klitzing and coworkers (1980) in fact found that when strong uniform magnetic fields in the z-direction were applied to a two-dimensional xy-plane of an extremely clean, cold semiconductor, the two-dimensional Hall conductivity σ_{xy}, which gives the current density in the x-direction due to an electric field in the y-direction, was given very exactly by an integral multiple of $e^2/(2\pi\hbar)$ at just these fields. The fields at which such a circumstance arises are calculated in Problem 22.1. At these same fields the two-dimensional resistivity measured by the field in the x-direction went to essentially zero. The values of Hall conductivity, $\sigma_{xy} = ne^2/(2\pi\hbar)$ with integral n, are so accurately given that this *Quantum Hall Effect* can be used as a standard for determining this combination of the fundamental constants, or the fine-structure constant $e^2/\hbar c = 1/137$ using the accurately known speed of light.

It was subsequently found by Tsui, Störmer, and Gossard (1982) that there was also this singular behavior at odd fractions, e. g., 1/3, 1/5, of the fields at which this filling occurs. Laughlin (1983) explained this behavior in terms of a collective state, describable by the Laughlin wavefunction,

$$\Psi(\{z_j\}) = A \prod_j z_j^n e^{-\alpha z_j^2}. \tag{22.10}$$

for fraction $1/n$, with z_j related to the coordinates of the jth electron by $z_j = x_j + iy_j$. For further discussion of integral and fractional Quantum Hall Effects, see Prange and Girvin (1987).

22.2 Magnetism of Atoms

We turn next to spherically-symmetric systems in a uniform magnetic field. For such a system a more convenient gauge for the vector potential is $\mathbf{A} = {}^1\!/_2\mathbf{H}{\times}\mathbf{r}$, which is $A_y = {}^1\!/_2Hx$, $A_x = -{}^1\!/_2Hy$ for a magnetic field H in the z-direction as illustrated in Fig. 22.2. The kinetic-energy term in the Hamiltonian is again given by Eq. (22.1). We add the potential $V(r)$ and now we may expand $\mathbf{A}$ for small r to obtain

$$H = \frac{1}{2m_e}\left(p^2 + 2e\frac{\mathbf{p}\cdot\mathbf{H}{\times}\mathbf{r}}{2c} + \frac{H^2e^2}{4c^2}(x^2 + y^2)\right) + V(r)\,. \tag{22.11}$$

Recalling that $\mathbf{A}\cdot\mathbf{B}{\times}\mathbf{C}$ is the volume of a parallelepiped with edges **A**, **B**, and **C**, we know that the second term is unchanged by rotating the vectors as $\mathbf{p}\cdot\mathbf{H}{\times}\mathbf{r} = \mathbf{H}\cdot\mathbf{r}{\times}\mathbf{p} = \mathbf{H}{\cdot}\mathbf{L} = HL_z$, with **L** the angular-momentum operator as described in Section 16.3. Thus, this term in the Hamiltonian linear in the magnetic-field strength is

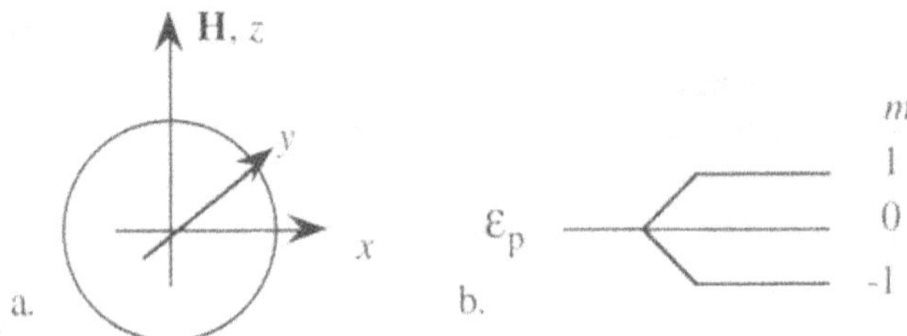

Fig. 22.2. a. A uniform magnetic field parallel to the z-axis is applied to a spherically-symmetric system, such as an atom. b. The three electronic p-states in such an atom are split into three levels, ε_p and $\varepsilon_p \pm {}^1\!/_2\hbar\omega_c$, called the Zeeman spitting.

$$H_{mag.} = \frac{eH}{2m_ec} L_z = \frac{1}{2} \omega_c L_z , \qquad (22.12)$$

with ω_c again the cyclotron frequency given in Eq. (22.7). This is illustrated for atomic p-states in Fig. 22.2b, and for d-states in the very simple Problem 22.2. It is striking that this splitting is into equal steps in energy, as was the motion of a free electron in a magnetic field in Eq. (22.8), but *the steps are just half as large* as for the free electron. We could understand this factor of two in detail for the case of the spherical harmonic oscillator discussed in Section 2.5, but it is in fact a very general result.

We have obtained this splitting as arising from a modification of the electron dynamics by the electron's deflection in the magnetic field. We can also think of it as from an interaction $-\mu\cdot\mathbf{H}$ between the magnetic field $\mathbf{H}$ and the magnetic moment μ arising from the orbit, analogous to the interaction between a magnetic field and an ordinary permanent magnet. From the first form in Eq. (22.12) we see that the magnetic moment to be associated with the orbit is

$$\mu = \frac{-e}{2m_ec} \mathbf{L} . \qquad (22.13)$$

The ratio $-e/(2m_ec)$ is called the *gyromagnetic ratio*. For this orbital interaction there is no moment for an s-state, and one *Bohr magneton*

$$\mu_B = \frac{e\hbar}{2m_ec} \qquad (22.14)$$

for a p-state. It is quantized as the L_z is quantized, being related by the gyromagnetic ratio. The d-state has two Bohr magnetons of magnetic moment.

We expect a similar interaction between a magnetic moment arising from the electron spin angular momentum, and might have expected the same gyromagnetic ratio, giving half the splitting for the spin-half electron. However, the spitting is the same, corresponding to twice as large a gyromagnetic ratio, $-e/(m_ec)$. Thus the Bohr magneton of Eq. (22.14) is also the magnetic moment of the electron due to its spin of one half. It follows that the energy of an electron in an s-state, as in the hydrogen atom, will also be split in a magnetic field into levels at

$$\varepsilon_{mag.} = \pm \mu_B H, \qquad (22.15)$$

relative to the degenerate level with no field, and there is splitting of the levels in the p-state in addition to those shown in Fig. 22.2b.

This all follows from the relativistic theory of the electron, Dirac theory, where the splitting of levels in the magnetic field comes automatically. It is possible to rationalize the result by thinking of the electron charge as being distributed on the surface of a sphere while the mass is distributed uniformly through the bulk of the sphere, but it is not clear what this means for a point particle.

The magnitude of the gyromagnetic ratio for any system divided by the value $e/(2m_ec)$ from Eq. (22.14) is called the *g-value*. Our discussion here would indicate a g-value of 2 for the electron spin, but the measured value can be slightly different. This arises from the interaction between the electron spin and its environment such that a rotation of the spin causes also rotation of orbital moment.

These various splittings can be directly detected by observing the absorption of electromagnetic radiation by the corresponding systems in a magnetic field. We may substitute values from the beginning of the chapter into Eq. (22.14) to see that μ_BH is 5.9×10^{-6} eV for a field of one kilogauss, of the order of easily attainable fields. We may convert this to a frequency by equating it to $\hbar\omega$ and divide it into the speed of light to see that it corresponds to electromagnetic wavelengths of a few centimeters, microwave frequencies. Thus the experiment can be done with specimens in a microwave cavity, varying the magnetic field and looking for fluctuations in the microwave absorption. These are called *spin-resonance* experiments. If a level is occupied by electrons of both spins, as for the 1s-states in helium, no energy can be absorbed by flipping a spin. That is the usual circumstance in solids, where all bond-states are occupied and antibonding states empty in a semiconductor, or all chlorine states filled and sodium states empty in rock salt. However, in a solid with a defect, such as rock salt with a chlorine atom missing, resonance can be observed for an electron attached to that defect. Spin resonance experiments usually ignore the bulk of the crystal and give direct information about the defects, or about surfaces if there are unpaired electrons (as opposed for example to electron pairs in a bond) at the surface. Spin resonance can also be observed it metals, and it involves only those electrons at the Fermi energy, which can be excited by flipping their spins.

The gyromagnetic ratio formulae apply also to nucleons, but the mass which enters is the nucleon mass, larger by a factor of order 10^5 than the m_e of the electron. Thus the spin-splitting of the nuclear levels is very tiny in comparison to that for electrons. Correspondingly any resonance experiments on the nuclei such as hydrogen are carried out with radiation of wavelength some 10^5 times larger, radio waves. This is called *Nuclear*

Magnetic Resonance (NMR) and has turned out to be much more important than electron spin resonance. It is used in magnetic resonance imaging for medical purposes by observing with one radio frequency and a magnetic field which brings protons into resonance only over a plane. The reflected waves can be used to image the proton density. It is also heavily used in chemistry, observing subtle differences in the environment of individual nuclei when their atoms are in different bonding sites.

22.3 Magnetic Susceptibility

We already noted that the diamagnetic susceptibility of a free-electron gas could be obtained by equating the shift in the energy of an electron gas, as illustrated in Fig. 22.1, to $-{}^1/_2\chi H^2$. This was a positive shift in the energy, corresponding to diamagnetism so the susceptibility was written χ_d and given in Eq. (22.9) as a negative number. There is another contribution to the susceptibility which arises from the electron spin moment. It is considerably simpler to calculate. We found in Eq. (22.15) that the energies for two different spin orientations were shifted up and down by $\mu_B H$. Thus we may imagine free-electron energies as a function of wavenumber (or energy bands in solids) separately for spin-up and spin-down electrons, as illustrated in Fig. 22.3. With a magnetic field parallel to the z-axis the spin-down electrons may have their energies uniformly lowered by $\mu_B H$ and the spin-up electrons raised by the same amount, as shown. In equilibrium the same Fermi energy ε_F applies to both spins, each with a density of states at the Fermi energy of $n(\varepsilon_F)/2$, so there is now an extra density of spin-down electrons $n(\varepsilon_F)\mu_B H/2$ and the density is reduced by the same amount for the spin-up electrons. With each having a magnetic moment along the field of $\pm\mu_B$, the magnetic moment density $\mathbf{M} = \chi_p H$ is $n(\varepsilon_F)\mu_B{}^2 H$. It is parallel to the field, lowering the energy, corresponding to a paramagnetic susceptibility, positive, and we have written it χ_p. This is also called the Pauli susceptibility and we have found it to be given by

$$\chi_p = n(\varepsilon_F)\mu_B{}^2 = \frac{3Ne^2}{4k_F{}^2 m_e c^2}, \tag{22.16}$$

where in the final form we wrote the density of states as $3N/(2\varepsilon_F)$ with N the electron density and substituted for μ_B from Eq. (22.14). This is three times as large and of opposite sign to the diamagnetic contribution which we gave in Eq. (22.9). The combination, two thirds of Eq. (22.16), is in rough accord with measured values for simple metals. However, there are corrections to both contributions, a g-value different from two for the spin moment, and any change from m_e in the dynamic mass which enters the diamagnetic

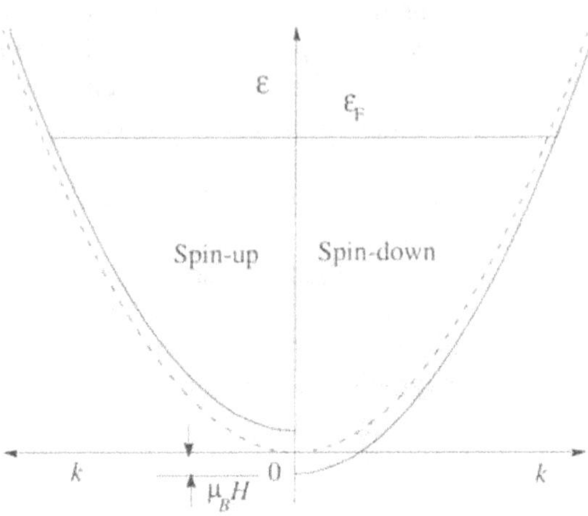

Fig. 22.3. A plot of spin-up and spin-down bands, shifted up and down by the interaction of the electron spin with an applied magnetic field. With the same Fermi energy for both spins, $n(\varepsilon_F)\mu_B H/2$ per unit volume are shifted to the spin-down band to give the magnetic susceptibility of Eq. (22.16)

contribution. In transition metals both corrections are large.

To calculate the magnetic polarizability of atoms, or the resulting susceptibility, we return to the Hamiltonian of Eq. (22.11). We calculate the energy to second-order in the magnetic field for each atom, multiply by the density of atoms, and equate the shift to $-{}^1/_2\chi H^2$ to obtain the susceptibility. Again there are both diamagnetic and paramagnetic contributions. The diamagnetic contribution comes from the final term in the kinetic-energy operator, which is already quadratic in the field and therefore enters only in first-order perturbation theory for the electronic state ψ. The shift in energy is

$$\delta\varepsilon = <\psi| \frac{H^2e^2}{8m_ec^2}(x^2 + y^2)|\psi>. \tag{22.17}$$

For a spherical system the expectation value for $<x^2>$ and for $<y^2>$ will be ${}^1/_3<r^2>$, of the order of an atomic radius squared, or internuclear distance squared, so we multiply by the density N of such electronic states (the atom density times the number of orbitals per atom of each type) to obtain

$$\chi_d = \frac{Ne^2}{6m_ec^2}<r^2> , \qquad (22.18)$$

quite similar in form to that for the free-electron gas. A remarkable feature of this contribution is that it arises even if all states are occupied with both spins. There is even a contribution from the core electrons in an atom, though the corresponding $<r^2>$ is very small. The contribution of the filled-d-shell electrons in noble metals is large enough to lead to a net diamagnetism, in contrast to other simple metals. Such contributes also arise for molecules, though then the $<r^2>$ must be evaluated for bond states. The calculation for semiconductors also involves a sum over contributions from bond states (see for example Harrison (1999), 159ff).

The paramagnetic contribution to the susceptibility of atoms and molecules arises from the term in the Hamiltonian linear in magnetic field, which we wrote in Eq. (22.12), as $eHL_z/(2m_ec)$, proportional to L_z. Most systems in the absence of a field will have no net angular momentum so the expectation value of L_z will vanish and there will be no first-order term, linear in H. We shall return to systems which *do* have net angular momentum, such as the O_2 molecule, shortly. In the more usual case, the shift in the energy must be obtained in second-order perturbation theory and will again be quadratic in H. Substituting from Eq. (22.12) for the matrix elements between some starting state $|i>$ and the states $|j>$ to which it is coupled, we obtain a second-order shift in energy of the i'th state of

$$\delta\varepsilon_i = \sum_j \frac{e^2H^2<j|L_z|i>^2}{4m_e^2c^2(\varepsilon_i - \varepsilon_j)} . \qquad (22.19)$$

For an atom, spherically symmetric, for which the energy eigenstates can be taken as eigenstates of L_z , the orthogonality of $|j>$ and $|i>$ guarantees that the matrix elements are all zero and there is no contribution. Similarly, if the magnetic field is applied along the axis of a cylindrically-symmetric molecule, there will be no contribution. However, in a semiconductor, with bonds along all cube-diagonal directions, all magnetic fields are guaranteed to be skewed with respect to some bond axes. States can again be chosen to be angular momentum eigenstates with respect to the bond axis, but then the magnetic-field interaction in the Hamiltonian, proportional to $\mathbf{H}\cdot\mathbf{L}$ as indicated just before Eq. (22.12), must be written in terms of L_x L_y, and L_z relative to these bond axes, which can in turn be written in terms of raising and lowering operators as discussed in Section 16.3. Then there is coupling between the bonding and antibonding states through the matrix elements of the raising and lowering operators which couple the s- to the p-states on the same atom. Note that it always gives a lowering in the energy of the occupied states and is therefore a paramagnetic contribution, called the Van Vleck term in the susceptibility. The needed matrix element is an intra-atomic one, which can be estimated, and gives a reasonable account of paramagnetism in semiconductors. In particular, the Van Vleck term is seen to decrease for polar semiconductors with larger gaps, while the diamagnetic term, called the Langevin term, is quite insensitive, as would be guessed from Eq. (22.18). (See, for example, Harrison (1999), 159ff.)

We return finally to systems, such as O_2, which have net angular momentum in the absence of fields. If we construct molecular orbitals for the oxygen molecule, we obtain a set which is qualitatively just like those we found for N_2 and showed in Fig. 5.7. The difference is that each oxygen atom contributes one less electron than a nitrogen atom so the bonding π-state, which was the highest-energy occupied state in N_2, contains only two electrons though there are two π-states (x-oriented and y-oriented) and each could accommodate electrons of two spins. We noted in Section 4.2 that when a set of degenerate orbitals are only partly occupied, the exchange interaction favors occupying them with parallel spin to the extent possible (Hund's rule). Thus one of the π-electrons will be in the x-oriented π-state and the other will be in the y-oriented π-state with the same spin. The net moment of the molecule is two Bohr magnetons from these spins, and there is no net orbital angular momentum. This is called a paramagnetic molecule because in an applied magnetic field the moment will tend to align with the field to lower its energy, a paramagnetic response. In this case the tendency is inhibited only by statistics. There are states for each molecule with the spin-angular-momentum components along the magnetic field of $\hbar$, 0, and $-\hbar$. Following just the procedure we used in Chapter 10 we see that the relative probability of the three states is $e^{-2\mu_B H/k_B T}$, 1, and $e^{2\mu_B H/k_B T}$. This leads to a dipole which is proportional to H at small fields, where the exponentials can be expanded, and corresponds to a susceptibility of

$$\chi = \frac{8\mu_B{}^2 N_{O_2}{}^2}{3k_B T}\ , \tag{22.20}$$

with N_{O_2} the number of oxygen molecules per unit volume.

22.4 Ferromagnetism

This same exchange interaction which was responsible for the paramagnetism of oxygen molecules is responsible for ferromagnetism in solids. It is in fact rather easily understandable for metals in terms of Fig. 22.3. We noted in Eq. (4.15) that the exchange energy of a free-electron gas, arising from the correlated motion of each electron with that of the electrons of the same spin, is

$$E_{ex} = -\frac{3e^2 k_F}{4\pi} \text{ per electron,} \tag{22.21}$$

If we were to arbitrarily shift some electrons from the spin-up band to the spin-down band, the exchange energy would become more negative for the spin-down band, and less negative for the spin-up band, as shown in the figure, and with more electrons in the spin-down band the total exchange energy would definitely be more negative. We may ask if this lowering could ever be greater than the increase in kinetic energy which this transfer causes. If so we have found an instability of the free-electron gas which would lead to a spin-polarized - that is to say, a ferromagnetic - state.

We gave in Eq. (2.12) the kinetic energy for electrons in a free-electron gas as

$$E_{kin} = \frac{3\hbar^2 k_F^2}{10m_e} \text{ per electron.} \tag{22.22}$$

We now separate the electrons by spin and Eqs. (22.21) and (22.22) apply to each set. We gave in Eq. (2.10) k_F in terms of the electron density, which we can rewrite for the number of electrons $N_\pm = N/2$ of a given spin

$$k_{F\pm}^3 = 6\pi^2 \frac{N_\pm}{\Omega} . \tag{22.23}$$

We can substitute for k_F in Eq. (22.21) and (22.22) and multiply by $N_\pm/\Omega$ to obtain the total energy density for the electrons of each spin,

$$E_\pm = -\frac{3e^2}{4\pi}(6\pi^2)^{1/3}\left(\frac{N_\pm}{\Omega}\right)^{4/3} + \frac{3\hbar^2}{10m_e}(6\pi^2)^{2/3}\left(\frac{N_\pm}{\Omega}\right)^{5/3} . \tag{22.24}$$

If we increase the spin-down density by δn , and decrease the spin-up density by the same amount, the change in energy to first order in δn is $\partial E_\pm/\partial(N_\pm/\Omega)\,[\delta n - \delta n] = 0$, but to second order is $^1/_2\, \partial^2 E_\pm/\partial(N_\pm/\Omega)^2 \delta n^2$, so if the second derivative is negative, the system is unstable against spin alignment. The condition for an instability is

$$-\frac{4}{3}\frac{1}{3}\frac{3e^2}{4\pi}(6\pi^2)^{1/3}\left(\frac{N_\pm}{\Omega}\right)^{-2/3} + \frac{5}{3}\frac{2}{3}\frac{3\hbar^2}{10m_e}(6\pi^2)^{2/3}\left(\frac{N_\pm}{\Omega}\right)^{-1/3} < 0 , \tag{22.25}$$

which can be written as

$$k_F < \frac{e^2 m_e}{\pi \hbar^2} = 0.60 \text{ Å}^{-1}. \qquad (22.26)$$

When this condition is satisfied, the energy continues to drop as more electrons are shifted, so it should proceed until all spins are aligned, as in Hund's rule. The k_F of Eq. (22.26) is only slightly less than the smallest Fermi wavenumber for the simple metals, 0.65 Å^{-1} for cesium. Thus it correctly suggests that all simple metals should be stable with equal numbers of both spins. In the transition metals, the density of states at the Fermi energy is higher, which decreases the kinetic energy term without changing the exchange term, as if m_e were increased in Eq. (22.26). Thus it is to be expected that ferromagnetism could occur in transition metals.

The conclusions are correct, and the physical origin is correct, but the picture can be misleading. First, it is not certain that this is the greatest instability. At low density the spin-density wave discussed in Section 21.2 may also be favored. Another instability at low density, the formation of a lattice of localized electrons, called a *Wigner crystal*, can also be favored. To see this we may construct a variational state in which each electron has a wavefunction which is a Gaussian around an individual lattice site. Then the Coulomb plus kinetic energy can be minimized with respect to the spread of the Gaussian. The resulting energy of that crystallized state can then be compared with the Coulomb plus kinetic energy for the uniform electron gas. At a spacing similar to that in Eq. (22.26) the Wigner crystal is found to be favored. There is another complication in our description of the ferromagnetic state in that the density of states is complicated in a transition metal, not simply an increased mass. The problem is more appropriately addressed in tight-binding theory, where it semiquantitatively accounts for the observed occurrence of ferromagnetism in the transition metals (Harrison (1999), 589ff).

Another aspect of the free-electron picture is misleading. It would seem to suggest that as we increased temperature and excited electrons into the reversed-spin band, the tendency to form the ferromagnetic state would weaken, as we discussed for the antiferromagnetic insulator in Section 21.2, and that ferromagnetism would disappear, as is found experimentally. In fact when ferromagnetism disappears experimentally, at the Curie temperature, there remains the same magnetic moment on each atom, and it is only the parallel orientation of the moments on different atoms which disappears. Thus the magnetic properties above the Curie temperature are those of a paramagnetic crystal, as we discussed for the O_2 crystal in the preceding section. The ferromagnetic transition metals are more easily understood in the atomic picture, with d-state spins aligned according to Hund's rule, and then weakly coupled between atoms to form the ferromagnetic state at low temperatures.

22.5 Spin-Orbit Coupling

There are important one-electron effects, as well as many-body effects, from the coupling between spin moments and orbital magnetic moments. This is an appropriate place to discuss that coupling. The origin of the interaction between the electron spin and its orbital motion arises from the magnetic field due to the relative motion of the electron and the nucleus, seen then by the electron spin magnetic moment. The interaction follows directly from the Dirac relativistic theory of the electron. It is not so easy to derive it from the nonrelativistic theory we have used, but we can understand the form it takes. If the electron were at rest and a particle of charge Q passed by, its current, which is proportional to Q and its velocity $\mathbf{v}$, would produce a magnetic field at the electron proportional to the current, inversely proportional to the square of the distance r, and to the sine of the angle θ between $\mathbf{r}$ and $\mathbf{v}$. The magnetic field would in fact be given by $\mathbf{H} = Q\mathbf{r}\times\mathbf{v}/cr^3 = \mathbf{E}\times\mathbf{v}/c$ where $\mathbf{E}$ is the electric field arising from the charge Q. The factor of the speed of light c makes the units of $\mathbf{E}$ and $\mathbf{H}$ the same. We can then make a transformation to the coordinate system of the nucleus (this should be done using relativistic equations). For a spherical system, the electric field $\mathbf{E}$ can be replaced by $(\mathbf{r}/er)\, dV(r)/dr$ with $V(r)$ the potential energy of the electron (charge $-e$) due to the nucleus and other electrons present. Then with $\mathbf{r}\times\mathbf{v}$ equal to the angular momentum $\mathbf{L}$ divided by m_e, the magnetic field is found to be

$$\mathbf{H} = \frac{1}{2em_e cr}\frac{dV(r)}{dr}\mathbf{L}\,. \qquad (22\text{-}27)$$

(See, for example Schiff (1968), p. 433.) The factor 1/2 is called the *Thomas precession factor* and comes from the use of a relativistic transformation.

The important point is that the magnetic field is proportional to the angular momentum of the orbit $\mathbf{L}$, and the magnetic moment of the electron is given by the spin angular momentum $\mathbf{S}$ times its gyromagnetic ratio of $-e/(m_e c)$, given just before Eq. (22.15). Thus the interaction of the spin moment with the orbital field is given by $\mathbf{L}\cdot\mathbf{S}$ times a factor, depending only upon r,

$$H_{SO} = \frac{1}{2m_e^2c^2r}\frac{\partial V}{\partial r}\mathbf{L}\cdot\mathbf{S}. \qquad (22.28)$$

For a Coulomb potential the interaction is proportional to Ze^2 as we would expect.

If only the interaction of a single electron with its spin is of interest, H_{SO} does not couple states of different l , so electronic orbitals of only a single l are of interest. We shall always use a basis of atomic states, with specific values of l and m and the matrix elements between them arising from spin-orbit coupling are written

$$<l',m'|H_{SO}|l,m> = \lambda_l \, \delta_{l',l} \, \frac{2<l\,m'|\mathbf{L}\cdot\mathbf{S}|l\,m>}{\hbar^2} \tag{22.29}$$

with

$$\lambda_l \equiv \frac{\hbar^2}{4m_e{}^2c^2} \int r^2 R_l(r)^2 \frac{\partial V}{r\,\partial r}\,\mathrm{d}r \tag{22.30}$$

having the units of energy. They have been calculated by Chadi (1977) for a number of elements which are important in semiconductors and are listed in Table 22.1. [In some studies λ is defined differently by a factor of two, but we used Chadi's definition.] We note that they grow rapidly with atomic number because they are dominated by the potential near the nucleus, proportional to the nuclear charge, and in fact they grow much more rapidly due to additional changes in $R_l(r)$.

How we now proceed depends upon the system we consider. For isolated atoms, spin-orbit coupling modifies the energies of the states, giving *fine structure* to the simple spectra we discussed in Chapter 4. We consider that first for hydrogen, or an alkali metal where only a single electron is involved in an important way.

When there is an interaction between two contributions to the angular momentum in a spherically-symmetric system, it is physically clear that

Table 22.1. Spin-orbit coupling parameters λ (in eV) for valence p-states, renormalized for use in solids, compiled by Chadi (1977). The spin-orbit splitting at the top of the valence band is 3λ in elemental semiconductors.

	Al 0.008	Si 0.015	P 0.022	S 0.025
Zn 0.025	Ga 0.058	Ge 0.097	As 0.140	Se 0.160
Cd 0.076	In 0.131	Sn 0.267	Sb 0.324	Te 0.367

there will not be eigenstates of each individual contribution, but there are eigenstates of the total angular momentum

$$\mathbf{J} = \mathbf{L} + \mathbf{S}. \tag{22.31}$$

For the isolated atom, the eigenstates of the total angular momentum $\mathbf{J}$ can be described just as were the eigenstates of $\mathbf{L}$ in Section 16.3. There are eigenstates of the squared total angular momentum $\mathbf{J}^2$ with eigenvalues $j(j+1)\hbar^2$. The quantum numbers j can only be $l \pm s$ depending upon whether the spin of $^1/_2$ is parallel or antiparallel with respect to the orbital angular momentum. Each eigenstate of $\mathbf{J}^2$ can be chosen to also be an eigenstate of the z-component of total angular momentum J_z with eigenvalues $\hbar j_z$ and with j_z taking values $j, j-1, j-2, \ldots -j$.

Further, we may write

$$\begin{aligned} \mathbf{J}^2|j,j_z\rangle = j(j+1)\hbar^2|j,j_z\rangle &= (\mathbf{L}^2 + 2\mathbf{L}\cdot\mathbf{S} + \mathbf{S}^2)|j,j_z\rangle \\ &= l(l+1)\hbar^2|j,j_z\rangle + 2\mathbf{L}\cdot\mathbf{S}|j,j_z\rangle + s(s+1)\hbar^2|j,j_z\rangle \end{aligned} \tag{22.32}$$

so that the $|j,j_z\rangle$ are also eigenstates of $\mathbf{L}\cdot\mathbf{S}$ with eigenvalues obtained by solving Eq. (22.32) for $\mathbf{L}\cdot\mathbf{S}|j,j_z\rangle$. They are given by $[j(j+1) - l(l+1) - s(s+1)]\hbar^2/2$. Thus, using Eqs. (22.28) through (22.31) a one-electron eigenstate of energy ε_l in the absence of spin-orbit coupling, would be split into levels of quantum number $j = l \pm s$ with energies given by

$$\varepsilon_j = \varepsilon_l + \langle j|H_{SO}|j\rangle = \varepsilon_l + [j(j+1) - l(l+1) - s(s+1)]\lambda_l. \tag{22.33}$$

For p-states, $[j(j+1) - l(l+1) - s(s+1)]/2$ is $^1/_2$ for $j = {}^3/_2$ and is -1 for $j = {}^1/_2$ so there are four states ($j_z = {}^3/_2, {}^1/_2, -{}^1/_2, -{}^3/_2$) at $\varepsilon_p + \lambda$ and two ($j_z = \pm{}^1/_2$) at $\varepsilon_p - 2\lambda$, rather than three states, of two spin orientations each, without spin-orbit coupling. This splitting is illustrated in Fig. 22.4.

The values appropriate to the free atom are typically $^2/_3$ of those renormalized for the solid and given in Table 22.1. The atomic spectra allow measurement of the energy differences between various spin-orbit-

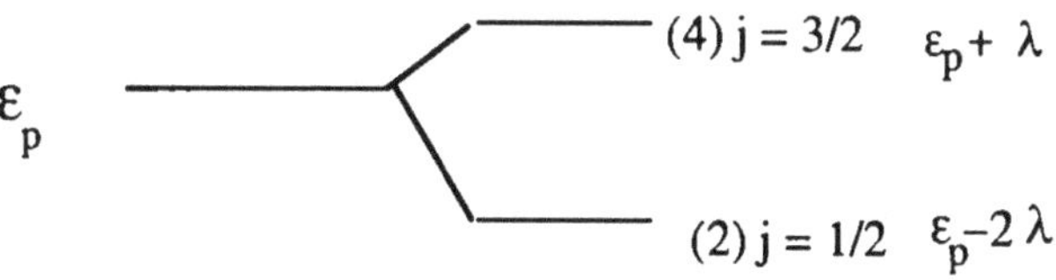

Fig. 22.4. The splitting of the hydrogen p-state by spin-orbit coupling, Eq. (22.33)

split states for various l-values. We see from that table that the splittings are quite small on the scale of the difference between different atomic term values so the splitting is indeed small, and appropriately called fine structure. In Problem 22.4, this same analysis is carried out for d-states.

The analysis is more intricate when more than one electron is included, so that there is spin-orbit coupling between different electronic states, but the principle is the same. For example, for carbon, for which we start with a configuration $1s^2 2s^2 2p^2$, the lowest state (by Hund's Rule) will be with the two p-states of parallel spin, so we imagine a two-electron state with spin quantum number $S = 1$, using a capital S for the total-spin quantum number. If we take the total orbital momentum to be approximately conserved, which is often reasonable since the interaction with the spins is weak, that total orbital-angular-momentum quantum number can be $L =$ 0, 1, or 2. These different states will have different Coulomb energy from the electron-electron interaction. For each choice one can then evaluate the $\mathbf{L}\cdot\mathbf{S}$ energy. Other approaches can be taken when the terms which dominate are different from what was assumed here.

We note briefly an important effect of spin-orbit coupling for the band structure of semiconductors (Chadi, (1977)). We noted in Section 13.5 that the states at the top of the valence band are made purely of p-states. We can expect those bands to be split into two (doubly-degenerate, since there were four total) bands at slightly higher energy and one (doubly-degenerate) band at slightly lower energy, and that is the case. The spin-orbit coupling can be directly included in a tight-binding band calculation, doubling the size of the Hamiltonian matrix since now the spin-up and spin-down orbitals are distinguished. Spin-orbit coupling provides intra-atomic matrix elements from Eq. (22.29), in addition to the interatomic matrix elements arising from the $V_{ss\sigma}$, $V_{sp\sigma}$, etc. The intra-atomic matrix elements are evaluated by writing

$$\mathbf{L}\cdot\mathbf{S} = L_x S_x + L_y S_y + L_z S_z = \frac{L_+ S_- + L_- L_+}{2} + L_z S_z \tag{22.34}$$

using Eq. (22.32). Then the procedure is straight-forward and leads to the expected splitting .

How the bands go away from $\mathbf{k} = 0$ is less obvious, but is shown in Fig. 22.5. In Part a are seen the doubly-degenerate (four-fold, including spin) heavy-hole bands and light-hole band. The heavy-hole bands can be written in terms of states which have $\pm\hbar$ angular momentum around the wavenumber of the state. These are split up and down depending upon whether the spin is parallel or antiparallel to this orbital angular momentum.

The light-hole band must match with the upper band since it is the one with the higher degeneracy as seen in Fig. 22.4. The result would be that the light-hole band drops rapidly with wavenumber and would cross the lower heavy-hole band. However, bands can only cross if they are uncoupled to each other. Any coupling will cause the bands to move apart, sometimes called an *anticrossing*. This is precisely what occurred in the formation of the antiferromagnetic insulator discussed in Section 21.2 . Before the symmetry was broken we could draw free-electron bands as in Fig. 21.3 in a reduced zone ($0 < kd < \pi/2$), or in an extended zone ($0 < kd < \pi$), or both, and they cross because there is no coupling between them. Once the symmetry is broken, the two crossing bands are coupled and the crossing is avoided as shown there. In Part b we see that exactly this happens here and the lowest band switches from heavy to light in the region where they would have crossed. These are important bands in an important region of energy so the result is of considerable consequence.

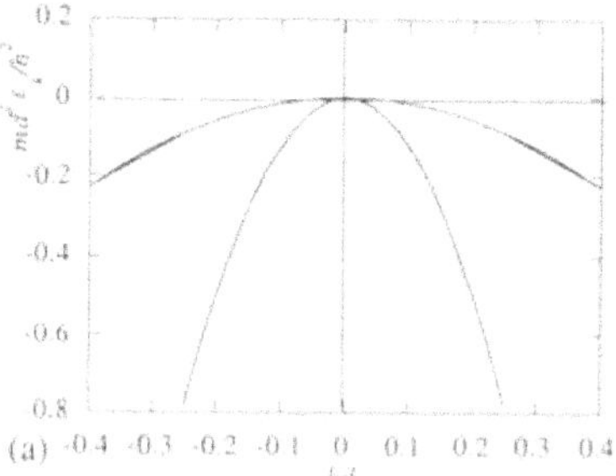

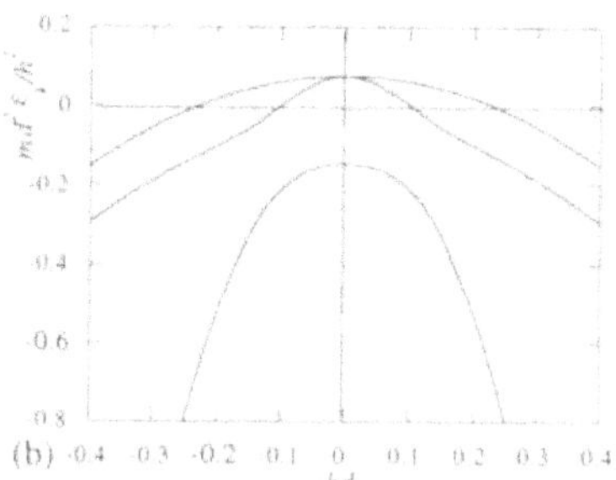

Fig. 22.5. The top of the valence bands of germanium (a) without spin-orbit coupling and (b) with spin-orbit coupling. The energies are in units of $\hbar^2/md^2$. Note that the light-hole band is at the upper energy at $\boldsymbol{k} = 0$ in Part (b). [After Harrison (1999)].

Chapter 23. Shake-Off Excitations

Shake-off excitations would seem to be a specialized topic, but they have turned out to be so central to a wide variety of quantum-mechanical properties, and so central to a number of insights, that they definitely deserve at least a chapter. They also are very suitable as a concluding chapter. They concern the behavior of "spectator" systems, such as an electron in an atomic state, when a second electron is removed. By way of introduction we discuss two limiting approximations for time-dependent problems.

23.1 Adiabatic and Sudden Approximations

One problem we have not discussed is boundary conditions which change with time. There are many situations related to this, but we illustrate it for an electron in a one-dimensional quantum well, as in Section 2.1 and as shown in Fig. 23.1. The Hamiltonian contains only the kinetic energy, but the positions at which the vanishing boundary conditions are applied change with time.

There are two limiting cases for which the answer seems obvious. The energy levels are discrete and if the boundary conditions change very slowly it is clear that an electron in the lowest state has no chance to jump to a higher state. An electron in this state, or in another state, is expected to stay in the expanding state, called following the state adiabatically, in analogy with the slow expansion of a gas against a piston, cooling the gas. On the other hand, if the boundary were very quickly to be expanded, it is clear that the initial state $|\psi_0(0)\rangle$ cannot change quickly; it changes only according to the Schroedinger Equation, Eq. (1.16). Thus we may neglect its change

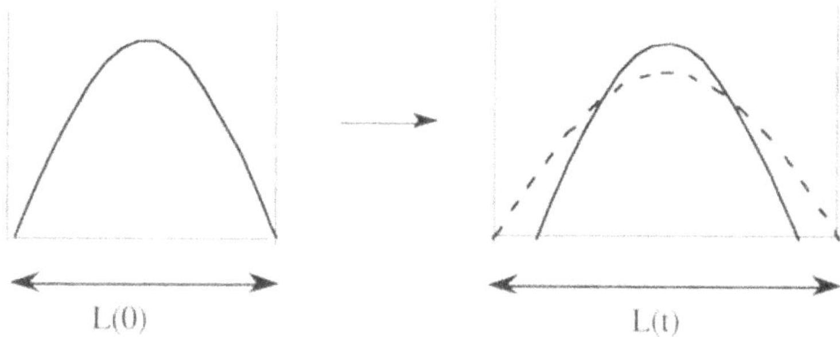

Fig. 23.1. If the well in which an electron is bound is increased in size slowly, the state will follow adiabatically as to the dashed line to the right. If it is increased suddenly, the state will initially remain as it was, as in the solid line to the right.

during the boundary shift, expand the original wavefunction in the eigenstates $|\psi_j(t)>$ appropriate to the new boundary, and can then easily follow the time evolution of each state. In fact, the squared expansion coefficients $<\psi_0(0)|\psi_j(t)>$ of the original state in each of the new states is exactly the probability of a transition occurring to that state. This is called the *sudden* approximation and is analogous to expanding the chamber holding a gas so rapidly that the atoms have no chance to do work against the moving walls.

It is not immediately obvious for the system in Fig. 22.3 what the criterion for fast or slow is, though we shall show that if the time taken to change the potential is small compared to $\hbar$ divided by the energy difference to some excited state, we may regard the change as fast, and if it is large, we may regard the change as slow. We do this by expanding the wavefunction in the eigenstates which depend upon the length $L(t)$ which in turn depends upon time, very much as we did in treating time-dependent perturbations in Sections 9.3 and 19.3,

$$|\psi(t)> = \sum_j u_j(t)e^{-i\omega_j(t)t}\,|j> \; . \tag{23.1}$$

Here

$$|j> = \sqrt{2/L(t)}\,\cos[\pi(2j+1)x/L(t)], \tag{23.2}$$

with x measured from the center of the well and energy

$$\varepsilon_j(t) = \hbar^2[\pi(2j+1)/L(t)]^2/(2m) \equiv \hbar\omega_j(t)\,. \tag{23.3}$$

There are also states with energy $\hbar^2[2\pi j/L(t)]^2/(2m)$, which are odd around $x = 0$, but these are not coupled to the states we consider. They are a separate problem. We may think of the adiabatic limit where the u_0 is one initially, and remains near one. We may see how any other u_j grows with time by substituting Eq. (23.1) into the Schroedinger Equation, multiply on the left by a particular $\langle j|$, dropping the zero-order terms, and keeping only the largest first-order term, that with $u_0 \approx 1$. This leads to

$$-\frac{\hbar}{i}\frac{\partial u_j}{\partial t}\, e^{-i\omega_j(t)t} \;-\frac{\hbar}{i}\langle j|\frac{\partial}{\partial t}|0\rangle\, e^{-i\omega_0(t)t} = 0\,. \tag{23.4}$$

Note that a term $\hbar t\ \partial\omega_0/\partial t\, \langle j|0\rangle$ was dropped because $\langle j|0\rangle=0$. To evaluate $\langle j|\partial/\partial t\,|0\rangle$ we take the derivative of $|0\rangle$ with respect to t, using Eq. (23.2). The term from $\partial/\partial t\,\sqrt{2/L(t)}$ is again zero because $\langle j|0\rangle = 0$, but the other term gives $\langle j|\partial/\partial t|0\rangle = (2/L(t))(\pi/L(t)^2)(\partial L(t)/\partial t) \int \cos[\pi(2j+1)x/L(t)]x \sin[\pi x/L(t)]\,dx$. We write

$$\begin{aligned}\lambda_j &= \frac{2}{L(t)}\int \cos[\pi(2j+1)x/L(t)](x/L(t))\sin[\pi x/L(t)]\,dx\\ &= 2\int_{-1/2,1/2}\cos[\pi(2j+1)u]u\sin[\pi u]\,du,\end{aligned} \tag{23.5}$$

which is independent of t and of order one. Note again that there is no coupling to the states which are odd around $x = 0$. With this form we have

$$\frac{\partial u_j}{\partial t} = -\lambda_j\frac{\pi}{L(t)}\frac{\partial L(t)}{\partial t}\,e^{i(\omega_j(t)-\omega_0(t))t}\,. \tag{23.6}$$

In order to proceed further we must specify $L(t)$. One way to represent a small change in length ΔL taking place in a time t_0 is to take $L(t) = L + \Delta L(e^{-t/t_0} - 1)$. Then $(1/L(t))\partial L(t)/\partial t \approx -(\Delta L/L)e^{-t/t_0}/t_0$. We may substitute into Eq. (23.6), integrate from zero to a large t , now taking $\omega_j - \omega_0$ independent of time since changes are of higher order in $\Delta L/L$. We obtain the probability of a transition to the state $|j\rangle$ as

$$u_j^*u_j = \left(\frac{\pi\lambda_j\,\Delta L}{L}\right)^2\frac{1}{1+(\omega_j-\omega_0)^2t_0^2}\,. \tag{23.7}$$

If the time t_0 taken to change the boundary is small compared to $1/(\omega_j - \omega_0) = \hbar/(\varepsilon_j - \varepsilon_0)$, the final factor is near one and we have a form for the transition rate which we may associate with abrupt changes. It will tend to decrease with increasing j because λ_j from Eq. (23.5) will tend to drop as $1/j$. If t_0 is long compared to $\hbar/(\varepsilon_j - \varepsilon_0)$, the final factor will become small and we expect few transitions, associated with an adiabatic following of a slowly-changing boundary. In Problem 23.1 we treat the other limit, where the change is rapid, and find explicitly the probability of particular transitions.

Similar criteria can be derived for a harmonic oscillator when the parameters, such as the spring constant, change with time. In that case transitions tend not to occur if the changes in parameters occur over a time long compared to the period of the oscillator, physically a very natural criterion. It is in fact closely related to the criterion we just found for the system in Fig. 23.1 since the energy to the excited states in the oscillator is of order $\hbar\omega$, and that energy divided into $\hbar$ is the period of the oscillator.

23.2 Vibrational Excitations

We turn to the question of vibrational shake-off excitations which may arise when an electronic transition occurs. Perhaps the simplest case conceptually is illustrated in Fig. 23.2. We imagine a particle bound in a state $|0\rangle$ on a platform, with an energy which varies as λu with the displacement u of the platform. λ would be mg for a particle of mass m in a gravitational field, or it could be the shift in energy level in a molecule as the internuclear distance is changed. The Hamiltonian will also contain a harmonic-oscillator kinetic and potential energy associated with the platform displacement u. Finally, there will be a set of freely propagating electronic states $|k\rangle$, coupled to the local state by a matrix element $V_{0k} = \langle 0|H(\mathbf{r})|k\rangle$ with H involving only the electronic coordinates. The term λu is absent from the Hamiltonian when the electron is in a state $|k\rangle$. This Hamiltonian

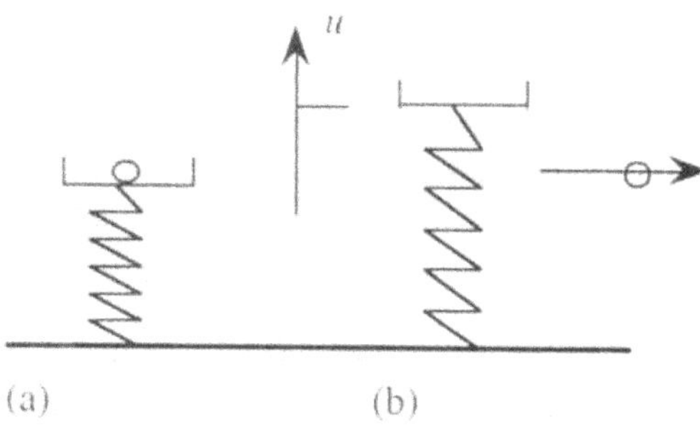

Fig. 23.2. In Part (a) a particle is bound in a local state, in which its energy is shifted by a displacement u of a harmonic oscillator. That state is coupled to freely propagating states, Part (b), in which there is no coupling with the displacement.

can be written, before we treat the vibrational terms quantum-mechanically, as

$$H = (\varepsilon_0 + \lambda u)c_0^\dagger c_0 + \varepsilon_k c_k^\dagger c_k + V_{k0} c_k^\dagger c_0 + V_{0k} c_0^\dagger c_k + {}^1\!/_{2M} P^2 + {}^\kappa\!/_2 u^2, \tag{23.8}$$

with M and κ the mass and spring constant for the platform oscillator.

Now we imagine the particle bound to the platform, and the spring compressed to its shifted equilibrium position. If the particle makes a transition off of the platform, the platform will return to its position, unshifted by the weight of the particle. The question might arise whether we should make the sudden, or the adiabatic, approximation: is the platform left vibrating (called a shake-off excitation), or does it adiabatically shift to its new equilibrium question. We are left with the question: "How fast does the transition occur?" The answer is far from obvious, and most people would guess incorrectly. The correct approach is simply to do the calculation, *including the platform in the quantum-mechanical problem*, and we can interpret the result afterward if we wish. The model captures the physics of a wide variety of problems in which we feel we should ask how long some process takes and often the result can be guessed by generalizing the result we obtain here. In other cases, one can redo the new problem.

We shall use the Golden Rule, with the perturbation V_{0k}, and the principal task is finding the initial and final states. In the initial state, with the particle on the platform, the equilibrium position is shifted to u_0 such that $(\partial/\partial u)({}^1\!/_2\kappa u^2 + \lambda u) = 0$, or $u_0 = -\lambda/\kappa$. It is essential to treat the platform quantum-mechanically (it is always essential to include all parts of the system in the quantum theory if it makes a difference, or if we cannot see how to proceed otherwise) so we might let the oscillator be in its ground state $\phi_0(u - u_0)$, with ϕ_0 the ground-state harmonic-oscillator eigenstate given in Section 2.5. Then the initial state is $|0,0\rangle = \phi_0(u - u_0)|0\rangle$, with the first zero in $|0,0\rangle$ referring to the electronic state and the second referring to the vibrational state. The final state $|k,n\rangle$ might also be in a ground state of the unshifted harmonic oscillator $\phi_0(u)$, or it might be an excited state $\phi_n(u)$, with an energy higher by $n\,\hbar\omega$. That is what we wish to learn. It is important that we use *unshifted* final states here so that the energy can be specified.

For the excited state, the matrix element for the transition becomes $\langle k,n|H(\mathbf{r})|0,0\rangle = V_{0k} \langle \phi_n(u)|\phi_0(u-u_0)\rangle$, with the *overlap integrals* given by

$$\langle \phi_n(u)|\phi_0(u-u_0)\rangle = \int du\, \phi_n(u)\, \phi_0(u - u_0). \tag{23.9}$$

(The harmonic-oscillator functions are real.) The same forms apply if the final state of the oscillator is the ground state, $n = 0$. We may immediately write the transition rate for each excitation level of the final state from Eq. (7.9),

$$P_{0n} = \frac{2\pi}{\hbar} \sum_k V_{0k} V_{k0} < \phi_0(u-u_0)|\phi_n u)>< \phi_n(u)|\phi_0(u-u_0)> \delta(\varepsilon_k + n\,\hbar\omega - \varepsilon_0). \tag{23.10}$$

All of the interesting questions can be answered by looking at this result. First, if there is no coupling, $\lambda = 0$, then $u_0 = 0$ and all of the overlap integrals are zero, except $n = 0$, for which the overlap is one. The oscillator is never excited and the usual formula, $(2\pi/\hbar) \Sigma_k V_{0k} V_{k0}\ \delta(\varepsilon_k - \varepsilon_0)$ applies, as it should. When the coupling is nonzero, the probability of a transition with no excitation is reduced by a factor $< \phi_0(u)|\phi_0(u-u_0)>^2$. This reduction is compensated for by some probability of transition to a state $|n>$, with probability proportional to $< \phi_n(u)|\phi_0(u-u_0)>^2$, with shake-off excitations of $n\,\hbar\omega$. In fact this compensation is in some sense complete since

$$\sum_n <\phi_0(u-u_0)|\phi_n(u)>< \phi_n(u)|\phi_0(u-u_0)> = 1, \tag{23.11}$$

if the sum includes $n = 0$. This is called a *sum rule* and it follows from the fact that the states $|\phi_n(u)>$ are a complete set. Thus we may expand $\phi_0(u - u_0)$ in the $\phi_n(u)$ as $|\phi_0(u-u_0)> = \Sigma_n |\phi_n(u)>< \phi_n(u)|\phi_0(u-u_0)>$ and then Eq. (23.11) is just the normalization condition on $|\phi_0(u-u_0)>$ There may be slight differences in the V_{0k} which enter, and the density of states will be different for different n , so the compensation of lost probability to $n = 0$ by probability to other n is only approximate. In that approximation, $<\phi_0(u-u_0)|\phi_n(u)><\phi_n(u)|\phi_0(u-u_0)>$ is the conditional probability that the particle will leave n quanta of excitation when it leaves the platform.

This $<\phi_0(u-u_0)|\phi_n(u)>< \phi_n(u)|\phi_0(u-u_0)>$ is in fact just the formula for the sudden approximation, the squared expansion coefficient of the initial state in the possible final states. However, now that the oscillator has been quantized, it describes also the probability of remaining in the ground state - appropriate to the adiabatic approximation.

We have indeed answered the question as to the probability of different events occurring and we may learn something by looking at the results. A convenient way is to plot the probability of the electron emerging from the system with different energies, $\varepsilon_0 - n\hbar\omega$, illustrated in Fig. 23.3. If the shift in the equilibrium position u_0 is small compared to the zero-point fluctuation a_0 of the oscillator (defined after Eq. (2.43)) we have a weak-coupling limit.

The two Gaussian states will only be shifted slightly from each other and $<\phi_0(u)|\phi_0(u-u_0)>$ will be near one. The values of $<\phi_n(u)|\phi_0(u-u_0)>$ will be small for other n, and decrease rapidly with n as shown in Fig. 23.3 (a). We may think of the slight reduction in P_0 as the probability that the oscillator makes a large enough zero-point fluctuation to bring the oscillator into the unshifted position.

If the shift in the equilibrium position u_0 is *large* compared to the zero-point fluctuation a_0 of the oscillator, the shifted Gaussian will barely overlap the unshifted Gaussian and there will be almost no chance of emission with $n = 0$. This is also true of small n but once we go to large enough n that the peaks in the harmonic-oscillator wavefunctions are large near $u = u_0$, states of energy $n\hbar\omega$ near the classical vibrational energy $^1/_2\kappa u_0^2$ (as we saw in Problem 2.9) the probabilities become large, as illustrated in Part (b) of Fig. 23.3. We may think of this *strong-coupling limit* as the classical limit and it is exactly what is expected in classical physics for the abrupt removal of the particle. The transition of the particle leaves the oscillator in its original displaced position, with energy $^1/_2\kappa u_0^2$ which then appears as vibrational energy. In fact, the state of the oscillator is a coherent packet representing a harmonic oscillator with displacement approximately equal to u_0, and that packet will oscillate as a classical oscillator. Many systems correspond to this strong-coupling, or classical, limit in which the behavior is as if the electronic transition were very fast and the harmonic oscillator, or atomic system, is slow. This statement is called the *Franck-Condon Principle*, but it is only true for some systems.

We may now go back and ask what time we should have assumed it took the transition to occur if we wished to guess whether the result corresponded to an abrupt or an adiabatic transition. The criterion for a "fast transition",

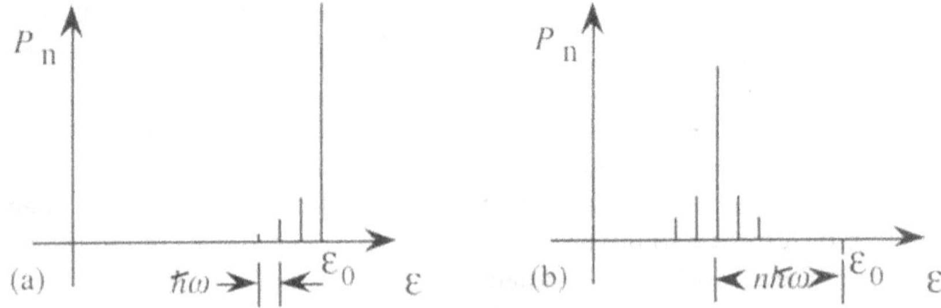

Fig. 23.3. The emission spectrum of a particle leaving a state coupled to an oscillator, as in Fig. 23.2. Part (a) is the weak-coupling limit in which the probability is high of leaving with the original electronic energy and leaving no shake-off excitations. There is a small probability of leaving a few quanta of energy behind. Part (b) is the strong-coupling limit in which the particle leaves approximately $n\hbar\omega = {}^1/_2\kappa u_0^2$ behind in shake-off excitation.

for the Franck-Condon Principle to apply, is that the equilibrium shift u_0 be large compared to the zero-point fluctuation a_0, or $a_0{}^2 << u_0{}^2$. We may multiply both sides by $^1/_2\kappa$, and $^1/_2\kappa a_0{}^2$ was seen in Eq. (2.44) to be $\hbar\omega/4$. Further , $^1/_2\kappa u_0{}^2$ is the spring energy in the relaxed state, twice the energy gained E_{relax}, by letting the spring relax as in Fig. 23.2 (a). Thus the condition for a fast transition is

$$\hbar\omega << 8E_{relax}. \tag{23.12}$$

This is the condition that $\hbar/E_{relax}$ is short compared to $8/\omega = 8T/2\pi$, with T the period of the oscillator. We would have guessed correctly when the sudden approximation is appropriate if we had said that the time for the particle to make a transition from the platform was $\hbar/E_{relax}$. This was the transition time which we suggested most people would guess incorrectly. It has nothing to do with the lifetime of the state on the platform, nor the matrix element V_{0k} with the external states, but only upon the parameters of the oscillator which we are using as the instrument to "measure" the transition time. This is a particularly important message in a time when it is often the style to guess answers by asking what are the relevant time scales, or distance scales, in order to avoid making a model and doing a calculation. The latter is much safer.

Electron tunneling is an important example which contains the same physics as this example, Fig. 23.2. If we think of the transition from the platform as a tunneling event, and ask whether the tunneling should be considered fast or slow compared to an oscillator period, we find that the tunneling time is again $\hbar/E_{relax}$ and has nothing to do with the thickness of the barrier. One place where this comes up is in the tunneling of electrons in a polarizable medium such as a semiconductor. We saw in Section 17.3 that a carrier polarizes the medium, reducing its energy to form a polaron. If that carrier tunnels to a different place, we might ask if it leaves vibrational energy behind, or does the polarization of the lattice disappear during the process. The answer is obviously that it will tend to leave vibrational energy behind if $\hbar/E_{relax}$ (with E_{relax} the energy gain in forming the polaron) is small compared to $1/\omega$. Probably a better, and equivalent, criterion is that it will tend to leave vibrational energy behind if E_{relax} is large compared to $\hbar\omega$. Such a case is frequently called a *small polaron*. It has distortional energy large compared to a phonon and usually well localized. This also means that if the E_{relax} is *small* compared to $\hbar\omega$ it is unlikely to emit a phonon, and it certainly cannot emit a part of a phonon. Another important example is electronic excitations in a molecule. Usually such a transition will shift the equilibrium spacing so that the molecule may, or may not, be left vibrating after the transition. We have seen how to learn which.

In our analysis we began with the oscillator in the ground state, but it is quite clear how to repeat the analysis with a starting excitation $n = n_0$. Then there are processes in which vibrational quanta are absorbed as the transition occurs, often called *phonon-assisted* transitions, as well as those with shake-off excitations.

A final closely related example is the Mössbauer effect (Mössbauer (1958)). When an atomic nucleus emits a gamma ray (Section 4.4) of momentum $\hbar\mathbf{q}$ of magnitude $\hbar\omega/c$, the atom must recoil with equal and opposite momentum, $-\hbar\mathbf{q}$. This conservation of momentum comes always from the matrix element and in this case it means that the terms in the Hamiltonian which describe the emission process within the nucleus must contain a dependence upon the center-of-mass coordinate **R** of the nucleus given by $e^{-i\mathbf{q}\cdot\mathbf{R}}$. Then if a free atom has a nucleus initially in a state $e^{i\mathbf{k}\cdot\mathbf{R}}/\sqrt{\Omega}$, it will go to a final state $e^{i(\mathbf{k}+\mathbf{q})\cdot\mathbf{R}}/\sqrt{\Omega}$. Now, if the atom in question is part of a solid, we may ask whether the nucleus is left vibrating, as if the gamma-ray emission were very rapid compared to a typical vibrational frequency ω_D of the solid, or if the momentum would be transmitted to the entire crystal with negligible recoil energy, as if the emission were very slow. The last situation is called the Mössbauer effect. It is quite easy to answer the question (see, for example, Harrison (1970)) much as we did above. We may write the component of **R** along the direction of **q** as u, and represent the binding of the atom in the solid as a harmonic oscillator with frequency ω_D. Then the matrix elements for the nuclear transition contain a factor $\langle n|e^{-iqu}|n'\rangle = \int du\ \phi_n(u)e^{-iqu}\phi_{n'}(u)$ with n' the level of excitation before, and n the level of excitation after, the transition. If prior to the transition the system is in the vibrational ground state, the conditional probability of finding it in the n'th vibrational state after is $|\langle n|e^{-iqu}|0\rangle|^2$. In particular, the probability of remaining in the ground state is readily evaluated using the harmonic-oscillator ground-state wavefunction (Eqs. (2.40) and (2.43))and found to be

$$|\langle 0|e^{-iqu}|0\rangle|^2 = e^{-(qa_0)^2}. \tag{23.13}$$

It is notable that we could solve this problem without knowing anything about the nuclear process other than the momentum transfer. It turns out that this formula applies also if the fluctuations come from thermal vibrations, with the zero-point a_0^2 in Eq. (23.13) replaced by the thermal $\langle u^2\rangle$. Thus, to increase the probability of the Mössbauer effect occurring, one goes to low temperatures so that it is suppressed only by the zero-point fluctuations. The effect is important since it leads to very sharply-defined gamma-ray energies, with negligible broadening from nuclear recoil.

Finally, we could ask what we should have assumed was the time required to emit a gamma ray if we wished to decide if the process was sudden or adiabatic. After the above analysis it may not be surprising that we should have taken the emission time as $\hbar$ divided by the classical recoil energy, $\hbar^2 q^2/(2M)$, with M the nuclear mass. (This form actually also illustrates the fact that if the recoil is by the entire crystal of mass NM the recoil energy is negligible.) As we expect, this "transition time" has nothing to do with half-life of the nuclear state.

23.3 Electronic and Auger Processes

The vibrational system in the preceding section has been a *spectator* to an event involving an electron, or other particle. In just the same way we can treat electronic states which are a spectator to another process. For example, in the beta-decay of a nucleus (discussed in Section 9.5), the emission of a beta ray increases the atomic number by one, and any electronic state on that atom will shift closer to the nucleus due to the extra nuclear charge. We may ask whether an electron in that state will follow the state, or whether it will be excited to a different final state. The answer is quite obvious by analogy with the results for vibrational excitations. If we write the electron state before the beta-decay as $|\psi_0'\rangle$, and the eigenstates after the decay as $|\psi_n\rangle$, we can define a conditional probability (probability, given that the beta-decay occurred) of finding the electron in the n'th state as

$$P_n = \langle\psi_0'|\psi_n\rangle\langle\psi_n|\psi_0'\rangle. \tag{23.14}$$

Again because of the different energies for the resulting beta-ray, there are corrections to this probability, usually not so important.

Another illustrative case is an electron in a donor state in a semiconductor. We found in Section 14.2 that the effective-mass wavefunction for such a state is hydrogenic,

$$\psi_\mu(r) = \sqrt{\frac{\mu^3}{\pi}}\, e^{-\mu r}\,, \tag{23.15}$$

with energy relative to the conduction-band minimum of

$$\varepsilon_D = -\frac{Z^2 m^* e^4}{2\hbar^2\varepsilon^2} = -\frac{\hbar^2\mu^2}{2m^*} \tag{23.16}$$

with $Z = 1$. If an electron is removed from the core of the donor atom, by an x-ray, the effective charge binding the donor state will increase from $Z = 1$ to

$Z = 2$. The energy drops by a factor of four and μ increases by a factor of two. Then the conditional probability of the electron shifting to the new ground state of the donor with $\mu = 2\mu'$ is

$$<\psi_\mu|\psi_{\mu'}>^2 = \frac{64\mu^3\mu'^3}{(\mu + \mu')^6} = \frac{512}{729} = 0.70. \qquad (23.17)$$

This is again consistent with a sudden approximation but gives the probability that the electron remains adiabatically in the ground state as Eq. (23.17).

When the excitations in question are electronic, and the change in potential is electronic, as in the case of the donor state, these are called *Auger transitions*. Another example is a transition in which a valence electron in an atom drops into an empty core state, which may shake off another valence electron, frequently ionizing the atom. When a valence electron in a metal drops to fill an empty core state, the screening of the resulting change in potential makes only a small change in each electronic state, very much like the small change in each state in a scattering, or tunneling, resonance (Section 8.4). Thus the probability of any one state making a shake-off transition is extremely small, but when summed over all states some shake-off is certain, as first showed by Anderson (1967). These transitions in the metal come at very low energies. The corresponding x-ray emission spectra would otherwise resemble the density of occupied states as a function of energy, but the spectra are modified near the highest-energy x-rays.

23.4 Inelastic Processes

We saw how a tunneling particle, as in Fig. 23.2, can leave vibrational energy behind, which is called inelastic tunneling. Similarly the tunneling electron can leave vibrational energy in the medium into which it tunnels, which is calculated the same way. A particularly interesting third case is in tunneling through an intermediate state as described in Section 9.1. The system is illustrated in Fig. 23.4, in analogy with Fig. 23.2. The transition rate is written as in Eq. (7.9),

$$P_{0f} = \frac{2\pi}{\hbar}\sum_f\left|\sum_i\frac{<f|H|i><i|H|0>}{E_0 - E_i}\right|^2\delta(E_0 - E_f)\,, \qquad (23.18)$$

but now the initial state is $c_1{}^\dagger|0>$, and we choose to put the harmonic oscillator in its ground state $\phi_0(u)$. (The analysis is the same, as in Section 23.2, if we choose an excited initial harmonic-oscillator state.) The intermediate states $|i>$ are all $c_2{}^\dagger|0>$, with the spring relaxed but with various

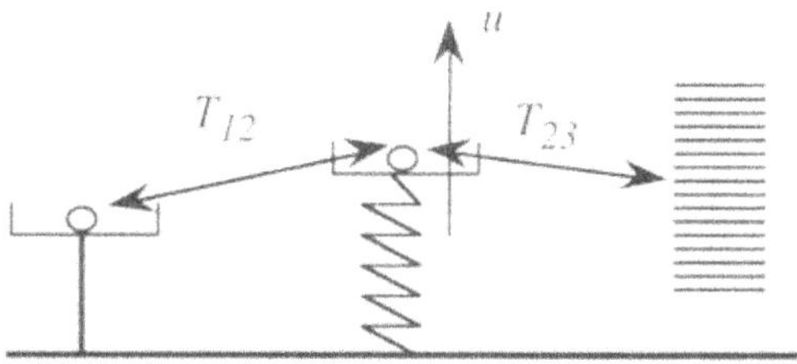

Fig. 23.4. An initial electron state $c_1^\dagger|0>$ is coupled to an intermediate state $c_2^\dagger|0>$ by T_{12}, which is in turn coupled to a set of final states $c_{k3}^\dagger|0>$ by T_{23}. There is a polarization term, $\lambda u\, c_2^\dagger c_2$, in the Hamiltonian, as in Fig. 23.2 and Eq. (23.8), so vibrations can be introduced in the intermediate system during tunneling.

vibrational states $\phi_{n'}(u - u_0)$. We write their energy $\varepsilon_2 - E_{relax} + n'\hbar\omega$. The final state is $c_{k3}^\dagger|0>$, with no relaxation since no particle is on the platform, and can have different vibrational states $\phi_n(u)$. Its energy is $\varepsilon_3 + n\hbar\omega$. Thus the sum over intermediate states, the second-order coupling of Eq. (9.2), becomes

$$<f|H^{2nd}|0> = \sum_i \frac{<f|H|i><i|H|0>}{E_0 - E_i}$$

$$= \sum_{n'} \frac{T_{12}T_{23}<\phi_0(u)|\phi_{n'}(u-u_0)><\phi_{n'}(u-u_0)|\phi_n(u)>}{\varepsilon_1 - (\varepsilon_2 - E_{relax} + n'\hbar\omega)} \quad (23.19)$$

and the energy delta function becomes $\delta(E_0 - E_f) = \delta(\varepsilon_1 - \varepsilon_3 - n\hbar\omega)$.

One interesting case has the final state, as well as the initial state, without excitations, $n = 0$. Then the second-order matrix element becomes

$$<f|H^{2nd}|0>= \sum_{n'} \frac{T_{12}T_{23}<\phi_0(u)|\phi_{n'}(u-u_0)>^2}{\varepsilon_1 - (\varepsilon_2 - E_{relax} + n'\hbar\omega)} \quad (23.20)$$

and the delta function is $\delta(E_0 - E_f) = \delta(\varepsilon_1 - \varepsilon_3)$. We have a sum of contributions from the intermediate states, of the same sign if $T_{12}T_{23}$ has simple behavior and $\varepsilon_2 > \varepsilon_1$. The transition goes through an intermediate, or *virtual*, state of higher energy. If $\varepsilon_2 < \varepsilon_1$ there may be intermediate states of the same energy as the initial states and the possibility of real transitions to the intermediate state.

The case in which there are excitations in the final state has the full factor $<\phi_0(u)|\phi_{n'}(u-u_0)><\phi_{n'}(u-u_0)|\phi_n(u)>$ from Eq. (23.19). It is interesting to note the two limits we discussed in Section 23.2. If the shift u_0 is large compared to the zero-point fluctuations a_0, the overlap $<\phi_0(u)|\phi_{n'}(u-u_0)>$ is small unless n' takes a large value, corresponding to a classical distortion of the spring from its relaxed position at u_0 to the initial position $u = 0$. Then for that large n' the overlap $\langle\phi_{n'}(u-u_0)|\phi_n(u)\rangle$ will only be large if n is near zero for the same reason and there will be no energy lost to the oscillator. This is the classical result, in which we may think of the platform as remaining at its initial position $u \approx 0$, with the electron darting through the intermediate state as in Section 9.1. This is in fact quantitatively what Eq. (23.19) gives for this limit with $n = 0$. With approximately the same elastic energy $n'\hbar\omega \approx {}^1\!/_2\kappa u_0^2 \approx E_{relax}$ added to each intermediate-state energy, the factor $1/[\varepsilon_1 - (\varepsilon_2 - E_{relax} + n\hbar\omega)] \approx 1/[\varepsilon_1 - \varepsilon_2]$ can be taken out from under the sum in Eq. (23.19), and the sum rule of Eq. (23.11) gives $T_{12}T_{23}$ for the sum over the numerator of Eq. (23.19) with $n = 0$. Thus we find $\langle f|H^{2nd}|0\rangle = T_{12}T_{23}/(\varepsilon_1 - \varepsilon_2)$ as if the platform never moved.

In the other limit, with the shift u_0 small compared to the zero-point fluctuations, the overlap $\langle\phi_0(u)|\phi_{n'}(u-u_0)\rangle$ is large for $n' = 0$ and small otherwise. Similarly then, with $n' = 0$, we see that $\langle\phi_{n'}(u-u_0)|\phi_n(u)\rangle$ will only be large if n is also equal to zero. Thus we again find no loss but in this case $\langle f|H^{2nd}|0\rangle = T_{12}T_{23}/(\varepsilon_1 - (\varepsilon_2 - E_{relax}))$ as if the platform shifted to its equilibrium position while the electron was present, with no phonons excited.

Loss to the harmonic oscillator arises only when the zero-point fluctuations are of a similar size to the shift u_0 in the equilibrium position. We may note that this was the condition that the tunneling time, $\hbar/E_{relax}$, is comparable to the period of the harmonic oscillator. This makes a very plausible intuitive picture. It is much like a person stepping briefly on a platform in passing over it. If he is on the platform for a small time compared to the period, it scarcely moves and no energy is lost. If he steps very slowly onto the platform and off, it will displace adiabatically down and up, again with no energy lost. Only if he steps on for a time of the order of the period of the oscillator will energy be transferred causing the platform to vibrate.

This same system can be used to address another interesting quantum-mechanical question, which is frequently discussed in terms of inelastic events causing a "loss of phase" of a wave packet. It is generally agreed that there is interference between two packets following different paths, but if an inelastic event (an energy loss or gain) occurs along one of the paths, the interference is lost. This is often interpreted as a randomization of the phase of that packet, which would indeed destroy the interference. However, if we consider a system for which we can distinguish a loss of phase from the elimination of one packet, a "collapse of the wavefunction", we find that the latter is the correct explanation (Harrison (1994)).

In order to address this question we need additional paths as shown in

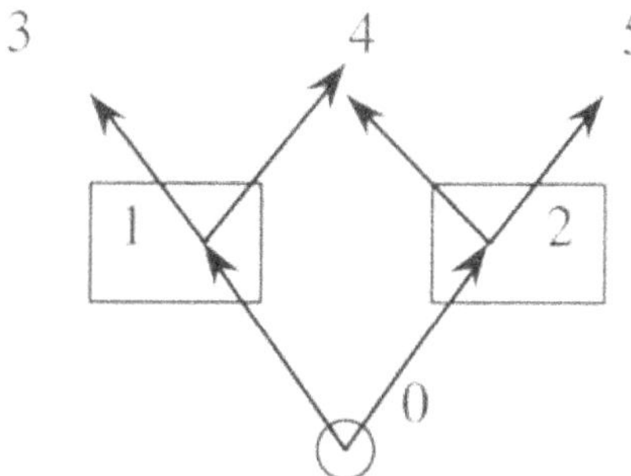

Fig. 23.5. An electron initially in the state |0> can tunnel through either of two intermediate states |1> and |2> into three final states as shown. With all matrix elements the same the amplitudes at |4> from the two paths add, so the probabilities of ending in |3>, |4>, or |5> are in the ratio 1:4:1. If an inelastic event occurs in |2>, the ratios are 0:1:1 (as expected), not 1:2:1 as would be predicted from a phase randomization by the inelastic process.

Fig. 23.5. We shall see that there is interference between the two paths leading to the states labeled |4>, so that if all matrix elements are the same there is four times the probability of that final state relative to that labeled |3> or that labeled |5>. We shall then allow an inelastic event of the type illustrated in Fig. 23.4 to occur in the intermediate state labeled |2>. We expect that this indicates that the electron took the path to the right, so that the probability of arriving at |3> is zero, and of arriving at |4> or at |5> is equal, and we shall find that to be true. If the inelastic event simply randomized the phase of the packet, it would indeed eliminate interference but would incorrectly predict equal probability for |3> and |5> and double probability for |4>. Such an assumption of randomized phases would give the correct *average* over equal numbers of inelastic events in both channels, but for the wrong reason. It is of interest to see how the inelastic event eliminates the second path in the correct theory.

We may again look at the second-order matrix element, which enters squared in the probability of each final state in Eq. (23.18). Without a coupling to vibrational states which could give inelastic processes these second-order matrix elements to the three final states are

$$\langle 3|H^{2nd}|0\rangle = \frac{T_{01}T_{13}}{\varepsilon_0 - \varepsilon_1} ,$$

$$\langle 4|H^{2nd}|0\rangle = \frac{T_{01}T_{14}}{\varepsilon_0 - \varepsilon_1} + \frac{T_{02}T_{24}}{\varepsilon_0 - \varepsilon_2} , \qquad (23.21)$$

$$\langle 5|H^{2nd}|0\rangle = \frac{T_{02}T_{25}}{\varepsilon_0 - \varepsilon_2} .$$

As we indicated, if all matrix elements entering are the same and $\varepsilon_2 = \varepsilon_1$, the second-order matrix element to the middle state |4> is double so the probability is four times. If we randomized the phase for the two contributions, the cross term would vanish and the probability of a transition to |4> would only double.

Now let the state |2> interact with an oscillator, as in Fig. 23.4, and let the corresponding oscillator be initially in the ground state $\phi_0(u)$ but excited $\phi_1(u)$ in the final state. [For purposes of this calculation we need consider only an oscillator for the state |2>.] Any vibrational state is allowed in the intermediate state, and if the intermediate state has the electron in the state |2> the intermediate states are $\phi_n(u - u_0)$ (we use an n rather than the n' of Eq. (23.19) for the intermediate state) with energy $\varepsilon_2 - E_{relax} + n\hbar\omega$. On the other hand if the intermediate state does not have the electron in the state |2>, but in the state |1>, the intermediate states are $\phi_n(u)$ with energy $\varepsilon_1 + n\hbar\omega$. We may rewrite the first line in Eq. (23.21) for the path through state |1> to |3> then as

$$<3|H^{2nd}|0>= \sum_n \frac{T_{01}T_{13} <\phi_0(u)|\phi_n(u)><\phi_n(u)|\phi_1(u)>}{\varepsilon_0 - \varepsilon_1 - n\hbar\omega} . \qquad (23.22)$$

We see that every term is exactly zero, either from the first overlap or from the second, or both. Thus there is no second-order matrix element, and no probability of the electron arriving in the state |3> if the final state has $\phi_1(u)$, exactly as we expected. If the final state had no phonons, the overlaps entering in Eq. (23.22) would be $<\phi_0(u)|\phi_n(u)><\phi_n(u)|\phi_0(u)>$ which equals one from the $n = 0$ term and transitions to the state |3> are allowed, again as they should be. In fact the oscillator coupled to the state |2> has no effect for this transition to the state |3> with no phonons excited.

The same factors, $<\phi_0(u)|\phi_n(u)><\phi_n(u)|\phi_1(u)>$ when there is an excitation to $n = 1$ in the final state, enter the first term in $<4|H^{2nd}|0>$ in Eq. (23.21) and it does not contribute. This orthogonality has eliminated the path through the state |1> from the results, as if the wavefunction had collapsed, though that is a clumsy way to describe a clear result. On the other hand, for the second term in $<4|H^{2nd}|0>$ with a single phonon in the final state the overlaps enter as $<\phi_0(u)|\phi_n(u-u_0)><\phi_n(u-u_0)|\phi_1(u)>$ which are nonzero for every n . The paths through the state |2>, with or without excitation in the final state, are calculated exactly as in the system in Fig. 23.4. All of the results of the calculation for this model are as we expect on physical grounds, and we have seen in detail how an excitation removes alternate paths.

Another system we should consider would again be that in Fig. 23.5, but now with a classical raising and lowering of the platform $u(t)$ randomly in time. This could be a representation of tunneling through a state which was shifted by thermal fluctuations. Then the corresponding matrix elements in Eq. (23.21) become time-dependent as in Section 9.3. If the fluctuations are sufficient, the growth of the coefficient of the final state through this term, calculated as in Section 9.3, becomes random in phase relative to contributions through the other path, and indeed a randomizing of the phase has occurred. Perhaps we can say that it corresponds to an array of inelastic processes as we described above, averaged over absorption and emission events, which then *does* give the same result as a randomized phase. This becomes more a matter of exactly what question is being asked than of how a particular physical system is being modeled.

This analysis sheds light on the question of how a particle, represented by a plane wave, can produce a string of droplets, a track, in a cloud chamber. Why cannot this plane wave generate a droplet far from the main track? The answer clearly can be given that the matrix elements needed for producing that droplet contain factors of the overlap of states associated with droplets in the main path. These overlaps are zero for the electron in the distant droplet.

This system represented in Fig. 23.5 also helps clarify the concept of "entangled states" which is often discussed in quantum theory, though we have largely avoided it here. We may introduce harmonic oscillators for both the intermediate states $|1>$ and $|2>$. Then in the course of transmission of the electron from $|0>$ to the final states there are many terms in the intermediate state, some with the harmonic oscillator for the state $|1>$ in an excited state. However, for those terms, the harmonic oscillator for the state $|2>$ is definitely in the ground state. When the oscillator for the state $|2>$ is in the excited state, that for state $|1>$ is always in the ground state. We cannot discuss the two oscillators separately because their states are entangled.

This entanglement feature is essential to quantum computing, which is currently under extensive discussion (e. g., Averin (1999), Nakamura, Pashkin, and Tsai, (1999)). It is usually based upon spin states, which we have been able to describe as spin-up or spin-down for almost all of our discussion, except in connection with spin-orbit coupling, because the spin did not appear in the Hamiltonian and these two eigenstates were degenerate. For computing in a binary system, each bit of information is a zero or a one, which we may represent as a spin up, $|\uparrow>$, or a spin down, $|\downarrow>$, called a *qubit*, a bit of quantum information. The quantum computer represents numbers with an array of such qubits. More generally the spin state of each qubit can be a mixture of spin up and spin down, $\cos\eta|\uparrow> + \sin\eta|\downarrow>$, and in fact such combinations arise automatically when the qubits interact with each other, and the interacting qubits are entangled in just the sense described for oscillators above. The lowering of one spin is inevitably accompanied by the raising of the other. It is this entanglement which may make it possible to greatly reduce computational time by following several scenarios (Section 1.5) at the same time. For this to succeed, it is also necessary to retain each scenario and to conserve the relative phases of the array of cubits, which may be the most difficult part.

This same mixture of spin-up and spin-down states is the basis of one method of "quantum cryptography". With an axis selected, binary information can be transmitted by sequentially sending particles of $|\uparrow>$ representing 1 and $|\downarrow>$ representing 0. If the receiver knows the axis direction, he can set his detector and read the message. If the message is intercepted by someone who does not know the direction, and guesses a direction off by some angle ϕ, he will read the wrong digit a fraction $^1/_2\sin^2\phi$ of the time, be unable to read the message, and have no way to correct his error. This same method can be used with photons with the polarization of the light playing the role of spin orientation. However, it is essential that the message come one photon at a time, or the interceptor can receive with multiple receivers, with multiple orientations, and sort it out afterward. It is the particle aspect which is essential, not the representation as polarization by $\cos\eta|\uparrow> + \sin\eta|\downarrow>$. That wave aspect is also there in the remarkable feature of classical light that crossed polarizers will prevent transmission of any light, but if a third polarizer is placed between them, at a 45° angle, 25% of the light will be transmitted. That result is immediate in the wave picture, but difficult to picture in terms of the spin-orientation of photons.

Epilogue

If we look back over the extraordinary range of topics we have discussed, we should remember that we have been exploring the consequences of a single idea, complementarity, which we introduced at the outset: *Everything* is at the same time a particle and a wave. That the consequences could be so pervasive, and yet not recognized before this century, is because the constant $\hbar$ which relates the particle and the wave is so small on the scale of everyday experience. It is however large enough that all modern engineers and scientists should understand this basic rule by which the world operates, and learn the approximations which allow them to apply it to their work.

Exercises

1.1 Compton scattering of an electron

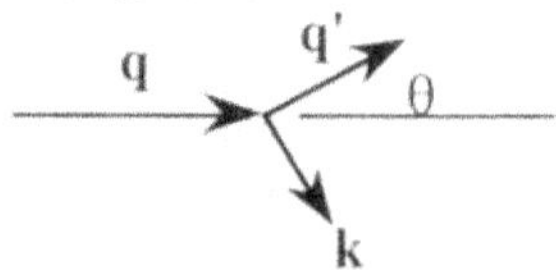

A photon of momentum $\mathbf{p} = \hbar\mathbf{q}$ is absorbed and another $\mathbf{p}' = \hbar\mathbf{q}'$ emitted at the same time by an electron initially at rest, called Compton scattering. The electron must pick up the missing momentum $\hbar\mathbf{k}$ (all components) and also pick up the missing energy. Using conservation of energy and momentum is enough to put a condition on the momentum gained by the electron in terms of the angle θ between $\mathbf{q}'$ and $\mathbf{q}$. One must use the relativistic form of the electron energy,

$$\varepsilon = \sqrt{(mc^2)^2 + (cp)^2} - mc^2 \,.$$

relative to the rest energy. Your result can be written in terms of the change in wavelength $\Delta\lambda$ of the light as

$$\Delta\lambda = 2\lambda_C \sin^2(\theta/2)$$

with $\lambda_C = 2\pi\hbar/(mc) = h/(mc) = 2.43\times10^{-12}$ meters, the Compton wavelength. Derive this result. This is one of the few derivations in this set of exercises.

2.1 π-states in a benzene ring

The benzene ring of six carbon atoms has a radius of about 1.53 Å. Imagine free electrons confined to such a circular ring.

a) Obtain the wavenumbers and energy of the free-electron states ($V(x) = 0$).

b) Actually we shall see that one electron per carbon lies in such a state. (These are called the π-electrons.) What is the lowest sum of electron energies, allowing one electron of each of two spins in each state?

c) What is the energy difference between the lowest empty and the highest occupied states? $\hbar\omega$ equal to this difference gives the optical absorption threshold.

2.2 Energy to break a benzene ring

Imagine breaking the benzene ring, keeping the length the same, and thus requiring the $\psi(0) = \psi(L) = 0$. (Note $\psi'(L) = -\psi'(0)$ is not ruled out now.)

a) Find the energy for these states.

b) Find the lowest sum of one-electron energies for the six electrons, and then the change from this sum in Problem 2.1. This is a contribution to the change in energy upon breaking the ring. (It might be called the energy of a resonant π-bond.)

2.3 Fermi energy in free-electron metals

Na, Mg, and Al have 1, 2, and 3 free electrons per atom, respectively. If the volumes per atom for the three elements are 39.3Å^3, 23.0 Å^3, and 16.6 Å^3, respectively, what is the Fermi energy $E_F = \hbar^2 k_F{}^2/2m$ to which the free-electron states are filled for each?

2.4 Density of states in a free-electron metal

Obtain the density of states $n(E) = dN/(Ad\varepsilon)$ per unit energy per unit area for free electrons ($E = \hbar^2k^2/2m$) moving in two dimensions (k_x, k_y) following the derivation in Section 2.2 for three dimensions. Note the difference in the dependence of $n(E)$ on energy E. [E is customarily used for energy when it is just kinetic energy, ε when it includes the potential and is therefore measured from some standard.]

2.5 Quantized conductance

We evaluated in Eq. (2.19) the net transmission of a channel with transmission $T_1 = T_2$ the same at each end. Reevaluate that sum with the two different (though of course at each end the transmission for flow to the left must be the same as that to the right). Note whether it can be asymmetric, corresponding to a rectifying channel.

2.6 A round quantum wire

Imagine a quantum wire, as in Fig. 2.4, but with a circular cross-section of radius r_0, and great length L. The energy eigenstates can be written in terms of cylindrical coordinates r, ϕ, and z. The radial solutions obtained from Eq. (2.23) with $V(r) = 0$ are Bessel functions of integral order $J_m(kr)$ as indicated in the text. These $J_m(kr)$ are similar in form to the j_l(kr) shown in Fig. 2.9, with $J_0(kr)$ going to 1 at $kr = 0$, and $J_1(kr)$ going to $kr/2$ at small kr. The only properties we shall need for this problem are the relation between k and the eigenvalue i Eq. (2.23), $\varepsilon = \hbar^2k^2/(2m_e)$, and the zeros of the first two Bessel functions,

$$J_0(x) = 0 \text{ at } x = 2.40, 5.52, 8.65...$$

and

$$J_1(x) = 0 \text{ at } x = 3.83, 7.02, 10.17...$$

obtained from Mathews and Walker (1964) p. 224.

What are the energies of the two lowest bands of $m = 0$ states, and of the lowest band of $= 1$ states? To what Fermi energy would we need to fill the $m = 0$ band to accommodate 1 electron per Å of length if r_0 is = 3 Å? Would this Fermi energy place electrons also in the lowest $m = 1$ band? In the second $m = 0$ band? [If it helps, you could take $L = 1000$Å.]

2.7 States in a large spherical cavity

When we consider the ionization of a (spherically symmetric) atom or impurity in a solid, we will want to treat free-electron states($V(\mathbf{r}) = 0$) using a large spherically symmetric boundary, rather than the parallelepiped we used in Section 2.2. The radius R of the boundary is very large. It is often convenient to work in terms of the full wavefunction $\psi(\mathbf{r})$, rather than the radial factor alone, $R(r)$, and we do that here.

a) Obtain the energies of the $l = 0$ free-electron states (s-states) by requiring $\psi(R) = Aj_0(kR) = 0$.

b) Normalize these states, $\int\psi^*(r)\psi(r)d^3r = 1$, by integrating over all space in the volume of radius R.

c) Obtain the number of s-states per unit energy in terms of R and energy $E = \hbar^2k^2/(2m)$, including the factor two for spin. [It will be proportional to R rather than to volume. If we were to add the density of p-, d-, etc., states, the total would eventually approach the density of states proportional to volume as we found for the parallelepiped.]

d) Similarly obtain the energies of the free-electron $l = 1$ (p-states) for very large R, using the $j_1(kr)$ given in Eq. (2.35). (Note that one term is smaller by a factor of order $1/(kR)$ and can be dropped when you set $j_1(kR) = 0$.)

[We will not do the normalization integral for p-states , which contains an integration of $\cos^2\theta$ and care would need to be taken with the radial integration for such higher l since individual terms diverge as r goes to zero.]

2.8 Existence of bound s-states

A deep spherical quantum well, of depth $-V_0$, and radius r_0 , will have bound s-states. However, for fixed r_0 , it will not have any bound states if V_0 is too small. How large must V_0 be, as a function of r_0, in order to have at least one bound s-state. [The solutions of Eq. (2.32) at $\varepsilon = 0$ will need to turn over for r less than r_0 to match to a decaying exponential.]

2.9 Harmonic-oscillator states

a) Integrate the Schroedinger Equation numerically,

$$-\frac{\hbar^2}{2m}\frac{\partial^2\psi(x)}{\partial x^2} + \frac{1}{2}\kappa x^2\psi(x) = \varepsilon\,\psi(x)\,, \qquad (1)$$

with $\hbar^2/m = 7.62$ eV-Å^2 (for m the electron mass) and $\kappa = 2$ eV/Å^2 (so $\hbar\omega = \hbar\sqrt{\kappa/m}$ = 3.90 eV) to obtain the n=7 solution. You might proceed as follows, or in some other way:

Start with $\varepsilon = 7^1/_2\,\hbar\omega$ and $\psi = x$ near $x = 0$ [appropriate for an odd n, and it does not matter that the resulting $\psi(x)$ is not normalized; for even n one could have used $\psi = 1$ near $x = 0$]. For odd n $\psi(0) = 0$, we can take $\psi'(0) = 1$, and from Eq. (1), $\psi''(0) = 0$. You can obtain $\psi(\Delta x)$ from

$$\psi(x + \Delta x) \approx \psi(x) + \psi'(x)\Delta x + {}^1/_2\psi''(x)\Delta x^2$$

and

$$\psi'(x + \Delta x)) \approx \psi'(x) + \psi''(x)\Delta x \qquad (2)$$

and again $\psi''(x + \Delta x)$ from Eq. (1) using everywhere $x + \Delta x$. The process is repeated interval by interval to a large x, and the energy adjusted up or down until it goes to zero at large x and has n nodes (for *all* x , including $x = 0$ and $x < 0$, not counting the zeros at $x = \pm\infty$). (You might use an interval $\Delta x = 0.1$Å, adjusting the energy to three significant figures to get $\psi(8\text{Å}) = 0$. It won't be quite the right energy since Δx is not infinitesimal and 8Å is not infinite.)

b) Plot the result. (You need not normalize and $x > 0$ is enough.)

c) How does the x-value for the largest peak agree with the classical x_{max} obtained from $\varepsilon = \frac{1}{2}\kappa x_{max}{}^2$? (Use your calculated ε, but it would not matter if you used the $7.5\hbar\omega$ since x_{max} is not so close at this small an n.)

d) Reestimate the energy using $\Delta x = 0.02$ Å, to see what sort of error was incurred from use of the finite interval.

3.1 Lagrangians and canonical momentum

Imagine a spherical bead of mass M sliding on a wire with position $0 < x \leq L$ with periodic boundary conditions at the ends. $V(x) = 0$.

a) What are the energy eigenvalues? [We have obtained exactly this before.]

b) If the bead can also freely rotate or spin (through angle θ) on the wire, its rotational energy would classically be $\frac{1}{2}I\dot{\theta}^2$ with the moment of inertia $I = \frac{2}{5}Mr_0{}^2$ for a homogeneous ball of radius r_0. What are now its energy eigenstates, allowing both spin and translation of the ball? [You may have to go through the Lagrangian, L, and the p conjugate to coordinate θ if it is not clear how to proceed. Then $p \to (\hbar/i)\ \partial/\partial\theta$.]

c) Letting $L = 2\pi r_0$ so that both motions depend upon the same parameters, what is the ground state energy and that of the next two lowest states, all in terms of M, r_0, and $\hbar$.

d) Now let this bead roll along a line without slipping, so the rotation speed $\dot{\theta} = \dot{x}/r_0$, and there is again only one independent coordinate. What are now the energy eigenvalues if we again apply periodic boundary conditions on $L = 2\pi r_0$? Note that L is chosen so that when the bead reaches the end the angle θ is exactly the same as at the start, so periodic boundary conditions are still satisfied at the ends. [It is an interesting possibility to take $L \neq 2\pi r_0$. It allows nonintegral angular momentum, when combined with the nonslipping condition, as we shall discuss for electrons in Section 10.5.]

e) What are now the lowest three energy eigenvalues, in terms of M, r_0, and $\hbar$?

3.2 Tumbling, translating, and vibrating Li_2

Lithium atoms have a mass M and form a molecule Li_2 with an equilibrium spacing d_0 2.67 Å. We describe the electronic structure in Problem 5.1, writing the electronic energy pl

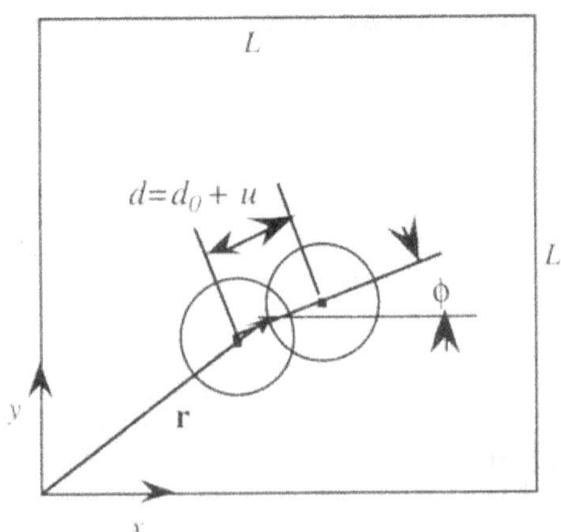

A lithium molecule, moving in a plane.

an overlap repulsion. This leads to a spring constant $\kappa = \partial^2 E_{tot}/\partial d^2|_{d_0} = \pi^2\hbar^2/(md_0^4)$ (with m the electron mass) giving a potential energy $V(d) = {}^1\!/_2\kappa(d-d_0)^2 \equiv {}^1\!/_2\kappa u^2$.

If we confine the atoms to a square plane, L on each edge, (confine with vanishing boundary conditions on the center of gravity) the molecule can translate, rotate, and vibrate. There are no potentials except the $V(d)$.

a) Describe the states of this molecule in terms of the quantum numbers which are needed to specify the state, and the parameters M, d_0, L, and $\hbar$.

b) Write an expression for the ground state energy, ground state with respect to the three or four motions .

Comments: We neglect any excitation of electrons to excited states (discussed in Problem 5.1 d). We also neglect any effect of stretching, $u = d - d_0$, on the moment of inertia for the rotating modes. [Note that a displacement of one atom by $u/2$, and the other in the opposite direction by $u/2$ produces the separation $d_0 + u$. There are many factors of 2.] You can obtain the kinetic energy, Lagrangian, and the momentum for each coordinate, p_u, etc., and go to the energy eigenvalue equation.

4.1 Variational state for hydrogen

To see how well variational solutions work, try $\psi(r) = A\exp(-\alpha r^2)$ as a variational solution for the ground state of hydrogen (which has a correct solution $\sqrt{\mu^3/\pi}\exp(-\mu r)$). What is the lowest energy ε you can get with this Gaussian form? You may evaluate

$$\varepsilon = \frac{\int 4\pi r^2\psi(r)\,(H\psi(r))\,dr}{\int 4\pi r^2\psi(r)^2\,dr}$$

analytically, with

$$H = -\frac{\hbar^2}{2m}\frac{1}{r^2}\frac{\partial}{\partial r}r^2\frac{\partial}{\partial r} - \frac{e^2}{r},$$

and tabulate it numerically as a function of α to obtain the minimum ε to a few hundredths of an eV. Compare with -13.6 eV.

4.2 Spherical systems

We gave in Fig. 4.1 the energy levels for hydrogen, including the two lowest $l = 0$ states and the lowest set of $l = 1$ states. Obtain the values for the corresponding states and sketch the levels for

a) a spherical cavity (infinitely high walls) of radius 2. Å . (You can use the $j_l(kr)$ given in Eq. (2.35) but need a numerical solution to get that energy.)

b) Repeat for a spherical harmonic oscillator, $V(r) = {}^1\!/_2\kappa r^2$, with κ chosen to give the same energy as in a) for the lowest $l = 0$ state. (2s-states can be made of a spherically-symmetric combination of states such as $n_x=2$, $n_y = n_z = 0$ harmonic oscillator states, similar to the way we made the p-states from $n=1$ and $n=0$ states in Section 2.5.)

Note: for both a) and b), all energy eigenvalues are positive. The shell model of the nucleus is usually used with the potential from part a), but sometimes with that from b).

4.3. Pseudopotentials and p-states

An empty-core pseudopotential radius, r_c, can be obtained for an alkali-metal atom by finding the value which will give the correct s-state energy by solution of the radial Schroedinger Equation based upon $\chi = rR(r)$, Eq. (4.2),

$$-\frac{\hbar^2}{2m}\frac{\partial^2}{\partial r^2}\chi + w(r)\chi = \varepsilon_s\chi \qquad (1)$$

with

$$w(r) = \begin{cases} 0 \text{ for } r<r_c \\ -e^2/r \text{ for } r>r_c. \end{cases} \qquad (2)$$

a) Obtain r_c for lithium (ε_s = -5.34 eV) and sodium (ε_s = -4.96 eV), integrating the Schroedinger Equation as described in detail in Problem 2.9 . Use a step in r of 0.01 Å and adjust r_c to within 0.01 Å so ψ(8Å)=0 .

b) Use the sodium r_c and find ε_p, the lowest p-state energy (within about 0.01 eV). It is interesting to compare this result with the empirical rule, Eq. (4.17).

Suggestions. It is helpful again to use $\hbar^2/m$ = 7.62 eV-Å^2 and e^2 = 14.4 eV-Å. Then for this s-state we can seek a solution of Eq. (1) above by numerical integration. It may save time to start with the Problem 2.9 program. At small r, you may take $\chi = r$ (not normalized) to set $\chi(0)$ and $\chi'(0)$, and obtain $\chi''(0) = 0$ from Eq. (1). Then proceed interval by interval as in Problem 2.9.

The value of r_c which gives a nodeless wavefunction that does not diverge at large r is correct. Using χ(8Å) rather than $\chi(\infty)$ is good to around 0.01 Å. You probably need to print out $\chi(r)$ values to see that there is no node between $r = 0$ and $r = 8$, but a tenth of the values of r is plenty. [If you like, you can run it for ε_s = -3.4 eV also to obtain the r_c used for the hydrogen 2s-state in Fig. 4.2.]

For Part b you need to add $\hbar^2 l(l+1)/(2mr^2)$ to $w(r)$ in Eq. (1), fix r_c and adjust the energy ε_p to get a nodeless solution with χ(8Å) = 0. At small r, you may take $\chi = r^2$ (not normalized).

The full free-atom calculation is a direct generalization of this procedure with $w(r)$ replaced by a more complete potential.

5.1. Molecular physics

The distance between the two nuclei in Li_2 is d = 2.67 Å. ε_s = -5.34 eV for the free atom from Hartree-Fock, Table 4.1. Based upon $V_{ss\sigma}$ from Eq. (5.6),

a) What is the energy of the bonding state and of the antibonding state?

b) What is the total change in energy of occupied one-electron states in formation of the molecule from free atoms, obtained as $\Sigma_{\text{electrons}}\,\delta\varepsilon_j$.

If this were the only contribution to the energy of Li_2, the molecule would collapse. The repulsion arising from the nonorthogonality of the two atomic s-states was approximated in Eq. (5.22) by an additional energy A/d^4. That is, the total energy becomes:

$$E_{TOT} = \frac{A}{d^4} + \Sigma_{\text{electrons}}\,\varepsilon_j\,.$$

c) Adjust A such that the correct equilibrium spacing is obtained. With this added repulsion, what is the formation energy (positive number) of the molecule?

Formation energies are typically overestimated by as much as a factor of two for first-row-element systems, though not for elements from lower rows. The experimental value is 1.07 eV for Li_2.

d) What energy of photon would you estimate is required to excite an electron into an excited electronic state of the molecule?

5.2 Polarizability of Li_2

Take an electric field **E** along the axis of the Li_2 molecule. If the s-states on the two atoms are separated by d, this will make an energy difference between the two lithium atomic states of eEd, making the molecule polar. This will also shift the average occupation of each atomic orbital from the original $u_i^2 = 1/2$ with no field.

Calculate the dipole which arises to first order in the field, (two electrons in the bonding state) to obtain the polarizability α such that the dipole is $p = \alpha E$. [α (not to be confused with polarity α_p) should have units of Å^3. Work through the polarity α_p to first order in the field E. Magnitudes are enough; clearly the electrons move in the direction they are pushed.]

5.3 Second-order perturbation theory

Another way (different from Problem 5.2) the polarizability α is calculated is by noting that a molecule has its energy shifted by $-\frac{1}{2}\alpha E^2$ by a small electric field E.

a) Calculate the shift in energy of the electrons in the bond by second-order perturbation theory, noting $\langle a|eEx|b\rangle = \frac{1}{2}(\langle 1| - \langle 2|)|eEx|(|1\rangle + |2\rangle) = -eEd/2$. Equate this to the $-\frac{1}{2}\alpha E^2$ to obtain another estimate of α.

b) Use this same perturbation theory to calculate the polarizability of a pair of electrons in a quantum-well state $\sqrt{2/L}\,\sin(\pi x/L)$ due to its coupling with the first excited state $\sqrt{2/L}\,\sin(2\pi x/L)$. [$L$ is the thickness, along the electric field, of the quantum well and the area does not matter. There are also contributions from coupling to higher states, $\sqrt{2/L}\,\sin(\pi nx/L)$ with n = 3,4, 5... (actually only even n contribute), which we do not include. The energies of the states are from $\hbar^2(\pi n/L)^2/2m$.]

Comments: There are different ways to obtain the matrix element, any of which are allowed:

One, you might look the integral up in the tables. [The author did not find it.]

Two, you might evaluate it analytically using $\partial/\partial a\int_{b,c}\sin(ax)dx = \int_{b,c}x\cos(ax)$. [The author did that.]

Three you might extract the L by changing variables and evaluate a remaining integral numerically. [The author did it as a check.]

Four, you might use *Mathematica*, or some such program.

In any case, you should end up with a formula which contains the dependences upon the parameters of the problem with a numerical factor. The dependence upon L of the result is of particular interest.

5.4. N_2 molecule

a) Using the molecular-orbital levels from the first column in Table 5.1, and the term values from Table 4.1, estimate the cohesive energy of the nitrogen molecule (including division by a factor of two as a rough estimate of the effect of overlap energy). The observed cohesion is much smaller at 9.8 eV per molecule.

b) Recalculate the cohesion using sp-hybrids $(|s\rangle \pm |p\rangle)/\sqrt{2}$ for the σ-states and keeping only the largest coupling V_2. (Again, $d = 1.09$ Å.) Compare the resulting cohesion with the value obtained in Part a).

5.5 sp-hybrids in CO

Consider the carbon monoxide molecule, with a spacing of 1.13 Å . Calculate the molecular orbital energies using sp-hybrids on the C and O to obtain the σ-states so that only a quadratic equation needs to be solved. You also need a quadratic equation to obtain the π-bonding states. See which states are occupied and obtain the cohesion as for N_2 in Problem 5.4, to be compared with the experimental 11.0 eV. Again the overestimate is greater than the usual factor of two for first-row compounds.

6.1. Tight-Binding benzene π-states

We redo Problems 2.1 and 2.2 for the benzene ring, using tight-binding π-states rather than free-electron states. Use $V_{pp\pi} = -(\pi^2/8)\hbar^2/(md^2)$ with $d = 1.53$ Å and we can measure energies from ε_p .

a) What are the energies of the six π-states (relative to ε_p)? What is the contribution of the occupied states to the cohesive energy? (Divide by two for the effect of the repulsion from nonorthogonality.)

b) If we break the ring but keep the other spacings the same, we saw in Fig. 6.4 that the states needed to go to zero at N+1 spacings. What is now the sum of energies of occupied states, divided by two?

c) How does the difference, the bond-breaking energy, compare with the energy per bond from Part a)?

6.2. Fitting interband coupling to free-electron bands

a) Given the formulae for $V_{ss\sigma}$, $V_{pp\pi}$, and $V_{pp\sigma}$ we have used, what would $\varepsilon_p - \varepsilon_s$ have to be to fit the free-electron bands *for a simple-cubic structure* at $\mathbf{k} = 0$?

b). No elements are simple cubic, but compare the free-atom term-value differences with the estimate based upon Part a) using simple-cubic spacings we estimate (if nearest-neighbor spacings vary with number of nearest-neighbor atoms X as $X^{1/4}$ (Harrison (1999)) for C (d=1.70Å), Si (d=2.60Å), Ge (d=2.70 Å) and Sn (d=3.10 Å). The term values do not vary nearly as much as this would suggest and are closest for systems where the band gaps are smallest (Sn).

c) The s-band $\varepsilon_k{}^s$ of Eq. (6.4) can be expanded for small k_z (or k_x or k_y) to get a term proportional to $k_z{}^2$. The coupling $V_k(sp) = 2iV_{sp\sigma}\sin k_z d$ entering Eq. (6.5) (good for a chain or for a simple-cubic structure) gives an additional term which can be calculated in second-order perturbation theory. It is proportional to $k_z{}^2$ at small k_z since $V_k(sp)$ is proportional to k_z and the energy denominator can be taken as constant. We fit that band difference as $[\hbar^2/(2m)][2\pi/d]^2$ (plus a term in $k_z{}^2$ which only gives fourth-order terms in the band energy we are studying, and can be dropped). What must $V_{sp\sigma}$ be to get the free-electron mass for the lowest band? Compare with the geometric mean of $V_{ss\sigma}$ and $V_{pp\sigma}$, and with our choice of $V_{sp\sigma} = (\pi/2)\hbar^2/(md^2)$. [The result could have been obtained by expanding Eq. (6.5), with $\varepsilon_k{}^s$ and $\varepsilon_k{}^p$ for the simple-cubic structure, for small k, but the way we did it may be good practice in the use of 2nd-order perturbation theory.]

7.1 Scattering by an impurity in two dimensions

We treated scattering by an impurity level in one and three dimensions for tight-binding s-bands, such that the perturbing matrix elements $\langle k'|H|k\rangle$ were equal to $(\delta\varepsilon_s/N)e^{-i(\mathbf{k}'-\mathbf{k})\cdot\mathbf{r}_i}$.

a) Use this same matrix element for a *two-dimensional* band to calculate the total scattering rate as a function of the energy of the electron. Use free-electron bands, as in the derivation of Eq. (7.12), for converting the sum to an integral over energy.

b) The scattering center could be represented by the diameter D of a scattering disk (rather than the cross-section in three dimensions), $vD/A = 1/\tau$ with v the electron speed, A the area of the system, and $1/\tau$ the scattering rate. What is that diameter (in Å) for a thermal ($\varepsilon = 0.025$ eV) free electron with the free-electron mass if $\delta\varepsilon_s = 2$ eV and the area per atom is 4 Å^2?

7.2 Transition from a local state

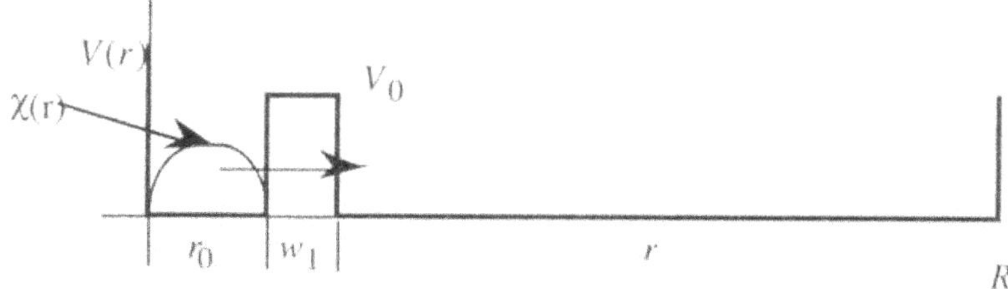

Imagine an electron bound in the lowest s-state $|0\rangle$ in a spherical quantum well (or *bowl*, since the minimum potentials are the same outside as inside) as above, with $|0\rangle = \sqrt{1/(2\pi r_0)}\,\sin kr/r$ for $r<r_0$, with $k = \pi/r_0$ and energy $\hbar^2k^2/(2m)$ We shall see in Chapter 8 that such a state is coupled to s-states $|k'\rangle$ outside the well, which if R is very large have wavenumbers k' equal to an integer times π/R, with energies $\hbar^2k'^2/(2m)$. We shall see in Chapter 8 how the matrix elements between the state inside and those outside is estimated, and Eqs. (8.15) and (8.20) will suggest squared matrix elements given by a form,

$$\langle 0|H|k'\rangle\langle k'|H|0\rangle = \frac{4\hbar^4k'^4e^{-2\kappa w}}{\kappa^2m^2r_0R},$$

with κ related to the energy and V_0 by $\hbar^2\kappa^2/(2m) = V_0 - \varepsilon \approx V_0$, so we take it independent of energy. Note that this has units of energy-squared and appropriate dependences upon the normalization distances r_0 and R.

Obtain a formula for the lifetime τ for this bound state, or the rate of transitions out $1/\tau$, using the Golden Rule. Note that the sum over final states is a sum over k' for spherically symmetric states vanishing at radius R.

8.1 Tunneling through a resonant state

We construct a program to calculate the transmission as a function of energy for a row of atoms (or a stack of atomic planes) for a one-band system with atomic levels of energy ε_j, coupled to nearest neighbors by the same $V_{ss\sigma}$ (here = -1 eV). Let all ε_j be the same ε_s for $j < 0$. Then the program begins with a transmitted wave to the left, $u_j = Te^{-ikdj}$ for $j \leq 0$, which satisfies Eq. (8.12) for these j with energy $\varepsilon = \varepsilon_s + 2V_{ss\sigma}\cos kd$. We can, for example, use this form to obtain the real and imaginary parts of u_j for $j = -2$ and $j = -1$. Then Eq. (8.12) can be used to obtain successive value of the real x_j and imaginary iy_j parts of u_j, keeping the same ε and inputting the successive values of ε_j for increasing j, as illustrated in Fig. 8.1, until we reach a j where all ε_j are the same (not necessarily the same as far to the left). Then we have passed the barrier and we may use Eq. (8.13) and any pair of neighboring u_j to obtain the transmission. (If the constant ε_s on the right is different from that on the left, as illustrated in Fig. 8.1, the wavenumber k must be determined using $\varepsilon = \varepsilon_s + 2V_{ss\sigma}\cos kd$ with the starting energy ε from the left, but with the ε_s for the right, and therefore a different k than used on the left. This is because Eq. (8.13) was obtained by evaluating the ratio of incident to reflected waves of the wavenumber appropriate to the right side. In the present exercise we treat the simpler case where ε_s is the same to the left and right of the barrier.)

For this problem we may again measure energies from the bulk ε_s appropriate on both sides of the barrier, and take $V_{ss\sigma} = -1$ eV throughout. That is, we take all ε_j equal to zero except near the central barrier. For this problem we take all $\varepsilon_j = 0$ except $\varepsilon_1 = \varepsilon_3 = 3$ eV. Since $\varepsilon_2 = 0$ this is a double barrier. The energies for propagating states run from $2V_{ss\sigma} = -2$eV to $-2V_{ss\sigma} = 2$eV. For each energy in this range kd for the transmitted wave is given in terms of the energy by $\varepsilon = 2V_{ss\sigma}\cos(kd)$ and for each energy the program will work through the barrier and calculate the transmission. Plot that transmission for the energy range of the entire band.

This double barrier has a resonance as described in Section 8.4 and you will find that the transmission goes to one at the corresponding resonance energy. Note that the same program, with different ε_j, and maybe different V_{ij}, will solve an extraordinary range of transmission problems. One example is Problem 8.3.

8.2 Scattering by a displaced atom

We think of the chain of atoms of length Nd as in Fig. 7.3 and Problem 8.1, but instead of changing the energy ε_s for one atom, we displace it to the right by δd. If the coupling $V_{ss\sigma}$ varies as $1/d^2$ the matrix element on the left decreases by a factor $d^2/(d + \delta d)^2$ and that on the right increases by a factor $d^2/(d - \delta d)^2$, but all ε_s remain at the energy which we took as zero.

Use the Golden Rule to calculate the reflectivity due to the displaced atom. Note that this could also be done numerically, and more accurately, using the program from Problem 8.1, but we wish here to see how the Golden Rule is used for such a case, and in particular how to calculate the needed matrix element.

The states in zero order are $|k\rangle = (1/\sqrt{N})\,\Sigma_j e^{ikx_j}$ and $\langle k'| = (1/\sqrt{N})\,\Sigma_i e^{-ik'x_i}$. The matrix element of the Hamiltonian without distortion is zero between states of different wavenumber, but there is coupling between the two from the changes in coupling just mentioned. If the position of the atom to be displaced is x_0, there is a change in coupling between the atomic state at x_0 and that at x_1 of $-2V_{ss\sigma}\,\delta d/d$, to first order in δd, which appears with a factor $(1/N)[\,e^{-ik'x_0}e^{ikx_1} + e^{-ik'x_1}e^{ikx_0}\,]$. There is also a change in coupling between atomic states at x_0 and x_{-1} of $+2V_{ss\sigma}\delta d/d$. This provides four terms in the matrix element,

$$\langle k'|H|k\rangle = -\frac{2V_{ss\sigma}\,\delta d}{Nd}\,[\,e^{-ik'x_0}e^{ikx_1} + e^{-ik'x_1}e^{ikx_0} - e^{-ik'x_{-1}}e^{ikx_0} - e^{-ik'x_0}e^{ikx_{-1}}\,]$$

$$= \frac{4iV_{ss\sigma}\,\delta d}{Nd}\, e^{i(k-k')x_0}\,[\sin kd\ -\sin k'd].$$

The calculation using the Golden Rule for $1/\tau$ and for the rate of reflection, for $k' \approx -k$, is now straightforward. a) Carry it out for k and k' small so that the sines in this matrix element can be taken equal to their arguments. The result will depend upon k .

b) Carry it out for states midband, where $kd = \pi/2$.

c) We may think of this electron with periodic boundary conditions as moving to the right, crossing the displaced atom once for every length Nd it traverses, and reflecting each time with probability R . Write R for the defect for electrons near the bottom of the band from the result in Part a. Had we calculated reflection directly as in Problem 8.1, the result , to lowest order in δd should be the same.

8.3 Tunneling through complex barriers

Problem 8.1 provides the basis for a wide range of interesting problems, simply by changing the parameters which enter. One such example is to calculate the transmission for a system as in Problem 8.1, with coupling $V_{ss\sigma} = -1$ eV and all $\varepsilon_j = 0$ except for, in this case,

$\varepsilon_1 = 3$ eV

$\varepsilon_2 = 0$ eV

$\varepsilon_3 = 1$ eV

$\varepsilon_4 = 0$ eV

$\varepsilon_5 = 3$ eV

[Once you have iterated through the chain from the transmitted side to the incident and reflected side, the reflectivity is obtainable as in Eq. (8.13).]

a) Plot the transmission for the energy range of the entire band.

b) How would you interpret the result?

9.1. Excitation from a quantum well

We consider carriers confined to the lowest subband of a quantum well of thickness d and area A as in the diagram to the left below,

$$\psi(\mathbf{r}) = \begin{cases} \sqrt{2/(dA)}\cos(\pi x/d)\, e^{i(k_y y+k_z z)} & \text{for } |x| < d/2 \\ 0 & \text{for } |x| > d/2\ . \end{cases}$$

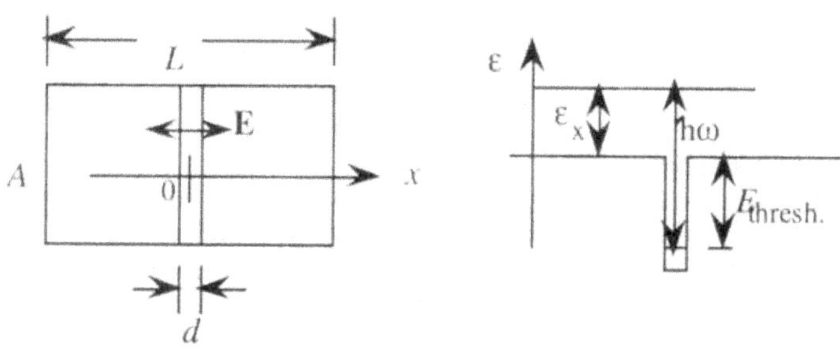

There is light on the system, which has its field **E** in the x-direction and wavelength large compared to the system so the perturbation can be taken to be eEx. Then, the quantum-well states are coupled to bulk states $\psi_k(\mathbf{r}) = \sqrt{2/(LA)}\,\sin(k_x x)e^{i(k_y y+k_z z)}$ of the same k_y and k_z with $k_x L/2 = n\pi$. See the energy-level diagram to the right for this problem (L is large compared to d but small compared to the wavelength of light.) The well states are not coupled to bulk states of the form $\cos(k_x x)$. [One could represent these bulk states more accurately, but this is a meaningful approximation.]

a) For fixed k_y and k_z obtain the matrix element using the first term in H_{el} from Eq. (9.11). (One way of doing the integrals is using $\cos A \cos B = [\cos(A+B) + \cos(A-B)]/2$.)

b) Use the Golden Rule and integrate over k_x, (for large L) to obtain the absorption rate (proportional to $u_{-q}u_q$) as a function $\hbar\omega$ and $k_x d$, related to each other and to the energy $\varepsilon_x = \hbar^2 k_x^2/2m$ above threshold as shown to the right above. We can use Eq. (9.10), and the relation between the vector potential and **E** above it to write $u_{-q}u_q$ as $\Omega c^2 E^2/(4\pi\omega^2)$ to obtain a result proportional to E^2, or the light intensity, with no factors of volume.

c) Plot the resulting absorption rate for an electron of given k_y and k_z, taking for simplicity d such that $E_{thresh.} = \hbar^2/(2md^2)$. Then in units of $E_{thres.}$, we have $\hbar\omega = 1 + (k_x d)^2$. For the plot do not worry about a leading factor, containing E^2 and fundamental constants, which is independent of $k_x d$ and ω.

[One way to proceed is to calculate both $\hbar\omega$ and the absorption in terms of $k_x d$ over the range $0 < k_x d < 3$ for the plot.]

10.1 Probabilities of defect charge states

We saw in Eq. (10.7) that the relative probability of occupation of a vibrational state of energy E_i is proportional to the Boltzmann factor $\exp(-E_i/k_B T)$ in equilibrium at temperature T. The same Boltzmann factor applies to a defect (maybe a vacant site in silicon) which can be neutral or which can have charge $+e$ by removing an electron of spin up, or a charge $+e$ by removing an electron of spin down, or can have a charge $+2e$ by removing both electrons. We call ε_0 the energy at which the first electron is removed, and define a reservoir energy μ at which it would be deposited (in equilibrium). It takes more energy, by U, to remove the second electron since it is from a positively charged defect. Thus the energy of each of the $+e$ states of the defect, relative to the neutral state of the defect, is $\mu - \varepsilon_0$ and that of the $+2e$ state, relative to the $+e$ states is $\mu - (\varepsilon_0 - U)$. The sum of the probabilities of the four defect states is one. Write formulae for the probability of each charge state.

Evaluate the probabilities numerically for $k_B T = 0.025$ eV (room temperature) and $\mu - \varepsilon_0 = 0.1$ eV. You may take $U = 0.5$ eV, but it does not affect the results appreciably. Fermi statistics could be derived in a similar way and this μ turns out to be the Fermi energy for the electrons. For this problem it enters only through the excitation energy $\mu - \varepsilon_0$ and drops out once we set the total probability equal to one for the defect. For the evaluation in this problem, one of the probabilities is so small as to be negligible.

10.2 Zero-point energy in solids

Estimate the zero-point energy, per atom, in a solid with a speed of longitudinal sound of $v_s = 6.4\times10^5$ cm/sec. and atomic volume $\Omega_0 = 16.5$ Å^3(aluminum). Use the Debye approximation of $\omega = v_s q$ up to a q_D such that there are as many modes in each of the three branches as there are atoms. Take the speed of the transverse modes to be smaller by a factor $1/\sqrt{2}$. (It's actually smaller than that, at about 3×10^5 cm/sec.)

The high-temperature vibrational energy is $3k_BT$, 0.075 eV per atom at room temperature, for comparison.

10.3 Carrier distribution in semiconductors

We obtained a formula for the density of electrons in the conduction band in terms of the energy difference Δ_c between the Fermi energy and the conduction band, the effective mass and the temperature, in Eq. (10.28). A similar formula applies for the number of holes in the light-hole band in GaAs and the upper heavy-hole band. We neglect the number of carriers in the lower heavy-hole band. Given the band gap of $E_g =$ 1.52 eV for GaAs and the masses: conduction band, $m_c/m = 0.067$; light hole band, $m_{lh}/m = 0.13$; and heavy-hole band, $m_{hh}/m = 0.62$,

a) where is the Fermi energy in the gap in intrinsic GaAs (equal number of electrons and holes) at room temperature ($k_BT = 0.025$ eV)?

b) What is then the density of electrons (per cm^3) in the conduction band?

c) What is the average kinetic energy (energy above E_c) of these electrons?

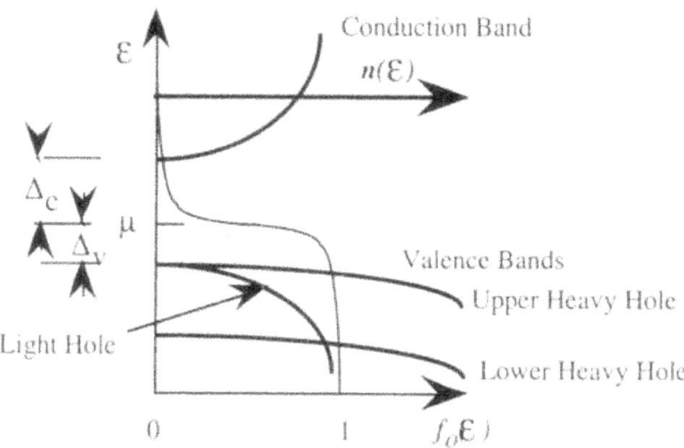

11.1 Time-dependent distribution functions

We solved the Boltzmann Equation for the case in which there was no dependence upon position and no dependence upon time to obtain Eq. (11.6). For the case of a uniform field turned on abruptly at $t = 0$, this dependence upon **p** remains correct for $t > 0$, but there will also be a dependence upon time. Try a solution $f_1(\mathbf{p},t) = f_1(\mathbf{p})g(t)$ in the Boltzmann Equation, Eq. (11.5) with $f_1(\mathbf{p})$ from Eq. (11.6). This will lead to a differential equation for $g(t)$. Obtain the general solution of the reduced equation, and a special solution of the full equation, and fit the sum to $g(0) = 0$ (when the field is first turned on) to obtain $f_1(\mathbf{p},t)$ for all $t > 0$.

12.1 Van-der-Waals interaction in 3D

We treated van-der-Waals interaction between two dipole oscillators oriented along the internuclear separation. Atomic dipole oscillators can also oscillate perpendicular to that separation. Transverse oscillators, of energy $\frac{1}{2}\kappa y_i^2 + \frac{1}{2}m\dot{y}_i^2$, will also contribute to the interaction. The transverse dipole $p = ey_1$ will produce a field in the y-direction , at an atom a distance r away in the x-direction, of $-p/r^3$, corresponding to an energy of interaction of $e^2y_1y_2/r^3$. Calculate the correction, to second order in this interaction, to the dipole frequencies $\omega_0 = \sqrt{\kappa/m}$ for these two coupled oscillators. Use this to obtain the change in the zero-point energy (to second order in the interaction) of the two oscillators in terms of the polarizability $\alpha = e^2/\kappa$. Add this lowering in energy, and that for polarization in the z-direction, to that we obtained for the x-direction, to obtain the full van-der-Waals interaction between two three-dimensional oscillators each of polarizability α. You could directly generalize this to two oscillators of different polarizabilities.

[It is an important point that this interaction applies directly to two *atoms* of polarizability α. Often a harmonic oscillator provides a valid model for an atom or molecule, with $\hbar\omega$ taken equal to the electronic excitation energy. Such an interaction is calculated more directly in the Problem 12.2.]

12.2 Many-body (van-der-Waals) interaction between atoms and molecules

For atoms or molecules the electron-electron interaction provides the counterpart of the $-2e^2x_1x_2/r^3$ between the dipoles shown in Fig. 12.1. Let the nuclei on two helium atoms be separated by $\mathbf{r}$. There is a first-order term in the energy from the interaction between each s-electron on one atom and an electron on the other atom, given by approximately e^2/r, which is not of interest. It balances an interaction energy between the electrons and a proton from the other atom. Now let the electron position on the first atom be $\mathbf{r}_1$ relative to its nucleus and that of an electron on the second atom be $\mathbf{r}_2$ relative to its nucleus. Then the electron-electron interaction is $e^2/|\mathbf{r}_2 + \mathbf{r} - \mathbf{r}_1|$. If $\mathbf{r}$ is large, and in an x-direction, this may be expanded in the components of $\mathbf{r}_1$ and $\mathbf{r}_2$ to obtain the leading term in x_1x_2 again as $-2e^2x_1x_2/r^3$. (Including other components, y_i, z_i, would be analogous to Problem 12.1.) The factor x_1 will have a matrix element between an atomic s-state $\psi_{s1}(\mathbf{r}_1)$ and an atomic p-state $\psi_{p1}(\mathbf{r}_1)$ (we consider the p-state which is proportional to x_1/r_1). We call that matrix element $X_{sp} = \int \psi_{p1}(\mathbf{r}_1)\, x_1\, \psi_{s1}(\mathbf{r}_1) d^3r_1$. Thus a two-electron state for the two up-spin electrons, for example, of two helium atoms, $\psi_{s1}(\mathbf{r}_1)\psi_{s2}(\mathbf{r}_2)$, will have a matrix element $-2e^2X_{sp}{}^2/r^3$ with the state with both atoms excited, $\psi_{p1}(\mathbf{r}_1)\psi_{p2}(\mathbf{r}_2)$. (This would also be true if we used antisymmetric combinations of states as in Eq. (4.14).)

a) Use this coupling to obtain the lowering in energy of the molecule in second-order perturbation theory in terms of X_{sp}, and $\varepsilon_p - \varepsilon_s$. Note that there are equal contribution for the excitation of both spin-down electrons, for the excitation of the spin-up electron on atom 1 and the spin-down electronic atom 2, and also for the spin-down electron on atom 1 and the spin-up electron on atom 2. The total is the van-der-Waals interaction between two helium atoms.

b) Similarly calculate the polarizability of the helium atom by calculating the shift in energy $-{}^1/_2|\langle\psi_{p1}|eE_xx_1|\psi_{s1}\rangle|^2/(\varepsilon_p - \varepsilon_s)$ of each electron by the perturbation eE_xx_1 due to the coupling with the same excited p-states, in analogy with Problem 5.3a. Equate the sum of the shifts of the two electrons to the $-{}^1/_2\alpha E_x{}^2$ of one helium atom to obtain the polarizability α of that atom.

c) Use Part b to write the van-der-Waals interaction of Part a in terms of polarizabilities for comparison with Eq. (12.2).

d) Follow through the reasoning to see the form of the result of Part c for two different atoms (e. g., He and Ne, where the excitation energy for Ne would be $\Delta\varepsilon = \varepsilon_s{}^* - \varepsilon_p$ with $\varepsilon_s{}^*$ the energy of the lowest excited s-state). Note again that there are only the two transitions, spin-up and spin down, to the s*-state on each atom.

13.1 Electron diffraction and Fermi surfaces

The nearest Bragg reflection planes in a simple-cubic lattice (which make up the faces of the cubic Brillouin Zone) are the planes which bisect wavenumbers $[2\pi/d]$ in cube directions. A divalent simple metal has a spherical Fermi surface with volume equal to that of the Brillouin Zone (a cube of edge $2\pi/d$ for a simple-cubic lattice) which is seen to intersect that cube.

a) Sketch a central cross-section (a (100) plane) of that cube and sphere. The portions of the sphere within the Brillouin Zone are Fermi surface in the first band.

b) If an electron has a wavenumber on one of those four segments from the first band while there is a magnetic field H present, perpendicular to the plane of the figure, that wavenumber will rotate with angular frequency $\omega_c = eH/mc$ (discussed in Section 14.1). As a free-electron it would complete a circular orbit, assume counterclockwise, in a time $T = 2\pi mc/eH$. However, due to the pseudopotential, when it reaches a Brillouin Zone face it will be diffracted (abruptly) to the opposite face and the wavenumber will continue to rotate counterclockwise until it again reaches a Zone face. Sketch the orbit which such an electron will follow in real space. (You will have the exact shape, in the diffraction limit, and do not need to work through the scale.)

c) In units of T, how long will it take the electron to complete the orbit?

13.2 Scattering by pseudopotentials, two dimensions

Redo the derivation of a formula for impurity scattering in two dimensions, starting with Eq. (13.15), leading to the two-dimensional counterpart of the second form in Eq. (13.17). The form is not as simple as in three dimensions.

14.1 Electron dynamics

Consider an electron moving in a simple-cubic tight-binding band

$$\varepsilon_{\mathbf{k}} = \varepsilon_s + 2V_{ss\sigma}[\cos k_x d + \cos k_y d + \cos k_z d\,]$$

with $V_{ss\sigma} = -(\pi^2/8)\hbar^2/(md^2)$ and $d = 2$ Å.

a) What is the maximum velocity (in cm/sec) such an electron can have in an x-direction?

b) In a field of 100 volts per centimeter in the x-direction, how long would it take an electron at rest at $\mathbf{k} = 0$ to again be at rest at $\mathbf{k} = 0$? (This assumes no scattering during this time, which is not realistic.)

c) Where would it end up after this time, relative to its starting point, and how far would it have traveled?

15.1 Phonon dispersion

Calculate the frequencies of transverse modes for $\mathbf{q}$ in a [100] direction in a face-centered-cubic structure, assuming a nearest-neighbor spring constant κ. The calculation follows closely the longitudinal calculation shown in Fig. 15.3, but the displacements are in different directions and the frequencies are not the same.

15.2 Vibrational specific heat

Write the sum (over wavenumbers and polarization $\lambda = 1,2,3$ for modes of frequency $\omega_{\mathbf{q}}^{\lambda}$) for the total vibrational energy for an fcc (face-centered-cubic) crystal in thermal equilibrium, using the average excitation number for each mode which we found in Eq. (10.11). Note that this does not assume a simple $\omega = qv$ which led to Eq. (10.14). However, it becomes simple in the limit of k_BT much greater than all $\hbar\omega_{\mathbf{q}}$. Obtain the leading term (in $k_BT/(\hbar\omega_{\mathbf{q}})$) in the vibrational energy per atom for that limit, and its derivative with respect to T, which is the specific heat per atom.

16.1 Second quantization manipulations

This is so simple it is hardly worth writing down, but follow the analysis in Section 16.1 to see that these are true:

a) If two states $|\Psi(\{\mathbf{r}_i\})\rangle$ and $|\Psi(\{\mathbf{r}_j\})'\rangle$ have different numbers of electrons, they are orthogonal, $\langle\Psi(\{\mathbf{r}_j\})'|\Psi(\{\mathbf{r}_i\})\rangle = 0$.

b) Show that if any state is occupied in $|\Psi(\{\mathbf{r}_j\})'\rangle$, but empty in $|\Psi(\{\mathbf{r}_i\})\rangle$, the two many-electron states are orthogonal.

16.2 Harmonic-oscillator number operator

Using the definitions

$$a^\dagger = \sqrt{\frac{M}{2\hbar\omega}}\left(\omega x - \frac{ip}{M}\right)$$

and

$$a = \sqrt{\frac{M}{2\hbar\omega}}\left(\omega x + \frac{ip}{M}\right),$$

evaluate $a^\dagger a$, noting also that $px - xp = \hbar/i$. Then manipulate the result to see that the Hamiltonian $(p^2/(2M) + kx^2/2)$ is $\hbar\omega(a^\dagger a + {}^1/_2)$.

17.1 Phonon emission by electrons

Complete the calculation of the rate at which an electron of wavenumber **k** spontaneously emits a phonon (no phonons present initially) outlined in Section 17.3, again taking $V_q = V_0 q_0/q$ and $\omega_q = \omega_0$, constant.

18.1 The field energy for photons

Use the expansion of the vector potential in Eq. (18.9) to write $A^2 = \mathbf{A}\cdot\mathbf{A}$ (you will need a sum over $\mathbf{q}$ and another over $\mathbf{q}'$) to evaluate the expectation value of the electric-field energy ${}^1/_{8\pi}\int d^3r\, E^2$ for a photon state with $n_{\mathbf{q}_0}{}^{\lambda_0}$ photons in a single mode of wave wavenumber $\mathbf{q}_0$, polarization λ_0, and no photons in any other mode. Note that $E^2 = A^2(\omega_0/c)^2 = A^2 q_0{}^2$ for this mode. Note also that there are four terms each with two $a_{\mathbf{q}}{}^\lambda$, $a_{\mathbf{q}}{}^{\lambda\dagger}$, $a_{-\mathbf{q}}{}^\lambda$, or $a_{-\mathbf{q}}{}^{\lambda\dagger}$, and that in the sum over $\mathbf{q}$, two terms contribute.

18.2. Optical matrix elements

We ordinarily get reasonable matrix elements taking atomic s-states as hydrogenic, $\psi(\mathbf{r}) = \sqrt{\mu^3/\pi}\, e^{-\mu r}$, which is normalized, with μ given by $\varepsilon_s = -\hbar^2\mu^2/(2m)$. p-states then are $\psi_{pz}(\mathbf{r}) = A'z\, e^{-\mu' r}$, etc., with $\varepsilon_p = -\hbar^2\mu'^2/2m$, and A' chosen for normalization.

a) Evaluate the A' which gives a normalized p-state wavefunction.

b) Evaluate the matrix element of $\partial/\partial z$ between the 1s-state of beryllium (ε_s = -8.42 eV) and the $2p_z$-state of beryllium (take ε_p = -5.81 eV). It should be given in $Å^{-1}$.

[You could check that the result is the same if you interchange left and right states, but you do not need to.]

18.3 Excitation of atoms

a) Take the absorption and emission rates from Eqs. (18.16) and (18.17) with the beryllium atom in a volume Ω, and the sum can be written as an integral, as indicated. If f_{2s} and f_{2p} are independent of spin, Σ_{spin} gives a factor 2. Substitute for $H_{\pm\mathbf{q}}{}^{\lambda}$ from Eq. (18.12), writing $<2p_z|\partial/\partial z|2s> = 1/r_a$ as in Eq. (18.14) and for every λ there is *some* p-state oriented along the polarization direction and f_{2p} applies to that p-state. Evaluate the integrals for absorption and emission, letting $n_{\mathbf{q}}{}^{\lambda}$ be a smooth function of q, (which would be $k_BT/(\hbar cq)$ if we had a thermal distribution at high temperature, but we do not make that assumption here). The results will depend upon $n_{\mathbf{q}}{}^{\lambda}$, the distribution functions and r_a.

b) For a single electron, $f_{2p} = 1 - f_{2s}$ and you can write the rate equation for df_{2p}/dt in terms of $n_q{}^{\lambda}$ evaluated at $\hbar\omega_q = \varepsilon_{2p}-\varepsilon_{2s}$. One could use this even if we applied a $n_q{}^{\lambda}$ which depended upon time.

c) What is f_{2p} (as a function of a steady n_q) in steady state? Sketch the result for f_{2p} as a function of $n_{\mathbf{q}}{}^{\lambda}$, which we may think of as $k_BT / \hbar\omega_q$.

19.1 2nd-order Stark Effect in a quantum well

The splitting of 2s and 2p-states in hydrogen due to a dc field, which we gave in Eq. (19.19), is called first-order Stark splitting, linear in field. The shift of the 1s-level due to coupling with the 2p-level is called a second-order Stark shift. Obtain the corresponding shift of the lowest quantum-well state ($\psi_1(z) = (\sqrt{2} / \sqrt{L})\sin(\pi z/L)$ for a well thickness L) due to coupling to the next-lowest state, $\psi_2(z) = (\sqrt{2} / \sqrt{L})\sin(2\pi z/L)$, by a uniform field E in the z-direction (a term in the Hamiltonian, $\delta H = eEz$). This will be a formula, proportional to E^2. [$\int dxdy$ would cancel and you don't need to include it. The result depends upon L.]

20.1 Interatomic interactions in metals

The interatomic interactions in a simple metal are given approximately following Eq. (20.11) as

$$V(r) = \frac{Z_1Z_2e^2\cosh\kappa r_{c1}\cosh\kappa r_{c2}e^{-\kappa r}}{r}.$$

Such a form can be used to estimate a wide range of properties of a metal. One simple property is the highest-frequency longitudinal vibrational mode propagating along a [100] direction in a simple-cubic metal, the mode in which nearest-neighbor atoms along the direction of propagation move in opposite directions, $\delta x = \pm u \cos\omega t$. Estimate the frequency for lithium, taking the interatomic distance for the simple-cubic structure as 2.75 Å, chosen to give the observed volume per atom in the real structure. You can evaluate k_F which will give a Fermi-sphere volume equal to half that of the Brillouin Zone (Z = 1 for Li), and then the Fermi-Thomas screening parameter from Eq. (20.9). We estimated the core radius for lithium in Problem 4.3, but we use the standard value of 0.92 Å (Harrison (1999), p. 453). This gives all parameters in $V(r)$, from which you can evaluate the net force from derivatives of $V(r)$, or numerically if the magnitudes of the displacements is for example $u = 0.01$ Å. Only two nearest neighbors contribute a force to first order in u_0 for displacements in a cube direction since the four nearest-neighbors in the lateral directions ($\pm d$ in the y- and z-directions) move with the central atom. We neglect contributions from more distant neighbors. You can then use force equals mass times acceleration, with the mass given by 1.67×10^{-24}g times the atomic weight of 6.94. [The Debye frequency for lithium, for comparison, is $\omega_D = 4.5\times10^{13}$/sec. (Kittel (1976), p. 126). Our estimate should be the same order of magnitude, but it is a different physical quantity and our estimate is approximate.]

21.1 Localized electrons in molecules

The Hamiltonian for a diatomic molecule, such as Li_2, can be approximated by three terms.

$$H_0 = \varepsilon_s(c_{1+}^\dagger c_{1+} + c_{1-}^\dagger c_{1-} + c_{2+}^\dagger c_{2+} + c_{2-}^\dagger c_{2-})$$
$$H_1 = V_{ss\sigma}(c_{1+}^\dagger c_{2+} + c_{1-}^\dagger c_{2-} + c_{2+}^\dagger c_{1+} + c_{2-}^\dagger c_{1-})$$
$$H_2 = U(c_{1+}^\dagger c_{1+} c_{1-}^\dagger c_{1-} + c_{2+}^\dagger c_{2+} c_{2-}^\dagger c_{2-})$$

with the subscript $\pm$ indicating spin. $H = H_0 + H_1 + H_2$ is the Hubbard Hamiltonian for a chain of two atoms. This can be solved exactly for two electrons present, but we look at approximate solutions.

a) If H_2 is negligible, it is the problem treated in Section 5.1, with ground state (Hartree-Fock) $|G_{HF}\rangle = \frac{1}{2}(c_{1+}^\dagger + c_{2+}^\dagger)(c_{1-}^\dagger + c_{2-}^\dagger)|0\rangle$. Evaluate the corresponding $E_0 = \langle G_{HF}|H|G_{HF}\rangle$, including the term H_2 in the Hamiltonian proportional to U. [It would also be possible to calculate the 2nd-order shift in energy due to H_2 using second-order perturbation theory, $\Sigma_j\langle G_{HF}|H_2|j\rangle\langle j|H_2|G_{HF}\rangle/(E_0 - E_j)$, but it would be rather intricate. That shift is called the "correlation energy".]

b) If H_1 is dropped, the ground state (called the Heitler-London state) is $|G_{HL}\rangle = c_{1+}^\dagger c_{2-}^\dagger|0\rangle$ (or any of three other states with the two electrons on different atoms). Evaluate $E_0 = \langle G_{HL}|H|G_{HL}\rangle$ for this state, including H_1. Aside from these four states there are only two other states, $|j\rangle$ each having both electrons on the same atom. Evaluate their energies from $\langle j|H_0 + H_1 + H_2|j\rangle$.

c) Evaluate the 2nd-order shift in energy of $|G_{HL}\rangle = c_{1+}^\dagger c_{2-}^\dagger|0\rangle$ due to H_1, which is $\Sigma_j\langle G_{HL}|H_1|j\rangle\langle j||H_1|G_{HL}\rangle/(E_0 - E_j)$. This lowering in energy would not arise if the two electrons were chosen to have parallel spin (e. g., $c_{1+}^\dagger c_{2+}^\dagger|0\rangle$) so it favors the spins of the ground state being antiferromagnetically aligned.

22.1 Quantum Hall Effect

The wavefunction for a free electron in a magnetic field, $\psi(\mathbf{r}) = \phi(x - x_0)\, e^{ik_y y}\, e^{ik_z z}$, which was introduced in Section 22.1 applies for a two-dimensional gas with the $e^{ik_z z}$ dropped. Let the system be a plane of dimensions L_x and L_y. Apply periodic boundary conditions on L_y and require the x_0 of $\phi(x-x_0)$ to lie in the region of length L_x.

a) Obtain a formula for the number of states (of a given spin) in each level ϕ_n for a given field.

b) For a density of electrons (number per area) of $1/A_0$ electrons of each spin, at what fields will the individual sets be exactly filled?

c) Relate this area per electron A_0 to an amplitude a_0 for a ground-state $\phi_0(x)$ such that $\frac{1}{2}\kappa a_0^2 = \frac{1}{4}\hbar\omega_c$ (potential energy equal to half the zero-point energy), with κ the effective spring constant for these orbits given in Section 22.1.

22.2 Zeeman splitting

Find the splitting of the different orbitals for a d-state ($l = 2$) in a magnetic field.

a) Allow also for the interaction of the two spin states with the field to sketch the new level diagram. (No spin-orbit coupling is included.)

b) Give the splitting in electron volts for $H = 1$ kilogauss.

22.3 Magnetic susceptibility

a) Obtain the Langevin shift in the energy, in eV, for a field of 1 kilogauss, of the two electrons in a helium atom (proportional to $\langle r_t^2\rangle = \langle r^2\rangle - \langle z^2\rangle$, using the 1s-wavefunctions obtained as in Problem 18.2, with ε_s taken from Table 4.1. The shift is very tiny.

b) There is no Van-Vleck term, Eq. (22.19), for the atom (since the s-states are eigenstates of L_z), but if these were two electrons in a molecule, the matrix element of L_z between each occupied state and an empty state (with energy such as the 2s-energy) would be of order $\langle 1|L_z|2\rangle \approx \hbar$. Using this value and taking the energy of the excited state to be of order $|\varepsilon_s|$ for helium, what would be the ratio of the Van Vleck to the Langevin term in the energy of the two electrons?

22.4 Spin-orbit coupling

Redo our analysis of the spin-orbit splitting of atomic p-states to obtain the corresponding result for d-states, with energies in terms of λ_2, and give the number of states at each energy.

23.1 Shake-off excitations

The s-states for an electron in a spherical quantum well of radius R are $A\sin(n\pi r/R)/r$, with A chosen so the states are normalized, illustrated below for the ground state, going to zero at $r/R = 1$. If an electron is in the ground state, and R is suddenly increased to $R' = 1.1R \equiv R/\nu$,

a) what is the probability (numerical) for the electron going to the new ground state? [This could be done by numerical integration or as a special case of the integral in Part b.]

b) Give a formula for the probability of its being excited to the new state of quantum-number n. (The needed integrals are quite simple.) Such an excited state in the expanded well is also shown below.

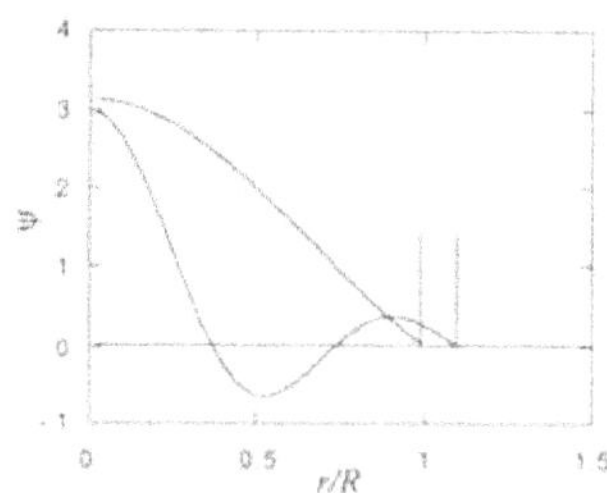

References

Adolph, B., V. I. Gavrilenko, K. Tenelsen, F. Bechstedt, and R. Del Sole, 1996, *Nonlocality and many-body effects in the optical properties of semiconductors,* Phys. Rev. **B53**, 9797. *262*

Aharonov, Y., and D. Bohm, 1959, Phys. Rev. II **115**, 485. Discussed also by Kroemer (1994), p. 235., *10ff*

Anderson, P. W., 1967, Phys. Rev. Letters **18**, 1049. See also Harrison (1970), p. 348, for discussion and a correction to the central formula provided by Anderson. *316*

Animalu, A. O. E., and V. Heine, 1965, Phil. Mag. **12**, 1249. Listed also in Harrison, 1966, 309ff for all the simple metals. *177*

Ashcroft, N. W., 1966, Physics Letters **23**, 48. *62*

Averin, D. V., 1999, *Solid-state qubits under control,* Nature **398**, 748. *321*

Baldereschi, A., 1973, *Mean Value Point in the Brillouin Zone*, Phys. Rev. **B7**, 5212. *280*

Bardeen, J., L. N. Cooper, and J. R. Schrieffer, 1957, Phys. Rev. **108**, 1175. *75, 282ff*

Bennett, W. R., W. Schwarzacher, and W. F. Egelhoff, Jr., 1990, Phys. Rev. Letters **65**, 3169. *26*

Bertino, M. F., A. L. Glebov, J. P. Toennies, F. Traeger, E. Piper, G. J. Kroes, and R. C. Mowrey, 1998, *Observation of Large Differences in the Diffraction of Normal- and Para-H_2 from LiF(001)*, Phys. Rev. Lett. **81**, 5608. *153*

Chadi, D. J., 1977, *Spin-orbit splitting in crystalline and compositionally disordered semiconductors*, Phys. Rev. **B16**, 790. *302, 304*

Cooper, L. N., 1956, Phys. Rev. **104**, 1189. *282*

Costa-Krämer, J. L., N. García, P. García Mochales, and P. A. Serena, 1995, Surf. Sci. **342**, L1144. *30*

Debye, P., 1912, Ann. d. Physik **39**, 789. *43, 146*

Dirac, P. A. M., 1926, Proc. Roy. Soc. (London) **A112**, 661. *8, 154*

Edwards, D. M., and J. Mathon, 1991, J. Magn. and Magn. Mater. **93** , 85. *26*

Edwards, D. M., J. Mathon, R. B. Muniz, and M. S. Phan, 1991, Phys. Rev. Letters **67**, 493. *26*

Einstein, A., 1907, Ann. d. Physik **22**, 180, 800. *146*

Einstein, A., 1911, Ann. d. Physik **34**, 170. *146*

Ensher, J. R., D. S. Jin, M. R. Matthews, C. E. Wieman, and E. A. Cornell, 1996, *Bose-Einstein Condensation in a Dilute Gas: Measurement of Energy and Ground-State Occupation*, Phys. Rev. Lett. **77**, 4984. For a much more complete discussion, see *Bose-Einstein Condensation in Atomic Gases,* edited by M. Inguscio, S., Stringari, and C. E. Wieman, IOS Press, Amsterdam, 1999. *149*

Fermi, E., 1934, Nuovo Cimento, **11**, 1. *135*

Giaever, I., 1960, Phys. Rev. Letters **5**, 147, 464. *121*

Hafner, J., 1987, *From Hamiltonians to Phase Diagrams,* Solid State Sciences **70**, Springer-Verlag, Heidelberg. *59*

Harris, S. E., 1997, *Electromagnetically Induced Transparency*, Physics Today, July, 1997, p. 36. *255*

Harrison, W. A., 1966, *Pseudopotentials in the Theory of Metals,* W. A. Benjamin, Inc., New York. *62, 174*

Harrison, W. A. , 1970, *Solid State Theory,* McGraw-Hill, New York; reprinted by Dover, New York, 1979. *163, 205, 221, 267, 268, 314*

Harrison, W. A., 1980, *Electronic Structure and the Properties of Solids*, W. H. Freeman, San Francisco; reprinted by Dover, New York, 1989. *78, 80*

Harrison, W. A., 1987, *Elementary theory of the properties of the cuprates* , in *Novel Superconductivity,* edited by Stuart A. Wolf and Vladimir Z. Kresin, Plenum Press, New York, p. 507. *281*

Harrison, W. A., 1994, *Inelastic events do not just randomize the phase*, Phys. Rev. **B50**, 8861. *318*

Harrison, W. A., 1999, *Elementary Electronic Structure* , World Scientific, Singapore. *26ff, 60, 97ff, 188, 194, 203, 221, 226, 242, 261ff, 267, 273, 274, 280, 297, 298, 300, 305, 330, 340*

Harrison, W. A., and Alexander Kozlov, 1992, *Matching conditions in effective mass theory,* Proceedings of the 21st International Conference on the Physics of Semiconductors, Beijing, China, August, 1992. Edited by Ping Jiang and Hou-Zhi Zheng, World Scientific, Singapore, Volume 1, p. 341. *120*

Harrison , W. A., and M. B. Webb, 1960, eds., *The Fermi Surface,* John Wiley, New York. Proceedings of an International Conference at Cooperstown, New York on August 22-24, 1960. *183, 191*

Hilborn, R., and C. Yuca, 1996, Phys. Rev. Letters **76**, 2844. *150*

Hoffmann, Roald, 1963, *An Extended Hückel Theory. I. Hydrocarbons*, J. Chem. Phys. **39**, 1397. *77*

Hybertsen, M. S., and S. G. Louie, 1985, *First-principles theory of quasiparticles: calculation of band gaps in semiconductors and insulators,* Phys. Rev. Letters **55**, 1418. *262*

Irvine, J. M., 1972, *Nuclear Structure Theory*, Pergamom Press, Oxford. *65*

Josephson, B. D., 1962, Phys. Letters (Netherlands), **1**, 25. *121*

Kittel, C., 1976, *Introduction to Solid State Physics,* Fifth Edition, Wiley, New York. *166, 340*

Kittel, C., and H. Kroemer, 1980, *Thermal Physics*, Second Edition, Freeman, New York. *18, 149*

Kogan, Sh., 1996, *Electronic Noise and Fluctuations in Solids*, Cambridge University Press, Cambridge. *164*

Kroemer, H., 1994, *Quantum Mechanics for Engineering, Materials Science, and Applied Physics,* Prentice Hall, New York. *1, 18, 41, 48*

Landau, L. D., and V. L. Ginsburg, 1950, JETP (USSR) **20**, 1064. *286*

Landauer, R., 1989, J. Phys. Cond. Matter **1**, 8099. *28*

Laughlin, R. B., 1983, *Anomalous Quantum Hall Effect: An Incompressible Quantum Fluid with Fractionally-Charged Excitations*, Phys. Rev. Lett. **50**, 1395. 292

Mann, J. B., 1967, *Atomic StructureCalculations, 1. Hartree-Fock Energy Results for Elements Hydrogen to Lawrencium.*. Distributed by Clearinghouse for Technical Information, Springfield, Virginia 22151, *60, 61*

Mathews, Jon, and Robert L. Walker, 1964, *Mathematical Methods of Physics* , Benjamin, New York. *33, 73, 324*

Moore, C. E., 1949, *Atomic Energy Levels,* Volume 1, National Bureau of Standards, Washington. *239*

Moore, C. E., 1952, *Atomic Energy Levels,* Volume 2, National Bureau of Standards, Washington. *239*

Mössbauer, R. L., 1958, Z. Physik **151**, 124. *314*

Nakamura, Y., Yu. A. Pashkin, and J. S. Tsai, *Coherent control of macroscopic quantum states in a single-Cooper-pair box,* Nature **398**, 786. *321*

Parkin, S. S. P. , R. Bhadra, and K. P. Roche, 1991, Phys. Rev. Lett. **66**, 2152. *26*

Prange, R. E., and S. M. Girvin, 1987, eds., *The Quantum Hall Effect*, Springer-Verlag, New York. *292*

Peierls, Rudolph, 1979, *Surprises in Theoretical Physics*, Princeton University Press, Princeton. *281*

Ransil, B. J., 1960, Rev. Mod. Phys. **32**, 245. *85*

Schiff, L. I., 1968, *Quantum Mechanics* , McGraw-Hill, New York. *36, 301*

Seitz, F., *Modern Theory of Solids*, 1940, McGraw-Hill, New York, *290*

Slater, J. C., and G. F. Koster, 1954, Phys. Rev. **94**, 1498. Also given in Harrison (1999), p. 544, and Harrison (1980), p. 481. *96*

Thornton, Marion, 1995, *Classical Dynamics of Particles and Systems,* Fourth Edition, Harcourt Brace & Co., Fort Worth, TX, *44*

Tsui, D. C., H. L. Störmer, and M. A. C. Gossard, 1982, *Two-Dimensional Magnetotransport in the Extreme Quantum Limit*, Phys. Rev. Lett. **48**, 1559. 292

Van Schilfgaarde, M., and W. A. Harrison, 1986, *Theory of the multicenter bond.* Phys. Rev. **B33**, 2653. *78*

Von Klitzing, K., G. Dorda, and M. Pepper, 1980, *New Method for High-Accuracy Determination of the Fine-Structure Constant Based on Quantized Ha Resistance.* Phys. Rev. Lett. **45**, 494. *291*

Weast, R. C., 1975, editor, *Handbook of Chemistry an Physics*, 56th Edition, The Chemical Rubber Company, Cleveland, *60*

Yukawa, H., 1935, Proc. Phys. Maths. Soc. Japan **17**, 48. *230, 231*

Subject Index

www.ingramcontent.com/pod-product-compliance
Lightning Source LLC
Chambersburg PA
CBHW061233220326
41599CB00028B/5416
9789810243944